The Sanitary City

HISTORY OF THE URBAN ENVIRONMENT

Martin V. Melosi and Joel A. Tarr, Editors

Martin V. Melosi

THE SANITARY CITY

Environmental Services in Urban America
from Colonial Times to the Present

ABRIDGED EDITION

UNIVERSITY OF PITTSBURGH PRESS

To Gina and Donna and to Adria and Steven, with love

Published by the University of Pittsburgh Press, Pittsburgh, Pa., 15260
Copyright © 2000, The Johns Hopkins University Press
Copyright © 2008, University of Pittsburgh Press
First University of Pittsburgh Press paperback, abridged edition, 2008
All rights reserved
Manufactured in the United States of America
Printed on acid-free paper
10 9 8 7 6 5 4 3 2 1

Library of Congress Cataloging-in-Publication Data

Melosi, Martin V., 1947–
  The sanitary city : environmental services in urban america from colonial times to
the present / Martin V. Melosi. — Abridged ed.
        p.   cm. — (History of the urban environment)
  Includes bibliographical references and index.
  ISBN-13: 978-0-8229-5983-0 (pbk. : alk. paper)
  ISBN-10: 0-8229-5983-6 (pbk. : alk. paper)
    1. Municipal water supply—United States—History.   2. Sanitary engineering—
United States—History.   3. Refuse and refuse disposal—United States—History.
4. Municipal services—United States—History.   I. Title.
  TD223.M45 2008
  363.6'10973—dc22   2007046614

CONTENTS

ILLUSTRATIONS

PREFACE

My intention in writing the original version of *The Sanitary City* was to provide a book that treated the development and impact of sanitary services in the United States over the entire span of the country's history. I saw the book as a synthesis of the historical work that had come before, but also a comprehensive study that would be a ready reference for anyone interested in the topic. I realized that even a book of such length merely scratched the surface of the topic. Yet its rather imposing size would not make it suitable for classroom use or for the reader who wanted something a little more manageable in length and focus. In writing the abridged edition, I have tried to retain the quality and character of the original without the length and detail. If you want more on the topic, let me refer you to the original published in 2000. In that edition, I wrote the following preface (with some necessary updating for this volume), which I believe still applies today.

Early in the evening of October 17, 1989, I settled into my recliner in front of the TV to watch the first game of the World Series between the San Francisco Giants and the Oakland A's. Having grown up in the Bay Area, I was particularly excited about the games to be played in my old backyard. Before things were much under way, the scene shifted to the Bay Bridge (connecting San Francisco with Oakland), which had been wrenched apart by what was apparently a serious earthquake. I immediately jumped to my feet and dialed my parents' home in San Jose, some fifty-five miles from the site of the disaster. Miraculously, I was able to get through. When my father answered, my folks were desperately trying to ride out the earthquake, which was at the moment tossing dishes and knickknacks from the shelves in a totally dark kitchen. They had no idea what was happening farther up the coast in the San Francisco–Oakland area. To say the least, this was a bizarre moment as I described the scene at the Bay Bridge to my father from 2,000 miles away in Houston.

My parents (who are no longer with us) survived the ordeal with only minor physical damage to their house, but with rattled nerves produced by the earthquake and its many aftershocks. As bad as the Loma Prieta quake was on that October evening—resulting in the deaths of fifty-five Bay Area residents and injuring hundreds more—it was not the major shaker experts had predicted. It was sufficiently bad, however, to collapse the upper deck of Interstate 880 near Oakland onto a lower deck, crushing cars and motorists; to plunge many communities into darkness because of damaged power lines; to rupture gas mains and set off fires; to destroy or damage hundreds of structures as far south as the Monterey Peninsula; to fracture water and sewer lines; to disrupt commuter traffic into San Francisco for many weeks; and to generally reshuffle the lives of thousands of people.

Coincidentally, I had left San Jose the day before the earthquake, returning to Houston from a research trip in California just in time also to miss Hurricane Jerry, which had hit Galveston. I was comforted in knowing that I had avoided both of these disasters and that my parents and my sister's family living in Redwood City were safe in the aftermath of the earthquake. Yet I could not forget that I had driven over that now-collapsed one-and-a-quarter-mile stretch of Interstate 880 several times only a few days earlier.

On October 18 or 19, I received a call from Bob Tutt at the *Houston Chronicle*, who wanted me to review the events surrounding Loma Prieta. As the resident urban historian at the University of Houston and as a specialist on public works and city services, I was not surprised that he had contacted me about the quake. It just so happened that I was in the early stages of research leading to the cloth edition of *The Sanitary City*, which accounted for my trip to California to do research in local depositories. With the book project on my mind and with a personal connection to the Bay Area, the interview (which appeared in the paper on October 22) became a meaningful discussion of the vulnerability of urban infrastructure subjected to natural disasters and how crucial public works and city services are to people's lives. Little did I know that there would be many more disasters—natural or otherwise—in the following years, especially the 9/11 attacks and Hurricanes Katrina and Rita, to make my point over and over again.

Unfortunately, it takes a major disruption like an earthquake—or a terrorist attack or a hurricane—to demonstrate graphically the importance of the streets and alleys, bridges, power and communications networks, water and sewer lines, and waste-disposal facilities to urban survival—let alone the countless lives that are lost or affected by the events.

How fitting it was that my original research for *The Sanitary City* began in earnest in the year of Loma Prieta.

ACKNOWLEDGMENTS

The original unabridged edition of *The Sanitary City* began in 1988 with a substantial grant from the Division of Research Programs of the National Endowment for the Humanities (NEH). The NEH, in general, and project officer Daniel P. Jones, in particular, demonstrated great faith in my work and patiently waited for the fruits of my labors. Without the NEH, this book would never have been written. I also was fortunate to receive significant supplemental funding from several other sources, including Pennzoil Company. The Smithsonian's National Museum of American History brought me to Washington, D.C., as a fellow, which allowed me to utilize some very useful manuscript and periodical collections. Curator Jeffrey Stine was mostly responsible for that opportunity. On my own campus, funding and release time were made available by the College of Humanities, Fine Arts, and Communication through a faculty development leave and a summer grant. The Environmental Institute of Houston provided crucial financial assistance for the last three chapters of the book. The University of Houston (UH) Energy Laboratory had been a longtime supporter of my work, including this project. I want to remember the former director, the late Alvin Hildebrandt, who not only shepherded several of my proposals through the system but who also took great personal interest in the substance of my research on cities.

Many people provided early critiques of my initial proposal and subsequent iterations. Jim Jones, then a colleague at UH, helped me think through what began as a hazy notion for a broad study on urban infrastructure, and put his crafty pen to work on my original grant proposal. Louis Cain of Loyola University of Chicago was the first to introduce me to the theory of path dependence, and helped me hone some of my ideas about city services. Engineer Bob Esterbrooks gave me excellent suggestions about the role of sanitary engineers. I also benefited from Harold Platts's thorough reading of the entire manuscript and his incisive comments, from Joel Tarr's vast knowledge of the subject, and from Josef Konvitz's efforts to make me look beyond the obvious. Many other colleagues commented on my original proposal or parts of the original manuscript, including Pete Andrews, Peter

Bishop, Blaine Brownell, John Clark, Mike Ebner, Bob Fisher, David Goldfield, Sam Hays, Suellen Hoy, Peter Hugill, Ken Jackson, Clayton Koppes, John Lienhard, Walter Nugent, Jon Peterson, Joe Pratt, Mark Rose, Chris Rosen, Bruce Seely, and Jim Smith. Two brief teaching opportunities at the University of Paris VIII and at the University of Helsinki allowed me to test some of the ideas in my work on colleagues and students from France and the various countries along the Baltic.

I was blessed with several hard-working students who ferreted out massive numbers of bibliographic references and photocopied thousands of pages of books and articles for me. Special thanks to Bruce Beaubouef, Thao Bui, Charles Closmann, Tom Kelly, Jack McClintic, Long Nguyen, Elisabeth O'Kane-Lipartito, and Bernadette Pruitt. To our former chief of staff in public history, Christine Womack, goes much gratitude for the hundreds of ways she made this book better. Library staffs at the Anderson Library at UH, the UH Law Library, Texas A&M's Evans Library, the Library of Congress in Washington, D.C., and the library of the American Public Works Association (especially Howard Rosen and Connie Hartline) provided very useful guidance. Special acknowledgment as well goes to Bill Ashley and his staff at University Media Services at UH for the excellent reproduction of the photographs in the original book as well as this current one.

Very early on in this project, George F. Thompson, president of the Center for American Places, pursued me to place *The Sanitary City* with him and to publish it with the Johns Hopkins University Press. I truly appreciate George's long-standing enthusiasm for my work. George's group also was instrumental in placing the abridged edition with the University of Pittsburgh Press. The director of Pittsburgh Press and my friend, Cynthia Miller, stood with me throughout the long process of crafting a shorter, more classroom friendly version of this book. And additional special thanks to Joel Tarr, colleague and friend, who nagged me incessantly on the importance of completing the abridged edition. He as much as anyone turned an idea into reality.

To my wife, Carolyn, goes my love for putting up with me again and again as I shuffled papers, grumbled about all of the work I had to do, and generally complained about everything and nothing. Fortunately, our grown daughters, Gina and Adria, and their partners, Donna and Steven, were spared most of my venting. I dedicate this book to them because of the joy that they bring Carolyn and me every day, even if we can't always be together.

In the end, the original book and the new abridged edition have probably given me greater satisfaction than anything I have written. I also was very fortunate to have others acknowledge the work, and I want to share my pleasure with them, including all those listed above.

The Sanitary City

Introduction

I n *Service Delivery in the City*, the authors stated that "delivering services is the primary function of municipal government. It occupies the vast bulk of the time and effort of most city employees, is the source of most contacts that citizens have with local government, occasionally becomes the subject of heated controversy, and is often surrounded by myth and misinformation. Yet, service delivery remains the 'hidden function' of local government."[1] Service delivery is a "hidden function" largely because it often blends so invisibly into the urban landscape; it is part of what we expect a city to be. While economic forces are essential to the formation of cities, urban growth depends on service systems that shape the infrastructure and define the quality of life.

I have chosen to focus on sanitary (or environmental) services—water supply, sewerage, and solid-waste disposal—because they have been and remain indispensable for the functioning and growth of cities. They provide water for domestic and commercial uses, eliminate wastes, protect the public health and safety, and help to control many forms of pollution.[2] Frederick Law Olmsted, the builder of New York City's Central Park, is credited with calling trees "the lungs of the city." Sanitary services are the circulatory system of the city.

Sanitary services also are important vehicles for revealing contemporary environmental thought as it relates to city development. They are linked inextricably to prevailing public health and ecological theories and practices,

1

which have played a large part in the timing of their implementation and in determining their form.

The Sanitary City is a broad history of water-supply, wastewater, and solid-waste disposal systems in American cities from colonial times to 2000 (or a little beyond) that analyzes their development, assesses their influence on urban growth, and evaluates their impact on the environment. While others have studied the history of these services individually, no one has attempted to integrate all three into one study covering such a long time period.

Like my *Garbage in the Cities*, this study takes a national perspective, rather than relying on case studies.[3] A major objective is to identify trends that characterize American approaches to sanitary service delivery and related environmental implications, drawing on the experiences of many cities. Also, the study gives attention to nineteenth-century English sanitation practices and their impact on the United States. The long time line allows for a detailed examination of changes in technology, in the evolution of regulatory authority, and in patterns of urban growth.

The Sanitary City is based on extensive research into contemporary periodicals, government reports, city publications, and proceedings of engineering and public health associations. It is organized chrono-thematically to show change over time but also to address several major issues: the influence of public health and ecological theory on sanitary services; the role of decision makers—sanitarians, engineers, physicians, and political leaders—in determining the choice of services to be provided; and the environmental implications of those choices.

The decision to choose between available technologies was informed by the prevailing environmental theory of the day. Prior to the twentieth century, when the initial technologies of sanitation were implemented, the miasmatic—or filth—theory of disease strongly influenced choice. Beginning in the 1880s through the end of World War II, bacteriological theory informed choice. Sometime after the war, new theories of ecology broadened the perspective of sanitary services beyond the narrower health outlook. These health and environmental theories were sufficiently widespread to constitute environmental paradigms. Consequently, the book is divided into three broad historical periods essential for examining the growth and evolution of sanitary services in the United States: the Age of Miasmas, colonial times to 1880; the Bacteriological Revolution, 1880 to 1945; and the New Ecology, 1945 to the 2000s.

Sanitary services are not organic entities, but specialized technical systems that help to shape the apparatus of modern cities.[4] Development of elaborate technical networks in the nineteenth century was a prime characteristic of the modern city. Whereas industrialization remained local or regional for many years, new technological innovations were quickly diffused nationally. This suggests that while American cities did not uniformly benefit—or suffer—from the direct economic impact of the Industrial Revolution, they were physically modernized as a result of new technical systems developing in the era. By the late nineteenth century, many cities in the United States entered a period of dynamic system building in areas such as energy, communication, transportation, and sanitation.[5]

The implementation of new urban technologies was not automatic, coincidental, or inadvertent, but intentional efforts by decision makers to confront existing problems faced by cities as they grew upward and outward in the nineteenth and twentieth centuries. Beginning in major cities of the industrial era, the decision to implement new technologies grew out of, as historian Jon Peterson argues, "struggles to surmount the limitations and failings of older urban arrangements that became apparent as a consequence of big-city growth." These struggles helped to define the environmental agenda in the delivery of urban services, and led to the building of technical systems that "assured the viability and even the special vitality of the new urbanism."[6]

Decision makers acted on choices for surmounting the limitations and failures of older urban arrangements and often promoted new technologies as a means to fulfill those choices.

City leaders responded to perceived immediate community needs often framed by rudimentary environmental considerations. For example, a pure and abundant water supply was vital for consumption by humans and fire protection; the diversion of sewage from homes and businesses and the removal of solid wastes improved health and combated disease.

A basis for defining the urban environmental agenda was built upon the common need for transportation, communication, energy, and sanitary services. The infrastructure, various technical systems, and sanitary services came to represent public goods, and thus required municipal—and later state and federal—commitments to increased public spending. Yet problems of equity and discrimination—in terms of who made the decisions and who received the services—persisted throughout the history of the American city. Favoritism, corruption, personal aggrandizement, and greed often thwarted rational decision making.

Sanitarians and engineers were the key conduits through which technical and scientific expertise about sanitary systems flowed from city to city. At the same time, they were receptors and disseminators of the prevailing environmental views that shaped those systems. The importance of sanitarians and engineers to municipal decision making rests on the premise that in the technologically networked city, changes made in the infrastructure and related services (barring natural disaster) are intentional efforts to shape or modify the physical city in order to accomplish social, political, or economic goals.

Decision making had short-term and long-term implications. In the short run, the implementation of technologies of sanitation often met the expectations of city leaders. Certain health problems decreased, delivery of precious water and disposal of wastes improved living conditions, and municipal reputations were enhanced. In the long run, the emphasis on project design rather than careful planning focused attention on immediate goals rather than on the potential resilience of the system or its capacity to adapt to pressures of growth.

It was common, even in the earliest days of the new systems, to emphasize centralization of function for the sake of efficiency and control, and to treat specific technologies as permanent solutions to the delivery of services, the eradication of disease, and the stimulus to urban growth. A commitment to permanence, however, often locked in specific technologies and limited choices for future generations—referred to as path dependence. Problems could arise if systems were either too well built or too poorly constructed. In the case of the former, an existing system could prove resistant to change; in the latter case, it might be in desperate need of replacement. Path dependency suggests that decisions made about sanitary systems in the nineteenth century had profound impact on cities more than a hundred years later.[7]

The story of sanitary services in the United States requires going beyond abstractions to reveal the intimate connection among technical systems, urban growth, and environmental impacts for more than 200 years. The first section of the book, "The Age of Miasmas," begins in seventeenth-century America, when cities faced poor sanitary conditions and suffered crippling effects of epidemic diseases with only a vague understanding of their cause. By the nineteenth century, a few larger cities developed community-wide water-supply systems with rudimentary distribution networks, but continued to regard waste disposal as an individual responsibility. The worldwide influence of the English "sanitary idea" in the mid-nineteenth century, however, linked filth with disease, and provided a clearer rationale and newer strategies for improving sanitary services. In his 1842 *Sanitary Report*, English lawyer and sanitarian Sir Edwin Chadwick took a bold stand on the need for an arterial system of pressurized water that would place house drainage, main drainage, paving, and street cleaning into a single sanitary process. While this remarkable hydraulic system was never implemented, nineteenth-century English sanitarians and engineers became leaders in setting standards for water and wastewater services throughout Europe and North America. More important, English theories of sanitation helped to provide the environmental context for developing new technical systems in the United States and elsewhere. Thus the development of North American water and wastewater systems in the mid- to late nineteenth century depended heavily on the expertise of English civil engineers and English public health leaders, the implementation or adaptation of English sanitation technology, and the absorption of English environmental values.

The earliest technologies of sanitation spread to North America in an era of rapid urban growth, especially after 1830, and were influenced by the English "sanitary idea" in the 1840s. Miasmas—filth or foul smells—were believed to be responsible for epidemics. While the miasmatic theory failed to uncover the root cause of disease, it placed great emphasis on the need for environmental sanitation to combat it. Primary attention was given to water supply and, to a lesser extent, sewerage. Citywide problems of solid waste were not clearly identified. The sanitary services of the time were designed to keep wastes from accumulating in the central cities, but could not prevent disease because of constraints of prevailing health theory. Despite their limits, modern water and wastewater systems owe their form and function to the "protosystems" of the nineteenth century that fostered the practice of environmental sanitation. The resulting American systems were not carbon copies of English models, but adaptations modified by local and national cultural, economic, technological, environmental, and political factors.

Part 2, "The Bacteriological Revolution," begins with the development of modern sanitary services from 1880 to 1920—a period of dynamic urban growth and development in the United States. The water-supply, wastewater, and solid-waste systems of this period were designed to provide permanent relief from threats to health along with convenience for city dwellers. The biological revolution of the late nineteenth century gave urban leaders new tools for combating epidemic disease, including bacteriological laboratories and use of inoculation and immunization. The desire to find permanent solutions to sanitary problems, however, placed reliance on systems designed in a previous era, not fundamentally redesigned within the new environmental context of the bacteriological age. In fact, the value of environmental sanitation as a primary tool for combating disease came into question as the goals of public health began to shift toward individual medicine.

From the end of World War I until the end of World War II, neither the quality nor character of sanitary services underwent substantial change compared to the previous period, with the possible exception of solid-waste collection and disposal. The view of solid waste as a more clearly defined form of pollution placed greater attention on developing new disposal options in particular.

A major challenge for municipal officials, engineers, planners, and sanitarians in the interwar years was to adapt sanitary services to growth characterized by metropolitanization and surburbanization, on the one hand,

and demand for such services in numerous small towns and rural communities, on the other. Decision making was complicated by two major disruptions—the Great Depression and World War II. From the fiscal perspective, the economic disorder of the late 1920s and 1930s changed the nature of city-federal relations and transformed what had been local service delivery into systems increasingly influenced by regional and national interests. Although the conflict between the germ and filth theories of disease had long been settled, from an environmental standpoint there continued to be a preoccupation with biological forms of pollution at the expense of a better understanding of chemical sources, especially industrial pollutants.

Part 3, "The New Ecology," suggests that external forces were central to influencing sanitary services after World War II. Urban sprawl placed increasingly stiff demands on the providers of water, wastewater, and refuse collection and disposal services. For the first time in American history, concern over a decaying urban infrastructure raised questions about the permanence of the sanitary systems devised and implemented in the Progressive Era and earlier. Dynamic peripheral growth and central-city deterioration were phenomena that characterized the postwar urban condition and no less affected the maintenance and development of sanitary services. In addition, an array of troubles soon characterized as a rising "urban crisis" increasingly shifted attention away from purely physical decay to social ills.

Sanitary services were situated not only within a new political and social context, but within a changing environmental context as well. The New Ecology and the modern environmental movement produced another paradigm in which sanitary services were being viewed with different eyes. New environmental awareness, emerging after the war and blossoming in the 1960s, had a powerful impact on the scientific and technical communities, including the sanitarians, public health professionals, and engineers who had been essential to the development of sanitary services in their original form. Of note was the changing focus from purely biological forms of pollution to chemical/industrial pollutants and pollution from municipal sewers. Solid waste became a national issue as a "third pollution," alongside water and air pollution.

After 1970, the so-called infrastructure crisis suggested massive physical deterioration of public works, as well as eroding financial resources for confronting a wide array of problems. A major question to be addressed was: To what degree did the potential breakdown of the technologies of sanitation, increasing awareness of the environmental implications of water and land pollution, and shrinking resources to confront the myriad troubles lead to a reevaluation of the premises upon which sanitary services had been based for more than two centuries? In addition, shifting attention from point-source to nonpoint pollution raised the uncomfortable possibility that exist-

---

ing technologies and the prevailing government regulatory apparatus were not designed to deal with newer and confounding environmental threats. Moreover, increasing quantities of solid wastes, tougher environmental legislation, and long-term dependence on sanitary landfills were leading to the unsettling feeling that a "garbage crisis" was on the horizon. The quest for the Sanitary City had yet to be fully achieved as the nation faced the new millennium.

I

THE AGE OF MIASMAS

From Colonial Times to 1880

Sanitation Practices in Pre-Chadwickian America

Prior to the 1830s, many American cities faced poor sanitary conditions and suffered crippling effects of epidemic disease. Few communities could boast of well-developed technologies of sanitation, and much of the responsibility for sanitation rested with the individual.

As England urbanized and industrialized in the eighteenth century, provincial urban communities only began to challenge the rural-dominated North American landscape. Colonial towns and cities grew in political, social, and economic importance, but only modestly in size and number.[1] The 1790 federal census showed that city dwellers represented less than 4 percent of the nation's population, and only two cities exceeded 25,000. Philadelphia (42,520) was the largest city in a country with just twenty-four urban places. By the end of the 1820s, the urban population had almost doubled, although fewer than 7 out of every 100 Americans lived in cities or towns.[2]

The limited scale of American urbanization does not mean that cities faced few health risks or that communal sanitary services were unnecessary. In the largest cities, the rate of growth was a key factor in stimulating concern about health and sanitation. New York, Philadelphia, Boston, and Baltimore experienced impressive rates of growth in each decade between 1790 and 1830.

Along with growth, European sanitation practices helped shape the initial American responses to water-supply, sewage, and refuse problems. Local circumstances affected the timing of new sanitary service delivery, but the forms and methods traveled across the Atlantic with the colonists or were borrowed directly from Europe.

While American urban communities seldom faced sanitation problems on a par with their European counterparts in these years, public and governmental perceptions and reactions were quite similar. Few people had an inkling about the causes of disease and illness. Individuals or private scavengers were usually responsible for disposing of wastes. And the role of government in protecting the community's health, guarding against the ravages of fire, cleaning streets, and providing pure water was obscure and untested in all but the largest cities.[3]

If American towns fared slightly better than European urban areas, it probably had more to do with less crowded conditions than with an enlightened outlook about sanitation. The "healthiness" of American towns was a matter of degree, however. Tolerance for nuisances and the almost serendipitous occurrence of epidemic disease played major roles in determining the sanitary quality of the communities. As late as the 1860s, Washingtonians dumped garbage and slop into alleys and streets. Pigs roamed freely, slaughterhouses spewed noxious fumes and effluent, and vermin infested dwellings—including the White House.[4] Few towns and cities were free of nuisances, and showed little resolve to move against the "noxious trades"— soapmakers, tanners, slaughterhouses, butchers, and blubber boilers—especially if they were located in the poorer areas.

Animals resident in urban communities were a part of preindustrial life. Horses for transportation; cattle, hogs, and chickens for food use; and dogs and cats as pets roamed freely through many vacant lots, streets, and alleys. Pigs and turkeys, in particular, were widely accepted as useful scavengers. Manure and dead animals were simply annoyances balanced against the value of sharing space with contributors to the town's welfare.[5]

Epidemic disease was taken much more seriously than sanitation, especially since many colonists feared it as the wrath of God.[6] While the relative isolation of North America limited the number of epidemics in colonial towns, they were no less ruinous than those in Europe once they spread. Transatlantic trade and urban growth in the seventeenth and eighteenth centuries led to an array of infectious disorders, including smallpox, malaria, yellow fever, cholera, typhoid, typhus, tuberculosis, diphtheria, scarlet fever, measles, mumps, and diarrheal disorders. Disruptive events, such as the Revolutionary War, reintroduced epidemics to major cities and towns.[7]

Smallpox was probably the worst of the early scourges, but while they were less frequent, yellow fever and diphtheria were as virulent. Yellow fever first attacked the Atlantic Coast in the 1690s, peaked around 1745, temporarily subsided, and then reappeared savagely in the 1790s in the port cities of Boston and New Orleans. In 1793, yellow fever took 5,000 lives in Philadelphia—one out of every ten residents. In 1798, the dread disease struck New York, killing 1,600 to 2,000 people out of a population of 80,000. By the 1820s, yellow fever virtually disappeared in northern states, but remained a chronic problem from Florida to Texas.[8]

The line between individual and governmental responsibility for responding to community needs was obscure through the early nineteenth century. Before then "the city was to be an environment for private money-making, and its government was to encourage private business."[9] In addition, at least until the political system was pressured to encourage wider participation, local government catered to the "better sort."[10]

Epidemics forced the government to deal with public health, at least from crisis to crisis, but the absence of regularized preventive action had to do with limited knowledge about contagious diseases. Previous experience became the best teacher. More affluent citizens, for example, could escape the city. In New York, about one-third of the 27,000 residents fled during the 1805 epidemic. The poor, who were unable to flee, usually suffered the most. To make matters worse, economic activity ground to a halt during epidemics as merchants and other business owners took flight, leaving workers at least temporarily unemployed. As the districts of the laboring poor grew, these became the focus of perceived threats to the city's health.[11]

Quarantining those who contracted or were suspected of contracting a disease was another method of reducing its spread. Massachusetts Bay Colony set up quarantining regulations as early as 1647 based on concern about the "great mortality" in the West Indies. Other communities lagged behind in passing such laws, and in many cases quarantine regulations were only temporary measures.[12]

Boston is often credited with authorizing the first permanent local board of health in the United States in 1797.[13] The threat of disease, especially yellow fever, stimulated some interest in permanent boards in other communities, but the few that were established often focused on nuisance abatement. Laypersons, especially the mayor and some council members, sat on the boards and rarely exerted much authority. Between 1800 and 1830, only five major cities established boards of health, and between the 1790s and 1830 all but Boston's were temporary.[14]

As late as 1875, many large urban communities had no health departments of any kind, partly because they did not have the authority to issue health regulations without approval from their state legislatures.[15] Equally important, until the mid-nineteenth century, was the peculiar relationship between epidemic diseases and the available means to control them. The propensity to regulate public health found greater sustained efforts in dealing with issues of nuisance. It was easier to visualize some sort of danger or inconvenience coming from noxious odors or putrefying wastes than from the mysterious appearance of yellow fever or smallpox.

Crude sanitary regulations were common in the American colonies by the late seventeenth century. In 1634, Boston officials prohibited residents from throwing fish or garbage near the common landing. Between 1647 and 1652, the local government passed other ordinances, including one that dealt with the construction of privies. In 1657, the burghers of New Amsterdam were among the first to pass laws against casting waste into streets.

Some effort was made to regulate the noxious trades by requiring butchers, tanners, and slaughterers to keep their property free of nuisances or by ordering the removal of slaughterhouses from the town limits. Between 1692 and 1708, Boston, Salem, and Charleston passed laws dealing with nuisances and trades deemed offensive or dangerous to the public. In New York City, the office of city inspector was established in 1804—the first permanent office concerned specifically with sanitation. Erratic enforcement of sanitary laws undermined the effort to protect the public health throughout colonial America, however, and continued to be a problem.[16]

New York City also was the first American city to establish a comprehensive public health code in 1866. In several ways, the courts provided an alternative remedy to municipal regulation. Much of the environmental law in the United States has been based on nuisance law derived from English common law. Nuisance law was primarily formulated from lawsuits relating to the use of land. Private nuisance action would lie where the defendant's unreasonable use of his or her property interfered with the reasonable use of the plaintiff's property. In the case of a public nuisance, an action could be brought against someone who obstructed or caused damage to the public in the exercise of the public's common rights. Private and public nuisance law had the capacity to abate specific sources of pollution instead of demanding a regulatory approach. Nuisance actions theoretically could be used to challenge individuals, municipalities, and industries relating to all manner of pollution.[17]

Throughout the nineteenth century, interpretations of nuisance were inconsistent. Before the Industrial Revolution—when the courts were more inclined to place economic concerns ahead of environmental issues—the application of nuisance doctrines could focus on strict property rights between individuals. Even in this period, the courts often were willing to protect the growth and economic expansion of towns and cities by invoking the public nuisance doctrine against individuals in violation of local ordinances and by protecting local governments from lawsuits.[18] Before the 1830s, the idea of nuisance played a more important role in efforts to dispose of liquid and solid wastes than the fear of epidemics.

For much of the country, dependence on wells or nearby watercourses for water supplies, the use of privy vaults and cesspools for human and household liquid wastes, collection of refuse by scavengers, and dumping or burning of garbage, ashes, and rubbish provided adequate sanitary services. In low-density areas, these methods resisted change or outright replacement. The practices often were publicly regulated but rarely publicly managed.

As populations increased, such approaches became less workable. The result was the development of the first technologies of sanitation—"protosystems"—that emphasized more sophisticated technologies, were increasingly capital intensive, were publicly regulated and often publicly operated, and removed the individual from direct responsibility. Prior to the mid-nineteenth century, almost all protosystems in America were devised for water supplies.[19]

More than any other sanitary service, an efficient water-supply system was a key factor in the well-being of urban populations. "In the United States, at least until the end of the nineteenth century, the presence of potable water was a major consideration in the location of towns."[20] City leaders increasingly devoted attention to the delivery of water supplies as urban growth accelerated in the early decades of the nineteenth century.[21]

What was to be learned from Europe about developing an effective water-supply system was unclear prior to the mid-nineteenth century because the urban context was so different. Centralized systems—or large systems dominated by private companies—had grown up with several of the major cities of Europe. Some dated from Roman times or were influenced by the great aqueduct construction practiced by the Romans.[22]

Two technical advances made delivery of water more practicable in eighteenth-century Europe and prompted the emergence of new private water companies: the application of steam power to water pumping and the wider use of cast-iron pipes. The first steam-driven pump was said to be installed in London in 1761. In 1776, a company was formed to furnish Paris with water from the Seine through the use of a steam pump.[23] By the nineteenth century, the steam pump provided reliable power to complement or replace gravity systems and offered a way to increase the volume of water from its source to consumers.

The delivery of water through cast-iron pipes provided a durable, cost-effective, and technically manageable way to improve the distribution of water supply to individual structures. Aqueducts could get water to a city, but provided no means to distribute it. Public wells were common, but did not solve the problems of transporting water to homes. Prior to the widespread use of pipes, cities relied on less-efficient means of distribution. Until the French Revolution, for example, the most common method for Parisians to obtain water was by dipping containers into the fountains in the public squares, or by "water carriers." Like Paris and other large cities, London relied on water carriers, especially for the houses of the affluent. The first significant use of cast-iron water pipe was developed to supply the Versailles Palace in 1685. In 1746, the Chelsea Water Works Company was probably the first to use cast-iron mains in London. Lead-pipe systems date back at least to thirteenth-century London, but lead was an inferior material with which to work, even without knowledge of its health hazards.[24]

Two events in late-sixteenth- and early-seventeenth-century London established England as a leader in citywide water supplies in Europe. First, in 1581 Dutch engineer Peter Morritz was granted a 500-year lease to construct waterwheel pumps on London Bridge to supply the city with water from the Thames. Some regarded this system as the first "modern" waterworks in London.[25]

A second key event was the incorporation in 1619 of the New River Company to supply water to individual houses. Water was brought to London from the River Lea and distributed through a network of wooden and then cast-iron pipes. The supply was superior to local, increasingly polluted sources. The success of the New River Company gave momentum to private enterprises organized to carry out public functions, and resulted in other companies.[26]

The early successes in providing water service in London were not sufficient to withstand some disagreeable impacts of the Industrial Revolution in the eighteenth century. The demographic shift in England at the time profoundly affected city growth and led to serious overcrowding in the major urban areas, with consequent health and pollution problems. As the world's first urbanized society, it was little wonder that England was the focal point for the development of good-quality water supplies to meet health and fire demands. In early-nineteenth-century London, rapid growth led to a frenzy among water companies scrambling to retain customers and to increase profits. Rapid construction in and around the city, and the refinement in pumping technology, made the business highly profitable for those who could capture the market.[27]

Between 1805 and 1811, five water companies were created by statute in London, and soon more joined them. Competition became so fierce that lines of rival companies were laid down in the same streets in populated districts, but not in sparsely populated areas. The battle for profits led to price wars, and by 1817 the eight water companies remaining in the competition teetered on the brink of insolvency. They survived by dividing the supply by district and by agreeing to raise prices.

The experience in delivering water in other English cities was different. Local governments regarded water management as "too vital a matter to be left entirely to individual initiatives or to profit-seekers."[28] Several town authorities took much more direct control of water service than in London. Leeds, Derby, Macclesfield, Huddersfield, and Manchester all fought to promote public projects. But as the towns grew into cities, the ability of local government to deal effectively with water management became problematic.

Until the 1840s, Parliament was more inclined to rely on market forces than state support to provide water service. The ability of private companies to raise capital under the terms of enabling legislation also worked against management by local authorities who lacked the power for long-term borrowing. Also, the flight of wealthy citizens from town centers made it difficult to modernize water systems through public control.[29]

Unlike cities such as Manchester, which had struggled to keep local government at the center of water management, London had become "a bastion of private enterprise in the water industry."[30] The water monopoly, however, faced criticism for raising rates, limiting distribution, and providing what was perceived as poor-quality water. In 1821, the House of Commons appointed a Select Committee to investigate the state of the water supply in London, the first occasion when the city's water supply was examined as a whole.

The findings did not satisfy the complainants since the report stated that there had been improvement in the supply, that the extension of the supply for private use and as a precaution against fire was satisfactory, and that the quality of the water in general was superior to every other European city. No action was taken, and many believed that the companies effectively were "whitewashed." The quality of the water supply continued to deteriorate in London. In some cases, the sewage outfall on the Thames River drained close to where water was drawn for the city. The river was rapidly becoming an open sewer.

A Royal Commission was appointed in 1827, again to inquire into the quality of London's water supply. Despite several shortcomings that were discovered, the report issued the next year essentially reiterated the findings of the previous study. But the commission did recommend that the city protect the quality of existing viable sources and obtain new sources. The commission questioned Parliament's support for market solutions to the water-supply problem, but Parliament failed to take action to regulate the private water industry.[31]

Soon an engineering solution emerged to provide a technical fix to an increasingly discredited system. In 1804, John Gibb built a successful slow sand filter bed at Paisley, Scotland, and in 1827 Scottish engineer and mill owner Robert Thom built a slow sand filter bed at Greenock, Scotland. Glasgow became the first town to have a piped supply of filtered water in Great Britain.[32] The Chelsea Water Works Company in London constructed a similar filter, and Thames water was run through the filter for the first time in January 1829.[33] The results were so good that the Chelsea filter beds became the prototype for what became known as the "English system." Soon they were adopted for the whole of London's water supply, and spread rapidly around the world. A primary aim of the filters was to reduce turbidity of the water to aid industries dependent on clear water, but the health value of filtration was unclear.[34]

With the completion of the Royal Commission's report and the introduction of slow sand filter beds, water supply as a major public issue temporarily dropped out of sight in England. The immediate impact of the English experience was to apply new technologies for more effectively delivering water to private homes and businesses. The quality of that supply, however, remained suspect.[35]

Not until the mid-nineteenth century did English experiences with water-supply systems begin to influence American cities. Prior to that time, most municipal officials did not detect problems that led them to seek alternatives to existing approaches. Many American towns and cities were on the cusp of change as the nineteenth century unfolded, but only under unique circumstances did water-supply protosystems begin to appear.

The fear of fire and epidemics was a great motivator for change. The old "bucket brigade" was grossly inadequate when whole blocks of homes and shops were endangered by fire. Prior to the completion of Philadelphia's system in 1801, it took the bucket brigade fifteen minutes to fill one fire engine with water; after the system was in place it took one and a half minutes.[36]

The hydrant became the modern symbol for fire protection, since it meant that water would be immediately available and abundant to fight a major conflagration. New York City, which earlier had taken a leadership role in fire protection, was slow in installing hydrants and did not do so before 1830. Hydrants made water quickly available for emergencies, but they also increased the use of water, making a large supply even more necessary.[37]

While the fear of fire always loomed, the startling impact of an epidemic increased public pressure for improved water supplies. Fear alone was insufficient to lead towns and cities to abandon traditional sources of water and familiar methods of acquiring it. A community needed a political commitment, fiscal resources, and access to new technology. Prior to the mid-nineteenth century, only about half of the major cities and towns had some type of waterworks. Most of them drew their supplies from wells, springs, or ponds, and did not have extensive distribution systems, if they had any at all. The great majority of the waterworks were located in the Northeast, with considerably fewer in the Old Northwest and Upper South.[38]

Before the turn of the century, most cities and towns depended on a combination of water carriers, wells, and cisterns to meet their needs. Even during the first several decades of the nineteenth century, many larger cities and smaller towns continued to rely on local sources of supply. Unless they hired water peddlers, each citizen used no more than three to five gallons per day.[39]

While community-wide water-supply systems developed slowly in American cities, in 1801 Philadelphia became the first to complete a waterworks and municipal distribution system sophisticated even by European standards. The necessary health, economic, and technical factors converged to produce what became a model for future systems. The Philadelphia waterworks, however, was an anomaly, since it did not spark an immediate nationwide trend.[40]

Concern for health prompted the campaign for a waterworks in Philadelphia. Despite uncertainty in determining disease causation, the correlation between pure water and good health was nevertheless a driving force in dealing with epidemics. *Scott's Geographical Dictionary* described the water in the densest areas of the city as having "become so corrupt by the multitude of sinks and other receptacles of impurity, as to be almost unfit to be drank."[41]

Although the issue of finding a new source of water had arisen earlier in Philadlephia, ravaging yellow fever attacks in 1793 and 1798 led political and business leaders to form a watering committee to deal with epidemics. The consensus was that polluted water from wells and cisterns caused the fever, and that the city's private wells should be replaced by a community-wide system. The waterworks also could provide needed water to clean the streets, fight fires, and add to the aesthetic quality of the city through public fountains.[42]

After examining various options, the committee accepted the proposal of Benjamin Henry Latrobe. The English-born engineer was also a practicing architect, who later worked on the U.S. Capitol from 1802 to 1817.[43] Latrobe recommended a system to pump water from the Schuylkill River, and to distribute it through mains made of bored logs. He proposed that water would be moved by a steam engine along the river up to a tunnel running under the streets and then by gravity to a pump house at Centre Square in the city. Another steam engine at Centre Square would pump the water to reservoir tanks at the top of the building, and then gravity would distribute water through the rest of the system. He began work on the system in 1799 and completed it in 1801.[44] Latrobe's aesthetics as an architect permeated the project. As one historian noted, "The Philadelphia Waterworks at Centre Square was an early example of Latrobe's influential neoclassical architectural style. The building was admired for its proportions and use of Greek protoypes."[45]

Even after full operation, the machinery never worked as planned. The cost of the system was high, the amount of water pumped was limited, and recurring yellow fever epidemics in 1802, 1803, and 1805 alarmed the citizens. In 1811, the Watering Committee replaced Centre Square with a larger plant in a different location. The plan of engineer Frederick Graff, Latrobe's former assistant, called for a pumping station along the Schuylkill at the foot of Fairmount rise (beyond the city limits), with construction of a reservoir on top of the hill in the city. The new facility was completed in 1815. Steam pumps again were employed, but the waterworks converted to more reliable water power in the 1820s. The Fairmount Waterworks served Philadelphia until 1911.[46]

In distributing the water, the new system in Philadelphia first relied on wooden pipes and eventually on iron pipes. From the seventeenth century until well into the nineteenth century, wooden mains were commonly used in American cities. The first wood conduits were probably laid in Boston in 1652. Winston-Salem, North Carolina, purportedly built the first citywide system with log pipes in 1776. While rotting and leaking were chronic problems, wooden mains had one advantage—in case of fire a hose could be connected directly to the main simply by drilling a hole in it. (After the crisis had passed, a wooden plug could be driven into the main. This practice probably was the origin of the term "fire plug.")[47] Soon after 1800, cast-iron pipe was introduced into the United States from England. By 1825, cast iron was half the price of lead, and only one-quarter the price in 1850.[48]

Not without its flaws, Philadelphia's waterworks was considered by many to be the most advanced engineering project of its time. Ultimately, Philadelphia had a system with a much greater capacity than existing demand, unlike comparable cities such as New York, Boston, and Baltimore.[49] To promote its use, citizens were initially offered free water for several years. Despite the fear of epidemics, many citizens had not been completely convinced to give up "their cold well water for the tepid Schuylkill water." By 1814, however, 2,850 dwellings were receiving water from the new system.[50]

The example of Philadelphia constructing a major waterworks was widely publicized, but a national trend in municipal, citywide waterworks was not evident until late in the century. Inexperience in dealing with such a major project, in part at least, helps to explain why urban population growth exceeded construction of waterworks for so many years. In 1800, there were seventeen waterworks for an urban population of 322,000; in 1830, there were forty-five waterworks for 1,127,000 urban Americans.[51] "Municipal governments in the nineteenth century were just emerging as effective governing bodies; often the decision to obtain a water system was the first major undertaking of a city government and the first which required a large initial outlay financed by bond issues."[52]

Rural-dominated state legislatures often attempted to check city growth by controlling services from the state capital or restricting the taxing and financing power of the city in its charter. Thus it was exceedingly difficult for cities to provide services, even when they accepted responsibility for them. "Home rule" for many cities did not become a reality until late in the century. Not surprisingly, almost every city and town initially turned to private agents or companies to supply water.

Private companies received franchises through the issuance of corporate charters, which was a typical way to generate public works activities in the eighteenth and early nineteenth centuries. Since few companies could meet the expectations of the cities for good service, plentiful and pure water, and low price, those who agreed to do so received franchises with substantial concessions. It was not unusual for a franchisee to get a long-term contract, exclusive rights to supply water, the right to acquire property by eminent domain, exemption from taxation, and other benefits.[53] In 1800, sixteen of seventeen (94.1 percent) American waterworks were private, and thirty-six of forty-five (80 percent) in 1830.[54]

In New York City, a freshwater pond had been the major source of supply even after 1800, while water from many private wells suffered saltwater infiltration and pollution from privy vaults, cesspools, and street drainage as early as 1750. In 1774, the Common Council contracted with an English engineer to build a municipal system using a steam engine to lift water into a central reservoir. The Revolutionary War derailed the project, and not until 1799—after a devastating yellow fever epidemic—was it renewed. City leaders realized that rivals Boston, Philadelphia, and Baltimore were building or were proposing to build waterworks.

Water quickly became the focus of a major political struggle. The council requested the legislature to provide it with special powers to establish a water system. Assemblyman Aaron Burr maneuvered to acquire a charter for a new private water company—the Manhattan Company—instead of supporting the development of a municipal system. The perpetual charter granted the company wide powers with few obligations, and Burr was intent on using the company to amass surplus capital in the hopes of building a banking business.

From the vantage point of water-supply service, the company was modestly successful. At its peak, it provided water for only one-third of the city, and was continually embroiled in controversy. Between 1801 and 1808, Burr faced staggering political setbacks, including losing control of the Manhattan Company. He was dropped from the board in 1802, making way for the rise of his political rival, DeWitt Clinton.

Clinton soon realized that meeting the demand for citywide distribution of water was impossible with the existing system. Discussions about selling the company to the city became more frequent, but in the short term the charter was revised, and the company continued its favored position.

The deteriorating quality of the water supply weakened the Manhattan Company's hold on the city's water service. In 1825, a bill granting a charter to the New York Water-Works Company was enacted. Controversy over its charter, the lack of good supplies of pure water in the immediate area, and the pressure from the Manhattan Company and other rivals ended the short-lived venture. The basic requirements for good, accessible sources of water were not met until the completion of the Old Croton Aqueduct in 1842, which provided for the first workable municipal system in New York City.[55]

Boston also endured a long water-supply debate before it developed an adequate protosystem in the 1840s. From 1630 to 1796, the city derived all of its water from wells and cisterns, and the quality was "hard, highly colored, often odorous, saline, bad-tasting, and sometimes polluted."[56]

In 1796, Govenor Samuel Adams approved an act creating the Aqueduct Corporation, which built a line from Jamaica Pond in Roxbury to the city. The distribution network was extended in 1803, but it did not provide service for the entire community. There was no further attempt to improve the existing system until 1825, but civic leaders chronically argued over the water supply through the mid-1840s.[57]

The pattern in the Midwest, South, and elsewhere was similar to experiences in the Northeast. Some of the larger cities made the transition to citywide systems early, with most cities and towns following more slowly. Cincinnati was the first "western" city with a waterworks. In 1813, community leaders contracted to drill "possibly 30" public wells in a single season. However, an 1817 ordinance chartered the Cincinnati Manufacturing Company to develop a system, one of the earliest such concessions granted. In 1839, the works was sold to the city. At the time of the purchase, the property only consisted of a pumping station and reservoir grounds. The company had run into financial trouble in the intervening years, and chronic problems in meeting its obligations stimulated distrust of the company and kept alive the possibility of public control.[58]

The St. Louis Water Works was built in 1830. In 1821, a general concern about fire hazards led to a demand for a better water supply. Finally, in 1829 the city council offered a $500 prize for the best plan. Within a short time the city signed a contract with Wilson and Company, and the work on the installation began in 1830, but water did not move through the pipes until the 1840s.[59]

Benjamin Latrobe brought his innovations from Philadelphia to New Orleans, which was on the brink of a major growth spurt. His plan was to secure a franchise for himself and his investors to turn a profit from the sale of water. The New Orleans waterworks was similar to Philadelphia's in several respects. A steam engine would pump water from the Mississippi River through a pipe into six elevated wood reservoirs. Gravity would carry the water through a combination of wooden and iron pipes. Benjamin's son, Henry, completed drawings of a fountain, which was never built, along the riverfront square.

Benjamin Latrobe was known for accepting projects that he was not immediately prepared to undertake. And since he was occupied in the Northeast, he sent Henry to the Crescent City in 1811 to begin the work. He himself did not arrive on the scene until 1819. Aside from the English engineer's absence, technical setbacks, and problems with investors, the Latrobes also had to contend with the disruption of the War of 1812. To his credit, Henry kept the project from unraveling. But in a major blow to his father and to the project, young Henry succumbed to yellow fever in 1817. Until Benjamin arrived in New Orleans, one of Henry's associates took charge. Upon completion of most of the work on the waterworks, Benjamin himself apparently contracted yellow fever and died. Within a year, the New Orleans Water Works Company, struggling to survive, was sold to the city. This project was Benjamin Latrobe's final engineering legacy.[60]

The technical achievements in developing early water-supply protosystems had some bearing on levels of consumption in the early nineteenth century. The new systems, however, did not provide for equity of service. In the mid-1820s, Cincinnati had more than 26,000 feet of wooden pipes, but served only 254 industrial and home users. At the time, the daily consumption of water probably averaged between three and five gallons per person, with higher consumption by those who could afford to purchase additional supplies.

For all the improvements begun by private companies through franchises, accessibility to water supply was still largely linked to class. Affluent neighborhoods and the central business district received the lion's share of water, while the working-class districts often relied on polluted wells and other potentially unhealthy local sources.[61] As historian Sam Bass Warner Jr. suggested, "Philadelphia, as the pioneer in waterworks, was the first to discover that to bring a water pipe to the sidewalk was still a long step from installing taps, toilets, or tubs inside the houses. For the urban poor, a generation and even longer elapsed before owners of slum properties installed plumbing."[62]

Despite the limitations of the new water systems, the few American cities that turned to community-wide approaches set patterns for modern sanitary services of the near future. Protosystems were precursors to more elaborate centralized systems adopted by the late nineteenth century. As in England, the application of these new technologies ran ahead of an effective understanding of the causes of disease and pollution, while they nevertheless attempted to enhance the healthfulness of the city and provide better protection against fire.

In the case of waste disposal, there was little or no linkage at this time between the search for a pure and plentiful water supply and methods of eliminating an array of rejectamenta and effluvia. Waste disposal had yet to rise much above the level of nuisance in the eyes of the public or city officials. In dealing with waste, Americans relied on approaches that had been commonly practiced in Europe for many years. Before the mid-nineteenth century, few large American or European cities constructed drainage systems and refuse-disposal facilities on a par with the great civilizations of Babylon, Mesopotamia, Carthage, or Rome. The ancient societies with the most highly developed sanitary systems offered large and dispensed services through hierarchical authority, but services were not equally distributed among the classes.[63]

Americans adopted Old World methods, focusing on individual responsibility for disposal of wastes as befit the circumstances of the country's urban centers prior to 1830. Cesspools, manure pits, and the pail system for removal of waste from privy vaults met local needs in much of Europe until the nineteenth century. Sewerage was primarily utilized for drainage rather than carrying wastewater. Sewers, if they existed at all, were mostly open ditches.[64]

The earliest mention of sewers in England dates back to the fourteenth century, but these were simply drainage ditches.[65] Prior to 1700, London had no sewers of any kind. In theory, cesspools or cesspits were viewed as the proper receptacles for excrement, and sewers were channels for surface water.[66] But in practice this did not occur. Until 1815, it was illegal to discharge waste other than kitchen slops into the drains of London. If deposited on land, much of the wastes could eventually flow through ditches along with rainwater or through covered and walled streams.

The introduction of the first water closets in England in 1810 offered city dwellers a more convenient—and seemingly more sanitary—method of disposing of human waste. This technical wonder encouraged greater use of water and, when linked to cesspools, reduced the effectiveness of the cesspools. Because of the volumes of water utilized, cesspool waste did not percolate into the soil but overflowed the cesspools and found its way into the streets and city drainage systems.[67]

Progress in refuse collection and disposal fared little better than sewerage before the nineteenth century. While the English Parliament banned waste disposal in public watercourses and ditches, the practice continued. Until the fourteenth century, Parisians were allowed to cast garbage out of their windows, and although several attempts were made at collection and disposal, the mounds of waste beyond the city gates were so high by 1400 that they obstructed the defense of the city. The plagues that invaded Europe between 1349 and 1750 provided some inducement for better sanitation, but responsibility largely remained an individual matter until the 1800s.

One significant improvement for better refuse collection and disposal was the practice of paving and cleaning streets, which began as early as the twelfth century. Paris started paving its streets in 1184 when, according to contemporary accounts, King Phillip II ordered the streets paved because he was annoyed by the offensive odors from the mud outside his palace. In 1415, Augsburg was the first city in Germany to pave its streets. Street cleaning at public expense came sometime later—in Paris, not until 1609. In the German principalities, street-cleaning work was frequently assigned to Jews and to servants of the public executioner. But street paving and cleaning, like other sanitary services, were often confined to commercial thoroughfares or affluent neighborhoods.[68]

Drainage and the disposal of liquid wastes fared similarly in America as they had in Europe. What historian Joel A. Tarr has called the "cesspool–privy vault–scavenger system" dealt adequately with the disposal of human and household liquid wastes in many communities until they experienced rapid growth or seriously altered the disposal system by introducing running water, which inundated the cesspools and privy vaults.

Human waste occasionally was deposited in leaching cesspools, but more often in privy vaults in cellars or close to the house. Privy vaults were relatively small, and were either covered with dirt when filled and replaced or emptied by the individual or by private scavengers. Most city ordinances required that the vaults be emptied at night, thus the term "night soil" became a euphemism for human waste. The privy vault disposal method operated reasonably well for many years, but the vaults were rarely watertight, required regular attention, and produced noxious smells.

Household liquids and wastewater found their way to on-site cesspools or dry wells in many communities, but too frequently were simply cast on the ground. Under the best circumstances, wastes were recycled on farmland or sold as fertilizers. The record of such uses, especially for night soil, was as erratic in the United States as it had been in England.[69]

More problematic was the impact of wastewater once it left private property and the flow of stormwater through the streets. While the cesspool–privy vault–scavenger system provided rudimentary handling of wastes, existing "sewers" offered increasingly little help in controlling drainage problems. By the end of the eighteenth century, major urban centers such as New York and Boston had sewers. A "sewer" in this early period was intended to carry off stormwater or to drain stagnant pools rather than to handle wastewater, and was most often a street gutter rather than an underground drain. In Boston, city authorities did not assume the maintenance of the drains or begin to build new ones until 1823. Only liquid wastes were allowed in the drains, and fecal matter was specifically excluded until 1833.[70]

As in England, many ordinances forbade placing any wastes in sewers in this period. Sometimes intentionally and sometimes quite inadvertently, the surface drains became open sewers carrying substantial waste matter that had been dumped there or came from overflowing cesspools and privy vaults.[71]

Unlike sewerage, street cleaning garnered serious attention because of the many functions that streets performed—transporting goods, allowing human and animal traffic, facilitating emergency fire service, and even offering a place for social encounters. Since streets were part of a community's "commons," street cleaning came to be regarded as a municipal responsibility before refuse collection. Individuals or scavengers carried the responsibility of disposing of refuse that they generated around their homes and businesses.

Pioneering sanitary engineer Samuel A. Greeley noted that "the beginnings of city cleaning were undoubtedly in street cleaning."[72] In many towns and cities, citizens frequently lodged complaints about the filthy state of the main thoroughfares and the neglect of conditions in alleys and on noncommercial streets. It was typical in Europe and America for city dwellers to use streets as a dumping ground for refuse. Horses and other animals contributed their share of wastes. Boston and New Amsterdam were first to pass ordinances prohibiting the most egregious practices, but these laws were difficult to enforce and rarely deterred citizens from tossing materials along almost every street and road.[73]

In some of the larger communities, scavengers removed clutter from streets and also carted away rubbish and garbage as early as the seventeenth century. Eventually, free-roaming swine and fowl were much less prevalent. In hiring scavengers, towns, especially those with moderate street use, could meet the street-cleaning needs of their citizens in nonresidential areas. Systematic street cleaning with paid crews became necessary in the mid-nineteenth century when greater vehicular traffic kicked up billows of dust and urban workhorses were more plentiful.[74] At best, many problems associated with liquid and solid wastes were dealt with casually in most towns and only with slightly more determination in larger cities until later in the nineteenth century.

There was little stimulus for American cities to alter their disposal practices prior to 1830. In the case of water supply, the fear of fire and epidemics, and eventually the experience of the English, produced some modest changes. The most significant was the construction of the Philadelphia protosystem. Yet this achievement was insufficient to set off a national trend. The English "sanitary idea" and the refining of the miasmatic theory of disease would provide the context in which elaboration in technologies of sanitation would take place in the mid- to late nineteenth century.

Bringing the Serpent's Tail into the Serpent's Mouth

EDWIN CHADWICK AND THE "SANITARY IDEA" IN ENGLAND

Mid-nineteenth-century England's "sanitary idea" made popular the notion that the physical environment exercised a profound influence over the well-being of the individual—that health depended upon sanitation. This concept reshaped thinking about the delivery of pure water, the removal of sewage, and the collection and disposal of refuse. As one writer put it, the greatest service of the sanitary idea was "in replacing . . . fatalism by a new faith in the power of scientific control of the physical environment."[1]

The advent of the sanitary idea offered a clearer rationale for improving sanitary services. Historian Ann F. La Berge, however, argues that late-eighteenth- and early-nineteenth-century France was first to provide a model for public health "theoretically, institutionally, and practically," and that there was "considerable cross-fertilization of ideas between public health advocates in Britain and France." By the 1850s, the British claimed preeminence in public health practices, such as sewerage and water supply, and leadership in the field passed to them.[2]

The ascendancy of British public health practices was a triumph for the theory of disease causation embedded in the sanitary idea. One of the key results was the transformation of Victorian cities from their Dickensian bleakness into more livable environments.[3] The greatest popularizer of the sanitary idea was barrister-turned-sanitarian Edwin Chadwick. Born on January 24, 1800, near Manchester, he had limited educational opportunities as a child and came to resent "the classically educated elite." Largely self-taught, Chadwick entered an attorney's office as an apprentice at age eighteen, and in 1823 was admitted to the Middle Temple to begin preparation to become a barrister. He supported himself by writing articles for newspapers, and developed associations with legal and medical students in London.

In 1824, Chadwick met Dr. Thomas Southwood Smith and later met John Stuart Mill—both Philosophical Radicals—and acquired his first knowledge of Benthamism.[4] The acknowledged leader of the Philosophical Radicals, Jeremy Bentham was a jurist, legal reformer, and utilitarian philosopher. He criticized traditional ideas of constitutional law and asserted the theory, as one writer stated, that "right actions are those which are most useful for the promotion of general happiness."[5]

From Bentham and economist David Ricardo, Chadwick acquired or reinforced his belief in an activist central government. As his reputation grew among the Benthamites, Chadwick was asked to open a debate on the Poor Laws at the London Debating Society in November 1829. Two years later, he became secretary for Bentham himself, assisting the aging utilitarian in the drafting of his *Constitutional Code*. Although they differed on several issues, Chadwick shared with Bentham a commitment to efficiency and the authoritarian nature of the state.[6]

The condition of the London slums became a major research interest for Chadwick, and while pursuing this work he contracted typhus, but fully recovered. In 1832, the year Bentham died, Chadwick was appointed to a commission inquiring into the English Poor Laws, resulting in the publication of the 1834 *Poor Law Report*. After more than two decades of public service, Chadwick returned to private life in 1854. Ostensibly for reasons of health, he resigned from the General Board of Health with the blessing of his many antagonists.[7]

Critics regarded Chadwick's social views as repressive, but his stature as an expert in Poor Law reform grew nonetheless. When three Poor Law commissionerships became available in 1834, Chadwick assumed he would receive one. But birth and wealth directed these appointments. His rigid stance on key issues also made him less attractive to a Whig government already operating on wobbly legs. Regarded as indispensable to the work, Chadwick was offered the position as secretary to the new Poor Law Commission. He initially balked, but reconsidered after being convinced that his power would be greater than his title. Although he played an integral role on the commission, Chadwick never acquired the decision-making influence he anticipated. Critics, however, continued to identify him with oppression of the poor.[8]

Throughout his career Chadwick battled with public officials, physicians, and engineers. Some considered him a martinet who did not work well with others. "No one ever accused Chadwick of having a heart," one observer noted.[9] He challenged those criticisms, not able to understand why his commitment to social change inspired such emotional reactions.

Chadwick ultimately turned to the field of public health, despite his image as insensitive to the poor and his disdain for physicians who cared little for preventive medicine. In the wake of the influenza epidemic of 1837–38, the commission inquired into the relationship between pauperism and sanitary conditions. Chadwick was given the responsibility to carry out the work, and he enlisted the aid of three doctors known for their devotion to environmental influences on health—James Kay, Neil Arnott, and Thomas Southwood Smith. It was not easy for Chadwick to share this opportunity to conduct a major national sanitary inquiry with physicians who were becoming prominent spokespersons for health reform in their own right.[10]

Chadwick brought great attention to the ravages of poverty and to the dismal health conditions of the industrial cities with the *Report on the Sanitary Condition of the Labouring Population of Great Britain* (1842). The document was widely disseminated, selling more copies than any previous government publication. It painted a vivid picture of urban blight, making emphatic the case for disease prevention.[11] As a good Benthamite, Chadwick also prescribed to the notion of "civic economy," which suggested that it was more expensive to create disease than to attempt to prevent it.[12]

The report was the culmination of an emerging movement in public health rather than the brainchild of any one reformer.[13] What made the report so radical was its denial of disease in fatalistic terms, and also its rejection of the view that poverty was the main cause of ill health.[14] It stated that ill health was a cause of poverty because disease had environmental roots. This was a forceful indictment of unsanitary living conditions in the industrial slums, as well as a severe criticism of physicians ignorant of the causes of contagion and of the moribund local health boards.[15]

The waves of cholera epidemics in England during the early nineteenth century underscored the powerful language of the report. In the late 1820s, many people accepted chronic dysentery and other endemic diseases as normal and disregarded warnings about the health problems mounting in the major cities. The cholera epidemic of 1831–32 changed all that, and reformers were taken more seriously. This first of several cholera epidemics to ravage Great Britain took 60,000 lives, many among the poor. Repeated cholera attacks struck the British Isles in 1848–49, 1854, and 1867. Tainted water was the medium of transmission for cholera. While its prevention was learned to be relatively simple, cholera terrified people because it hit suddenly and violently and was extremely contagious.[16]

English sanitarians of the period drew an immediate correlation between pollution and disease.[17] Bill Luckin has rightly characterized Chadwick's view as "proto-environmentalism," because it identified an environmental causation for disease, but without the understanding of the role of pathogenic organisms or a notion of pertinent ecological factors contributing to disease.[18] In addition, Chadwick was typical of the Benthamites and middle-class reformers who believed that "pauperism and disease were alike gratuitous and preventable."[19]

The so-called filth, or miasmatic, theory dominated the thinking of sanitarians until late into the century. Because disease was understood to arise from putrefying organic wastes, bad smells (miasmas), and sewer gases—and could not be transmitted from person to person—the filth theory is described as anticontagionist. Many of the diseases confronted were intestinal, thus environmental sanitation was credited with substantial success.[20]

Having provided a rudimentary environmental context for identifying the cause of disease in his report, Chadwick began to shape an administrative structure for implementing new methods of disease prevention. By emphasizing environmental aspects of hygiene, he envisioned doctors and other medical personnel extending their roles beyond specific treatment of sick individuals to a broader range of social action, especially inoculation programs and environmental sanitation. As a Benthamite, he concluded that the rights of the few were outweighed by the needs of the many. Local control of sanitary services, therefore, had to be supervised by a strong central authority and required the hiring of paid inspectors. This was precedent setting and predated modern civil service.[21]

For Chadwick, the appropriate response for dealing with unhealthy conditions was to be found in improved public works, including waterworks, sewers, paved streets, and ventilated buildings.[22] He proposed a hydraulic (or arterial-venous) system that would bring potable water into homes equipped with water closets, and then would carry effluent out to public sewer lines, ultimately to be deposited as "liquid manures" onto neighboring agricultural fields.

Chadwick was strongly influenced by John Roe, a railway and canal engineer who had been surveyor for the Holborn and Finsbury Sewers Commission. Roe introduced Chadwick to what one writer called "all of the evils of the sewers of the day." Roe took Chadwick down into the sewers themselves to show him the vermin and the crumbling brickwork of the old drains. One of Roe's solutions was to use a constant flow of water in concert with small egg-shaped sewers to increase the velocity of the flow and the sewer's carrying capacity. This concept fit well into Chadwick's arterial system, although it broke with the conventional thinking of the day.[23] With the addition of the fertilization phase, Chadwick noted, "we complete the circle, and realize the Egyptian type of eternity by bringing as it were the serpent's tail into the serpent's mouth."[24]

Many of Chadwick's contemporaries, including vested interests, claimed that his scheme was impractical. In the end, his comprehensive hydraulic system was not adopted. It fell to the increasing leverage of local authorities who were unwilling to allow the state to dictate the type of their public works programs, and to incrementalism because of budget limits or other constraints.

Chadwick's report and his attempts to implement the hydraulic system nonetheless marked a turning point in modern sanitary services. For the first time, four essential criteria for citywide service were united: a clear environmental context (health depended on sanitation), an administrative structure, a substantial technical response, and recognition of the need for breadth of service delivery. Not until the passage of the 1875 Public Health Law would these criteria converge.

The development of the new services occurred against a backdrop of important legislation in the mid- to late nineteenth century. The Public Health Act of 1848 was the culmination of Chadwick's sanitary work, although his influence on the movement itself had waned by the late 1840s. The act marked the first time that the British government took responsibility for protecting the health of its citizenry. Along with the Sanitary Act of 1866, the 1848 law was the departure point for developing modern legal machinery for dealing with sanitation.

In another sense, the 1848 act failed to meet Chadwick's own expectations about administering sanitation reform. Anticentralizers won concessions, since the act did not establish a national framework of local authorities. Also, boards were allowed (but not required) to appoint medical officers and were permitted (but not obliged) to undertake paving, sewerage, and water-supply programs. A relatively weak Central Board of Health was created, but its term initially was limited to five years.[25]

In some respects, Chadwick's own actions contributed to the success of the anticentralizers. The report set forth no clear idea on how central authority would be utilized to implement the sanitary plan, or how it was superior to local action. His first priority was to gain support for his analysis of public health conditions, then "there would follow as a matter of course an administrative machine that only Chadwick could properly direct." If this was his plan, it failed, and the anticentralizers won the day.[26]

Several pieces of legislation followed the 1848 act, many of which addressed the specific issues related to sanitary services, especially sewerage. The consummation of the work of the sanitary movement was the Public Health Act of 1875, which came two decades after Chadwick's withdrawal from public participation in the movement. Almost all earlier legislation was consolidated and extended in a remarkably comprehensive sanitary code. The new law was the broadest articulation of public health thinking prior to the emergence of bacteriology in the late nineteenth century, and provided impetus for the principles inherent in environmental sanitation.[27]

For more than a decade after the publication of the 1842 report, Chadwick's public life consisted of attempting to establish a new sanitary system, grappling with legislation, and generally trying to sustain his career. Christopher Hamlin has persuasively argued that the debate over Chadwick's arterial plan ended "not . . . with the triumph of Chadwick's system, but with an affirmation of the flexible, client-driven practice that characterized British engineering." In other words, it was the practice of British engineers to deal with problems defined by their clients—not to advocate their own technical visions—and to operate in a setting where there was no single technical solution.[28]

The debate over an integrated sanitary system, nonetheless, had a bearing on the development of English sanitary services and the diffusion of English public works concepts throughout the world. While an abundant, pure water supply was central to the Chadwickian system, primary attention turned to sewage disposal and utilization. Refuse disposal, more focused on public street cleaning and private responsibility for disposal, was never an integral part of Chadwick's sanitation plans. Garbage and other solid wastes were treated as nuisances rather than as health threats at this time.[29]

The principles of acquiring and distributing pure water had been essentially worked out earlier in the century. While few communities satisfied their water needs by 1850, the approaches for doing so seemed to be well understood. If the water-supply issue found a place in the debate over sewerage, it had to do with the changing circumstances for sewage disposal aggravated by increased water usage. The old system of sewage disposal simply was incapable of handling the amounts of water and effluent leaving the homes and businesses that had piped-in water.[30]

The Health of Towns Commission became the lightning rod for debate over sewage between 1842 and 1845, marking the beginning of the gradual adoption of the water-carriage system.[31] The problem of sewage flow had interested Chadwick primarily because existing public sewers and cesspools allowed surface water and what it carried to seep into the ground rather than depositing the material in some outfall. Also, Chadwick believed that the arterial system offered a way to capture the sewage for use on agricultural lands, providing possible revenue for a variety of urban improvements.[32]

Chadwick required the help of engineers to implement his system. Civil engineering had only recently emerged as a recognizable branch of engineering proper in Great Britain. What distinguished "civil" from other branches of engineering was the focus on infrastructure such as roads, bridges, and tunnels—and soon water-supply and wastewater systems. Commercial and industrial expansion of Great Britain in the eighteenth century stimulated the development of new projects outside of government. Until the 1750s, the state—particularly the military—was the major patron of engineering throughout Europe. Prior to the nineteenth century, however, British engineers engaged on "civilian" projects had begun to exchange ideas and develop a professional identity.[33]

Chadwick's and Roe's notions concerning the arterial system challenged engineering expertise. There were seven sewer commissions in Greater London (regarded by Chadwick, with some justification, as corrupt and inefficient). They were appointed bodies that often protected the interest of architects, builders, and surveyors who served on the commissions. They were also "quasi-judicial bodies" that administered the sewers in their districts. While the link between the commissions and civil engineers was tenuous, Chadwick criticized most engineers for not embracing his hydraulic system, as he had criticized physicians for not practicing preventive medicine.

The Metropolitan Sanitary Commission, established in 1847 to reform the sewer administration of London, became Chadwick's vehicle for attacking the more traditional engineers. His assault not only focused on the technical inferiority of his opponents' work but also on their failure to meet the moral obligations of sanitary improvement. He even abandoned some of his old allies, including Roe. Egg-shaped sewers were out, glazed earthenware pipes were in. Increasing the velocity of flow became Chadwick's chief obsession.

The 1849 Metropolitan Sewers Commission, also dominated by Chadwick, was established to build a system of sewers for Greater London. Roe was back in the fold by then, responsible for sewer-flow experiments. At the same time, the arterial system was under testing. But moving from design to construction proved difficult. Chadwick's engineers found it necessary to deviate from his principles to achieve a practical solution in construction. Chadwick's approach also was undertaken outside of London. The work was sanctioned by the General Board of Health, the body charged with building sanitary works and, according to Hamlin, "Chadwick's last stronghold." By 1852, when enough sewers were completed to evaluate the approach, the results were disappointing.

Chadwick returned to his proposal for strong administrative control, arguing that the way to keep the sewers clear was to educate the public on how to use the lines and to regularly inspect household connections. Most engineers and administrative bodies did not agree and refused to endorse his proposal. Hamlin argues that Chadwick tried to change the purpose of sewers "from removing surface and soil moisture to spiriting away wastes." While both functions were required, Chadwickians were "reluctant to acknowledge a need for separate sanitary and storm sewers."[34]

Criticizing Chadwick ignores the fact that the sewer-design debate was essential for determining how to cope with both liquid wastes and the increasing volume of water. Chadwick had made a strong case against viewing sewers only for drainage purposes, thus making sewerage a major issue in the battle for good sanitation. While his technical scheme, his assault on engineers, and his authoritarian approach worked against the free-flowing arterial system, he irreparably altered the perspective on the value of sewers. Nonetheless, an incremental approach to building sewer lines dominated public works practices for the remainder of the nineteenth century and into the twentieth. Chadwick's failure to complete a successful pilot program eroded his credibility. His plan's greatest shortcoming was its ambitiousness.

While most English towns eschewed the particulars of Chadwick's integrated approach, sewer development per se flourished as the era of water-carriage systems began by midcentury.[35] Water-carriage systems achieved importance especially after 1847, when Parliament gave local authorities power to discharge sewage directly into rivers or the sea.[36]

The spread of water-carriage systems demanded attention be given not only to drainage technology but also to outfalls, to the location of discharge points, and eventually to different forms of water pollution. Cities such as Birmingham and Manchester began to grapple with these issues relatively early.[37]

Following an epidemic of cholera, Parliament passed the Nuisance Removal Act of 1855, which established the Metropolitan Board of Works to develop an adequate sewerage system for London. Joseph William Bazalgette, chief engineer to the board, began the sewage project for London in 1859. The main drainage was virtually completed by 1865.[38]

Bazalgette had been a consulting engineer at Westminster, mainly involved with railways, but ultimately he became a staff member for London's Metropolitan Sewer Commission and soon was appointed engineer-in-chief.[39] Opposed to the Chadwickian system, Bazalgette proposed a series of main intercepting sewers, which ran east-west to catch discharges before they entered the Thames. The discharge would then be redirected into outfalls far downriver from the city. Pumps would be employed during high tide to force the sewage into the river.

The decision where to locate the outfall was controversial. While receiving support from the Metropolitan Board, Bazalgette's ambitious 1856 proposal was initially rejected. Two years later, the government reversed itself largely because of the so-called Great Stink of 1858. Hot weather and the use of thousands of water closets created an ungodly stench lasting two years, caused by putrefying sewage caught in the tidal reach of the river. Boat crews suffered from headaches and nausea, and sessions in Parliament were made bearable only by hanging sheets soaked in chloride of lime from each open window. Where engineering drawings and verbal persuasion had failed, an assault on the nostrils gave Bazalgette his victory. During the following two decades, approximately eighty-three miles of sewers were laid, draining 100 square miles of the city.[40]

Increasing use of water closets and implementation of the water-carriage system redirected pollution problems away from households and into rivers and streams. Concern over tainted water supplies of a generation earlier was revisited in Great Britain by the adoption of technologies meant to improve sanitary conditions in the cities. Piecemeal sewerage development left "end of the pipe" issues as an afterthought. The battles over sewerage did produce salutary results in reducing the death rates. A study of twelve large towns in Great Britain before and after the adoption of sewerage systems indicated a drop from 26 per 1,000 deaths to 17 per 1,000 deaths.[41]

The use of sewage as agricultural fertilizer never materialized to the extent that Chadwick envisioned. By 1880, about 100 towns had tried irrigation on sewage farms. The crops responding to repeated applications were limited, and land along the periphery of towns once only relegated to sewage irrigation often became more valuable for other purposes.[42] Several English farmers turned to new concentrated fertilizers, such as superphosphate or South American guano, rather than to the constant-flowing diluted sewage fertilizer that, as Hamlin stated, "kept coming whether it was needed or not."[43]

Not until the 1880s and 1890s did Victorians begin to understand sewage treatment as a biological process.[44] More often than not, sewage pollution became as much a political and jurisdictional issue as a technical problem. Pollution of streams and rivers was particularly critical in Great Britain because the land area of the country was so limited, the population along watercourses was dense, and sources of water were shared by several municipalities.[45] Discharging sewage into watercourses was cheap and practical for many towns that already invested the ratepayers' money in a water-carriage system. Once waste was flushed down a water closet or otherwise left a house through pipes, the interest of the citizenry dropped off sharply.

Concern over river pollution predated the commission's report by twenty-five or thirty years, but the most intense debates began in the 1850s. Luckin has drawn an interesting distinction between the debates over river pollution in London and those in northern industrial areas of England. The dominant economic and social structure in London at midcentury was nonindustrial. As a result, Londoners looked upon river pollution from the vantage point of commerce and consumption rather than production. The growing suburban ring also insulated the city from rural economic interests and values. Thus in London the most offensive source of pollution was human waste, not manufacturing waste. Attention turned primarily to the problem of water closets and household effluent. In the industrial North and the west Midlands, the struggle between the new industrial bourgeoisie and those in government wanting to limit the power of the manufacturing class focused on industrial pollution instead of human wastes.[46]

Because river pollution had national significance, inquiries, investigations, and new legislation were promoted to address the growing problem. A Royal Sewage Commission was appointed in 1857 to ascertain how to safeguard rivers and how to determine the best methods of disposing sewage. It stated that "the increasing pollution of the rivers and streams of the country is an evil of national importance, which urgently demands the application of remedial measures." The final report in 1865 recommended land treatment of sewage, but it was unenthusiastic about the profitability of sewage farming. The report asserted that towns causing pollution should cease to do so, and suggested that where cesspools were a health hazard, they should be replaced by a more modern approach. Such a conclusion did not effectively address the polluting capabilities of water-carriage systems.[47] The Rivers Pollution Prevention Act was passed eleven years later and became the basic water-pollution law for seventy-five years. The law contained several safeguards and reservations to protect industrial interests, however.[48]

The English experience with water-carriage systems had mixed results. Homes and commercial establishments with piped-in water acquired the means to discharge their effluent effectively. But the cost of this new inner-city efficiency was displaced pollution problems along almost every major watercourse, which threatened the purity of the water supply and aggravated relations between upstream and downstream communities.

Accompanying a cartoon of the filthy state of the Thames published in *Punch* in 1855 was a poem that read in part:

> King Thames was a rare old fellow,
> He lay in his bed of slime,
> And his face was disgustingly yellow,
> Except where 'twas black with slime.
> Hurrah! Hurrah! for the slush and slime![49]

A year earlier, the widely recorded Broad Street Pump episode took place. Dr. John Snow, a London physician, had been studying the causes of cholera. In 1849, he hypothesized that the disease was caused by an organic poison that could be discharged in human feces. If the infected feces entered the public water supply, an epidemic was sure to follow.

While investigating a severe outbreak of cholera near Broad Street, he learned that a workshop in the same area (which had its own well) reported no cases of the disease among its employees. This led Snow to seek out the polluted well. Simply by breaking the pump handle he ended the scourge. Snow had made the link between polluted water and epidemic disease.[50]

The government authorities learned from the cholera epidemics in the late 1840s and early 1850s. In 1852, Parliament passed the Metropolis Water Act, which required that all water drawn from the Thames (and other rivers supplying the city with water) must be filtered by January 1, 1856. In 1855, responding directly to Broad Street, municipal authorities in London required all water companies to supply filtered water. Within ten years, various British and European cities installed filters.[51] By the standards of the day, filtering water was the best way to ensure a "safe" water supply. Not until the 1880s and the advent of bacteriology would it become apparent that filtration, while valuable in combating many pollutants, could not in and of itself stave off waterborne disease.[52]

In grappling with efforts to secure a pure water supply and to provide effective sewage disposal, the English strongly influenced the development of sanitary systems. Improvements, such as water closets, pumps, sewage piping, and slow sand filters set the standard for the day. Legislative and court action established benchmarks or demonstrated the limits of governmental and judicial action in an era when laissez-faire notions were competing with demand for greater utilities regulation.

The development of new sanitary services in England in the mid-nineteenth century left an important legacy, one that profoundly influenced the United States. First, the correlation between the sanitary idea and the need for proper sanitation infrastructure and services became gospel. A pure water supply was not just a convenience, but a necessity for good health.

The second legacy was less definitive, but still significant: the manner in which new sanitary services were delivered. Chadwick's hydraulic system was clearly the most environmentally sophisticated notion of its time. Yet linking sanitary functions into a singular, closed system proved impractical. Many engineers were not comfortable with developing a system that fell outside their normal relationship with clients. They also were not convinced that Chadwick's system could work if implemented. Financial resources for such a massive construction project would have been extremely difficult to accumulate, let alone allocate.

Most significant, the kind of centralized authority that Chadwick envisioned for his system did not exist in a society possessing strong decentralizing tendencies, and where private companies played a significant role in the delivery of services. Because of the complex issues involved and the many vested interests, control and management of sanitary services in England were the product of a shared authority among local government, private companies, and Parliament.

Unfortunately, the failure to integrate or coordinate sanitary services may have delayed improvements in public health for some years.[53] At the very least, Chadwick's vision of bringing the serpent's tail into the serpent's mouth initiated a dialogue in which modern sanitary services could be evaluated. He helped to establish an environmental context for public health reform through the sanitary idea. He raised the important question of developing an administrative structure to implement change. And he linked the demand for a pure and abundant water supply with the necessity to devise a complementary wastewater evacuation system. While the English wrestled with converting the sanitary idea into a plan of action at home, they also helped to set in motion a sanitation revolution in other parts of the world, including the United States.

The "Sanitary Idea" Crosses the Atlantic

Beginning in the 1830s, urban growth in the United States and vague notions connecting waste with sickness led to several citywide technologies of sanitation, especially water-supply systems. In the following decade, the design and development of these systems were strongly influenced by the English "sanitary idea" pioneered by Edwin Chadwick. American cities underwent their first major sanitary awakening between 1830 and 1880, a change so profound as to establish a blueprint for environmental services for years to come.

The fundamental characteristics of the new technologies of sanitation were born in an era of miasmas. Sanitarians spread the word about environmental sanitation as essential to fighting epidemic disease. Civil engineers provided technical expertise for the design and construction of new systems. Private companies and local governments expended resources to implement the changes.

Modern sanitary services in the "Age of Miasmas" evolved against a backdrop of accelerating urban growth. Between 1830 and 1880, American cities rapidly multiplied, the rate of population growth soared, and urban physical plants grew upward and outward. The number of urban places with a popultion over 2,500 increased from 90 to 392 between 1830 and 1860. The concentration of cities in the Northeast continued, growth in the urban South lagged behind, but the western edge of the urban frontier pushed closer to California.[1]

The population of American cities grew by 552 percent to 6.2 million between 1830 and 1860—the fastest rate of urbanization the nation had ever experienced. Between 1820 and 1870, the urban population grew three times as fast as the national population. The enduring importance of commerce and the expansion of manufacturing were the major economic forces driving up the population of cities.[2]

While the total American urban population was small by the standards of western Europe, the United States was making significant strides by midcentury. In addition, American cities were becoming less like their European counterparts in physical appearance, with a tendency for greater suburban development. Most striking was the case of Philadelphia in the 1840s, where the central city grew at a modest 29.6 percent, but the suburbs expanded by almost 75 percent.[3]

Rapid growth and the proliferation of cities produced greater potential breeding grounds for disease and increased the need for improved health and sanitation measures. Subsequently, English sanitary reforms attracted a receptive audience in the United States. The sanitary idea was persuasive because it became easier to compare urban problems after 1830 than it had been in a previous era of limited urban development.

Charles E. Rosenberg, in his classic study *The Cholera Years*, recognized the transformation in American thinking about disease between the years 1832 and 1866. He observed that "cholera in 1866 was a social problem; in 1832, it had still been, to many Americans, a primarily moral dilemma. Disease had become a consequence of man's interaction with his environment; it was no longer an incident in a drama of moral choice and spiritual salvation."[4] Throughout much of the nineteenth century, it was typical to blame the poor, the infirm, or members of nonwhite races for the scourge of epidemic disease. Newly arriving immigrants raised the greatest fears, especially when they were crammed into filthy and dilapidated housing. Ironically, cholera—the "poor man's plague"—made victims of the very people accused of breeding the disease. In New York City, blacks and Irish immigrants were the most frequent casualties. In Philadelphia, the case rate among blacks was nearly twice as great as among whites.[5]

In southern cities, cholera was considered a race disease. In Richmond, Nashville, Atlanta, and other southern cities, cholera appeared first in black sections of town.[6] Local government in southern cities emphasized social cohesion as a major objective. Thus poor health conditions threatened all citizens, and public and private funds were intermingled in an attempt to develop an effective health-care system. Some historians have even argued that health and disease-control facilities were generally more advanced in southern cities. Disease was a "constant companion," since freezing temperatures that killed bacteria and viruses arrived so late. Nevertheless, fighting epidemics often was not successful, especially since an understanding of contagion was nonexistent.[7]

Yellow fever, unlike cholera, spared more black lives than white lives. People of West African extraction suffered least. In the great yellow fever epidemic of 1878, only 183 of 4,046 victims in New Orleans were black; in Memphis, only 946 of the more than 5,000 yellow fever deaths in the city came from the "colored population."[8]

While anticontagionism was eventually discredited, its widespread adoption in the nineteenth century was nonetheless a victory for empiricism and rationalism over sermonizing and moral outrage. Environmental sanitation appealed to simple logic and the senses, offering a way for people to participate directly in cleaning the cities, and ostensibly to eradicate disease. That it misrepresented the root cause of disease was a serious (and sometimes fatal) flaw, but its call for the removal of waste was a worthy objective.

The miasmatic theory "emerged into practical vitality" during the 1850s. Based on the apparent relation between filth and disease, the theory was crude at first, implying that organic decomposition per se caused disease. Eventually, filth was recognized as the medium for transmitting disease instead of the primary source of contagion. This provided a bridge to the later acceptance of the bacteriological—or germ—theory.[9]

While the germ theory was not firmly established until after 1880, the idea of contagion had been circulating at least since the sixteenth century. It was incorrect in detail until Louis Pasteur and Robert Koch clearly linked a specific organism with a specific disease. In 1871, an advocate of the theory was severely criticized in *Scientific American* for postulating that yellow fever was caused by a living organism. Only a few years later the germ theory was more widely accepted.[10]

Controversy over contagionism and major cholera pandemics in the mid-nineteenth century made the way easier for anticontagionism to win converts.[11] The emphasis on environment over personal habits of hygiene in Chadwick's report set the tone for the strategies to be employed in combating disease and improving sanitation for much of the remaining century. Few Americans, however, were inclined to share in its social views and had little taste for Chadwick's centralist ideas on government administration.[12]

Possibly the earliest graphic example of Chadwick's influence in the

United States was the 1845 study published by New York City inspector Dr. John H. Griscom, entitled *The Sanitary Condition of the Laboring Population of New York.* A New York City native, Griscom was a graduate of the University of Pennsylvania medical school and became a dispensary physician working with the poor. He had corresponded with Chadwick, and in the 1840s joined with other physicians to form the New York Academy of Medicine. He also was active in a wide range of social issues, including poor relief, prison reform, and immigration administration.

In his first report as city inspector in 1842, Griscom included a commentary with emphasis on the state of the poor. The Board of Aldermen was so displeased with Griscom's characterization of the city's sanitary condition that the aldermen chose not to reappoint him. Undaunted, he expanded the commentary into a book with a title reminiscent of Chadwick's report.[13] The first in-depth study of health problems in New York City, *The Sanitary Condition* ranged over many topics. It was infused with the environmental view of disease, and closer than most Americans to accepting Chadwick's social perspective.[14] Little immediately came of Griscom's efforts, however. New York City was infamous as an extraordinarily unhealthy city with problems so complex and political webs so entangled that one report could not change entrenched practices or reverse long-standing policies of neglect.[15]

Lemuel Shattuck, who also corresponded with Chadwick and was a friend of Griscom's, was more successful in gaining attention for the emerging public health movement in the United States. Shattuck, however, placed little stock in the social justification for sanitary reform. Through much of his life, he consistently exhibited an intense nativism that held immigrants responsible for the spread of disease and poverty.[16]

Born in 1793 in Massachusetts, Shattuck was raised in a farming community in New Hampshire. There his religious views were shaped by the Second Great Awakening, which imbued him with a piousness born out of the precariousness of life. With a strong sense of order and an eye for detail, he gained a reputation as a good amateur historian and genealogist. He eventually settled in Boston, where he attempted a career as a book publisher. His work in genealogy led him to a fascination with statistics, and in 1839 he was one of the founders of the American Statistical Association. He also served on the Boston City Council in 1837 and in the General Court in the following year.

At Shattuck's urging, the legislature passed a law in 1842 requiring the registration of births, deaths, and marriages. In 1845, he prepared the first comprehensive urban census of the century, which included an important sanitary survey of the city. Among the most disturbing data, Shattuck believed, was the fact that one-third of Boston's population was foreign-born or the children of foreign-born, and that the conditions under which they lived were detrimental to the city as a whole. He concluded that the municipal government should be compelled to remove every potential cause of disease.[17]

In 1849, Shattuck again was elected to the General Court, and as chair of a committee for the study of health and sanitary problems in the Commonwealth, he was the driving force behind a plan for a state sanitary survey.[18] The *Report of the Sanitary Commission of Massachusetts, 1850* reviewed public health progress in Europe, outlined the history of sanitation in the state, and called for a comprehensive public health administration. Lacking the vivid imagery and emotional punch of Chadwick's work, it focused on relationships that could be measured statistically. It also reinforced Shattuck's view that immigrants were primarily responsible for the degradation of the cities and the spread of disease. The conclusion stated that since a large portion of the population was not abiding by proper sanitary principles, the state must assume the responsibility to assure the public health.[19]

Despite the underlying assumptions, the report was a remarkable pioneering document, especially since no state or federal public health programs existed. To administer public health practices, it called for state and local boards and a system of sanitary inspectors.

The definitions of public health were stretched broadly to include sanitation programs for towns and buildings, public bathhouses and washhouses, maritime quarantine, vaccination against smallpox, promotion of health in infancy and childhood, and the control of food adulteration. The concept of public health also incorporated disparate issues such as control of smoke nuisances, control of intoxicating beverages, supervision of those with mental disease, the erection of model tenements, and emphasis on town planning. Public health was linked closely to the field of medicine through a call for the establishment of training schools for nurses, the teaching of sanitary science in medical schools, and the inclusion of preventive medicine in clinical practice.[20]

Whether Shattuck intended it or not, the report addressed many of the social issues of concern to Chadwick, Griscom, and others. It also strongly reflected the emphasis on prevention rather than cure of disease, which was at the heart of the filth theory. Despite its seminal nature, Shattuck's report failed to push the legislature to immediate action. It was simply too big a pie to eat at one sitting. It recommended fundamental changes in the way public health was confronted, and called for a vast commitment of will and dollars to address problems so often ignored.[21]

As many scholars concede, the political circumstances of the 1850s are important to understanding the response to public health reform. In Massachusetts, the Protestant elite that had controlled state politics for generations was having to confront large-scale Catholic immigration, and thus was "more concerned with voting behavior than hygienic habits."[22] On the national level, slavery in particular, but also women's rights, temperance, and prison reform were vying for public attention. From a business standpoint, proposals for extensive environmental sanitation threatened tenement landlords who loathed the idea of housing reform, and raised the specter of large private and public assessments for construction of new water systems, sewers, and drains.[23]

Yet Shattuck's plan was not summarily dismissed but merely delayed. As early as 1848, the American Medical Association (AMA) appointed a committee on public hygiene, which took particular note of several of Shattuck's objectives. But when the Massachusetts General Court failed to act on the report, momentum for public health reform in the AMA dimmed. Prior to the Civil War, Chadwick's, Griscom's, and Shattuck's ideas were kept alive in a series of sanitary conventions from 1857 to 1860, with the majority of the attendees being physicians. Interest in public health within the medical profession—with a shift in emphasis from quarantining to environmental sanitation—was rejuvenated by the conventions. Had not the Civil War intervened, the conventions may have been on the brink of effectively lobbying for a national public health organization.[24]

Several practical improvements in the public health field grew out of experiences on the battlefield, helping to make way for more formal governmental responses to the ideas generated at the sanitary conventions and in the Griscom and Shattuck reports. The reformist U.S. Sanitary Commission—an outgrowth of the Women's Central Association of Relief for the Sick and Wounded of the Army—pressured the army to become a semiofficial auxiliary, and pushed for good public health practices in installations in the North and in the military occupation zones in the South. Thousands of working-class and middle-class women looked to the health needs of soldiers in the army hospitals and in the field.[25]

By 1865, the sanitary idea was being assimilated into several public institutions. In that year, the Citizens' Association of New York investigated sanitary conditions in the city, echoing the conclusions of both Griscom and Shattuck. The following year the New York legislature passed the New York Metropolitan Health Law, which established a Metropolitan Board of Health with wide powers for New York City and Brooklyn. This proved to be the first effective health department in a major city, and quickly became a model for other municipal boards of health.[26]

Previous local boards had been notoriously political and ineffective. Without power of their own, they had to rely on the police to enforce local ordinances, which was rarely a high priority. In some cases, board members took payoffs to distribute contracts for waste removal or street maintenance.[27] However, sanitary reformers had higher hopes for the Metropolitan

Board. In some respects, the rise of sanitary science in America in this period helped to restore the sagging image of physicians in the wake of several devastating epidemics.[28]

The creation of the Massachusetts State Board of Health in 1869 clearly established the notion of the modern public health institution. There was a nineteen-year gap between the issuance of Shattuck's report and the establishment of the Massachusetts board, but in the intervening years the state maintained a reputation as a place of relatively good health. Dr. Edward Jarvis became a critical force in the public health community by collecting and analyzing data. Like Shattuck, he linked morality and personal health, asserting the primacy of personal responsibility in maintaining moral and physical order. To Jarvis, the role of the state was essentially that of a broker in protecting and promoting public health.

Through the work of Jarvis and others, support grew for sanitary programs based on the filth theory. The creation of the Board of Health coincided with the efforts of the Massachusetts legislature to reorganize state government after the Civil War. The board met for the first time in September 1869 and attempted to create effective local boards. In light of tepid local response, other issues were explored, including housing, condition of the poor, hygiene education, methods of slaughtering animals, sale of poisons, and investigation of various diseases. Changes did not come easy, but the board survived various onslaughts and helped to sustain the enthusiasm of sanitary reformers.[29]

The incorporation of the sanitary idea into American health institutions was well established by the 1870s, and the primacy of the filth theory was reaching its zenith. Despite the difficulties encountered by both New York and Massachusetts, the trend in larger cities was toward permanent boards of health. Chicago and Milwaukee established boards in 1867, Louisville in 1870, Indianapolis in 1872, and Boston in 1873.[30]

Professionalization of the sanitary community also was improved by the formation of the American Public Health Association (APHA) in 1872, an outgrowth of the prewar sanitary conventions.[31] The most important national expression of the spread of sanitary reform in the United States was the creation of the National Board of Health in 1879. By 1873, the federal government had two military agencies engaged in health work, the Marine Hospital Service and the Army Medical Corps. They were a source of research on epidemics and provided statistical studies, but their charge was not extensive enough to constitute a national health organization. The spread of sanitary reform activities brought attention to public health issues, but nothing was as dramatic as an epidemic to raise awareness of such matters. In 1873, a cholera epidemic in the Mississippi Valley killed approximately 3,000 people, and a yellow fever epidemic racked the cities of New

Orleans and Memphis. Soon after the passage of a national quarantine law in 1878, the South was struck with the worst yellow fever epidemic in the nation's history.[32]

Such dramatic events gave primacy to the idea of a National Board of Health, despite the widespread preoccupation with severe economic dislocations caused by the Panic of 1873. While the National Board lasted only four years, many of the services it provided were continued through other bodies. A precedent was established for the federal government to take some responsibility for the nation's health, as the Public Health Act had done in England in 1848.[33] Statistics indicated that sanitation methods were having the desired result. Between 1860 and 1880, the death rate in all but a few cities had fallen from approximately 25 to 40 per 1,000 to 16 to 26 per 1,000.[34]

The emerging profession of civil engineering played a dominant role in promoting and implementing environmental sanitation programs, particularly the new technologies of sanitation. As historian Terry Reynolds has stated, civil engineering was reaching "a vigorous adolescence" in the early to mid-nineteenth century through its role in constructing canals and railroads, and building much of the urban infrastructure. A new breed of domestically schooled engineers became available to cities, alongside the imported European engineers and American engineers who had received their training on the job.

Between 1820 and 1860, the U.S. Military Academy at West Point produced an increasingly larger percentage of the civil engineers working throughout the country. Since the "Point" graduated more engineers than the army could use for several years after 1820, many left the military for the civilian world. In pre–Civil War America, West Point may have trained as many as 15 percent of all civil engineers.[35]

The first civilian engineering school was founded at Norwich University in Vermont in 1820. Originally intended to train officers for the state militia, Norwich began offering engineering studies in 1825 and instituted a formal three-year course of study by 1834. In 1835, Rensselaer Polytechnic Institute in New York began a one-year course of study in civil engineering, and in 1850 a three-year program was created. Other schools attempted to develop engineering programs with little success, but after 1850 permanent programs appeared at the University of Michigan, Harvard, Yale, Dartmouth, Union College, and elsewhere.

The process of gaining professional respectability and standing for civil engineers was as gradual as the effort to develop a homegrown educational system. Because there were so few engineers in the United States before 1860, and because the country was so large, regional and national professional organizations took several years to take root. Political cleavages and rivalries, and differences in engineering philosophies, also kept the fledgling profession diffuse. In 1867, the American Society of Civil Engineers (ASCE) became the first successful national professional engineering society in the United States. Between 1850 and 1880, the number of engineers grew from 512 (including mechanical and civil) to 8,261 (civil only). Among these were numerous consultant engineers, who sold their expertise to clients across the nation.[36]

Changes in education and professional association contributed to the American "style" of engineering, which "placed more emphasis on reducing labor costs and on economy of construction than the parent traditions had and placed less emphasis on strength, permanency, aesthetic appeal and safety."[37] This definition, while rightly stressing the Americanization of engineering, is overly general. As will be seen in later chapters, an emphasis on "permanence" was at the very heart of the construction of new water and wastewater systems. American engineers also continued to borrow heavily from their European counterparts, although they were obliged to adapt to the conditions existing in their own cities.[38]

City politics was undergoing substantial change in the mid-nineteenth century, creating an uneasy setting for the advocates of good sanitation practices. Many governments continued to promote economic activity, with local business leaders in key positions of authority. Such governments showed a service preference for downtowns and elite residential neighborhoods.[39] With the vast influx of immigrants between the 1850s and 1890s, and the inability of cities to govern themselves effectively, some neighborhood groups, citywide machines, and state legislatures asserted their influence.[40]

The first ward machines, which developed in New York City in the 1850s, were bent on cultivating political power through neighborhood loyalty. In 1866, the archetype of city machines—the Tweed Ring—took control of Gotham. Since power was a commodity to be bought and sold, favors were traded for votes. Patronage was a favorite means of acquiring supporters, and public works, in particular, offered many such opportunities (for example, garbage collectors, fire inspectors, and waterworks supervisors), along with the issuing of permits, franchises, contracts, and licenses.

Amid the political shifts in city government in these years, state legislatures also found ways to control local services or usurp local authority not captured by ward or city politicians. Yet the result of the political transformation of the cities was not simply a deviation in priorities for service from one class to another or complete neglect for infrastructure development.[41]

As in any period, supporters of the new technologies of sanitation had to understand political realities to achieve their goals. Not all major cities produced well-oiled political machines. As urban historians have argued, city machines were often "fragile coalitions" or "makeshift alliances" that suffered internal conflict or were weakened by opposing parties and factions.[42]

Any municipal government calling itself reformist or otherwise was likely to use patronage or favoritism in some fashion in allotting contracts and franchises—if less likely to create jobs for supporters in the overt manner of the political machine. The increasing cohort of professionals employed in city government often assured more subtle approaches to negotiating the politics of city service implementation.[43]

The sanitary community in general, civil engineers included, embraced the anticontagionist views embedded in the sanitary idea to frame their understanding of municipal health needs and the type of services to deliver. They also used these views to persuade political leaders to support the objectives of environmental sanitation. Of course, environmental sanitation was an overly simplistic tool for raising the city above its current state of unhealthfulness and for assuring plentiful and safe quantities of water and efficient methods of waste disposal. Yet it placed responsibility for public improvement in human hands. That fact gave it great power and influence in reformist circles, and its imprint on modern sanitary services became indelible.

Pure and Plentiful

FROM PROTOSYSTEMS TO MODERN WATERWORKS, 1830–1880

Early in the nineteenth century, a few water-supply protosystems began to appear in major American cities. Philadelphia's project set the standard for these protosystems, although it did not spark an immediate national trend. By 1880, some water supplies were evolving into modern citywide systems. Not only did they deliver greater quantities of water over a larger area, but they also included rudimentary safeguards to ensure purity. A growing preoccupation with water quality—a direct result of the sanitary movement—was bringing attention to filtration techniques and new methods of water treatment. City leaders and sanitarians alike were demanding more from their water-supply service than convenience at the tap.

In absolute terms, the number of waterworks multiplied at an increasingly accelerated rate from 1830 to 1880 (see Table 4.1). During the 1850s and 1860s, however, the number did not keep pace with the chartering of new cities. Urban population increased at a faster rate than the number of waterworks until 1870, when the trend began to reverse itself.

Table 4.1. Percentage of American Cities with Waterworks

Year	Number of Works	Number of Cities*	Cities with Works (%)
1830	45	90	50
1840	65	131	50
1850	84	236	36
1860	137	392	35
1870	244	663	37
1880	599	939	64

*With 2,500 or more in population.
Source: U.S. Bureau of Census, *Census of Population: 1960*, vol. 1, *Characteristics of the Population* (Washington, D.C.: Department of Commerce, 1961), pt. A, 1-14–15, table 8; Earle Lytton Waterman, *Elements of Water Supply Engineering* (New York: Wiley and Sons, 1934), 6.

Momentum for developing new water-supply systems built gradually. Some communities experiencing modest growth continued to rely on wells and other local supplies or expected private companies under franchises to provide water service. Yet even cities undergoing rapid expansion were often leery of the capital investment required for citywide systems. By the 1870s, the trend toward more public water supplies was evident. There was a shift from private to public ownership in the period (9 of 45 public, 36 of 45 private in 1830) with relative parity by 1880 (293 of 599 public, 306 of 599 private).[1]

The crucialness of adequate supplies of water to meet the needs of citizens, commercial establishments, and industry—and the emerging mandate of cities to protect the public health—meant that authorities in the largest urban areas wanted centralized systems under their direct control. Boosterism was an additional motivation, since an effective water system was a powerful promotional tool to enhance a city's economic base. While many water companies had been profitable, capital investment in the more modern systems was steep, and operating costs were on the rise. Private service, therefore, was gradually phased out in several communities. In addition, public control of the water supply enhanced the authority of city government vis-à-vis the legislature or rival cities; thus private owners often were under pressure to sell out.

The desire of city leaders to convert private systems into public systems, or to build new public systems, rested on more than the will to do so. The central issue was the ability of cities to incur debt to fund major projects and to sustain the high costs of operating the new technologies of sanitation. As the nineteenth century unfolded, city finances underwent changes in scope and complexity that ultimately made the development of public sanitary systems achievable.

The impetus was blunted temporarily by the Panic of 1873, when retrenchment and conservative fiscal policy brought on a "pay as you go" philosophy. Also, until 1875, waterworks franchises were designed to be attractive in order to induce private companies to deliver water and to provide an adequate number of hydrants for fire protection.[2]

The power of taxation was central to a municipality's budgetary process. In cities throughout the United States, ownership rather than income became the basis for taxation. Fee simple land tenure set the pattern in a society that embraced private enterprise as a way of conducting business. Among other things, this view led away from assessments on business revenues to more general taxes. Before 1870, citizens expected that municipal services would be paid for on a "cost of service" basis without profit to the city. As services expanded, increases in taxes became inevitable.

The general property tax emerged as the basic mechanism for raising revenue. Unlike in England, where local taxation was based on the rental value of property, in the United States total property wealth became the standard. While the property tax was the most important source of local income by midcentury, taxing powers of the cities usually were controlled by the states. In some cases, the state imposed additional responsibilities on city government with no additional financial support. For example, epidemics of the period led cities to request state aid, but without much success.

Especially in the case of capital improvements like water-supply systems and sewerage, special assessments were an important financial tool to augment property taxes. As the range of city services expanded, demand for more revenue increased. The requirements of urban growth meant that reliance on taxes and special assessments was likely to satisfy neither the citizenry nor the local business establishment. Increased municipal debt became more common in the nineteenth century.

By 1870, the popular referendum was a tool for obtaining public support for bond issues. Many cities went deep into debt in the mid- to late nineteenth century to fund municipal improvements, but also to attract business. Most significant, methods of taxation and the allocation of municipal funds prevalent by 1870 set the pattern for urban finance for more than a century.[3]

Incentive to utilize new powers of taxation and debt grew out of the evolving role of municipal government in the nineteenth century, with an emphasis on greater service delivery. Major property holders were the primary supporters, and government most often catered to them. Also important was the booster role of local government. Municipalities increasingly sought funds for several public works projects and infrastructure—including canals, railroads, bridges, roads, and ports—and services such as public health, police, and fire protection. In the case of public health, the timing of these projects coincided with the advent of the sanitary idea and the growing faith in environmental sanitation.

The administrative structure of many preindustrial cities, built around city councils and weak mayors, rarely provided efficient and effective leadership. The councils had multiple duties, including serving as a lower court, but their power was often limited. On the one hand, in the ward system of representation, factionalism could undermine the cohesiveness of the council. On the other, rural-dominated legislatures were reticent to give cities expanded powers.[4]

Machine rule made it particularly difficult to develop large, capital-intensive projects. Ward bosses spent considerable time dealing with the daily concerns of their partisans. In the 1870s, for example, the Republican machine in Philadelphia left the water department with few funds for maintenance or expansion, since profits from the department, along with rake-offs from public works contracts, were used to run the machine.[5] Others used the water department directly to generate graft, such as the Tweed Ring, which required merchants to buy useless water meters for almost four times their value.[6]

In developing citywide water-supply systems in this era, substantial public investment proved difficult for all but the largest and most fiscally sound cities. If the legislature was not withholding extension of greater authority, the council was debating the wisdom of increasing the city's bonded indebtedness or was engaged in partisan debate. Going back at least to 1855, the percentage of public water systems tended to vary with the general financial health of the cities, at least until the 1880s, when other issues also influenced the decisions.[7] In addition, municipal indebtedness had steadily grown in order to finance several improvements, including water supply. By 1860, municipal debt was three times the federal debt and almost equal to the aggregate state debt.

Liberalization of charters and other fiscal changes also provided an opportunity for cities to finance water-supply systems and other public works, especially beginning in the 1860s. In most cases, a combination of local circumstances and the experience of other cities influenced the shift from private to public.[8] An 1834 act that created the Board of Water Commissioners provided for the legal and administrative tools that resulted in the completion of the Croton Aqueduct and Reservoir—a major step to alleviate New York City's water-supply needs—in 1842. This laid the foundation for the first workable municipal system in the city.[9]

Boston also suffered under many years of water-supply politics before it developed a system in the 1840s. In 1796, a General Court act created the Aqueduct Corporation, which constructed a line from Jamaica Pond in Roxbury to the city. The system was extended in 1803, but there was no further attempt to improve it until 1825. From 1825 (a year in which the city suffered a great fire) until 1846 (a period in which several epidemics rocked the city), civic leaders were embroiled in debate over the water supply.

The city council began inquiries that produced several plans for a new supply. Factions in the council continually skirmished over a favored approach, as private water companies looked on with grave concern about the growing mood for a public supply. Those in favor of a municipal supply were persistent, and a water referendum in December 1844 resulted in a major victory for proponents of a municipal system. Long Pond (later renamed Lake Cochituate) was selected as the city's source of supply, to be purchased at city expense. The city, however, did not gain the legal authority to establish a municipal supply until 1846. In that year the General Court passed Boston's Water Act, which provided for the development of Long Pond. The city was authorized to finance construction with municipal bonds.[10]

In October 1848, the Cochituate Aqueduct opened, ushering in an era of municipal control of water supply in Boston. The completion of the Cochituate system also changed the focus of water-supply debates. It placed monitoring of the water system in the hands of experts, under whose influence Boston was favored by avoiding future water shortages and escaping the devastation of epidemic disease. The success of the Boston system not only elevated the stature of technical experts but also reinforced the faith in environmental sanitation.[11]

While the Baltimore Water Company did not suffer by comparison with other works, it faced public criticism throughout the 1830s and 1840s. Extension of service was limited by profitability, typical in most cities. Citizens in outlying areas or in poorer districts around Baltimore were excluded from service under the company's conservative policy of extension. Continued population growth also spurred demands for public service, but the city did not acquire the Baltimore Water Works until 1854.[12]

The first waterworks in Chicago was not established until 1840, under the auspices of the Chicago City Hydraulic Company. It built the city's first pumping station and reservoir, and Lake Michigan was the primary source of water. The distribution lines only reached a small portion of the southern and western divisions, while four-fifths of the city's residents continued to obtain their water from the polluted Chicago River or from water carriers. With a cholera epidemic in 1852, city officials assumed control of the system. Since the epidemic was believed to have originated with the wells, the Lake Michigan supply gained greater significance.[13]

The St. Louis Water Works was built in the 1830s. In 1823, the mayor began promoting the idea for a citywide system, and in 1829 the city council offered a cash prize for the best plan. Within a short time, St. Louis officials agreed to a contract with Wilson and Company. The work began in 1830, but "no water flowed through the pipes until the 1840s."[14]

Outside of a core of emerging major cities in the industrial East and a few others sprinkled throughout the country, the transition from private companies to municipal service more typically occurred in the late 1860s and after. For example, in Buffalo, New York, the Jubilee Spring Water Company had distributed water through log pipes as early as 1826, but the city did not establish a municipal system until 1868.[15]

Milwaukee's first recognized waterworks was built in 1840 for the United States Hotel, while most citizens received their water from local springs and wells. In response to citizen pressure, the Common Council authorized the issuance of bonds in 1857 to finance a waterworks. The project was never completed, and a second attempt was sidetracked by the Civil War. Serious progress on a waterworks did not commence until 1868.[16]

In the South, municipal water systems were rare in this period. In Reconstruction-era Atlanta, plans moved ahead for a new waterworks, but its primary thrust was for fire protection and to serve business and industrial needs. Without a creditable municipal water supply, the more affluent turned to purchasing spring water or depended on personal wells. In black neighborhoods, drainage was poor, sewer outfalls often dumped wastes there, and wells were badly polluted. Likewise, in Memphis little attention was given to residential water service.[17]

While changes in the administration of waterworks were slowly evolving, far less subtle changes were occurring in water-supply technology. New sources of supply became necessary when old sources could no longer meet demand or became severely polluted. The only viable alternatives were digging new wells; pumping water from nearby lakes, rivers, and streams; seeking distant sources; or filtration (not until the 1870s). Good location was a significant advantage for cities forced to change or augment their water supplies. Filtration (and treatment), however, eventually helped to defy the limits of location.

For Chicago, location offered new sources in close proximity to population centers. When the town was founded in 1833, the water of the sluggish Chicago River was considered pure, with variation in quality from season to season. Water also was drawn from shallow wells. In the 1850s, as the Chicago River became an open sewer, the public water supply was pumped from an inlet basin on Lake Michigan, a distance of 3,000 feet from the mouth of the river. Lake Michigan offered a magnificent alternative as a water source, extending over 22,400 square miles.

As the city grew, and as the lake water close to shore became increasingly polluted, the intake pipe was moved farther out and deeper into the lake. In 1863, the Common Council approved a plan to construct a two-mile tunnel burrowed under the lake bottom connected to a new intake. The project proved to be much more difficult than anyone imagined. Despite the arduous task, this first lake tunnel only supplied the needs of the city until 1871. After the fire of that year, a new tunnel and pumping station needed to be built.[18]

Most major cities growing as rapidly as Chicago did not have the advantage of such a convenient water source, and thus consideration had to be given to distant sources. But like Chicago, they would need to confront skyrocketing capital costs and the sheer scale and complexity of the engineering task required to develop a new supply.

The Old Croton Aqueduct (1842) is regarded as a sublime engineering feat, and as a symbol of the conquest of nature in service to the urban population explosion. The Croton Aqueduct project is also an important example of changes in the scale and complexity of modern water-supply systems. Several attempts to solve New York City's water problem failed, but in 1835 the fortunes of the city changed. Citizens, frustrated by the poor state of well water and frightened by a recent outbreak of cholera, were ready to support a new plan. In a rare moment of political harmony, the voters, the state legislature, and the New York Common Council agreed to construct a forty-one-mile aqueduct from the Croton River in Westchester County to New York City.

The Croton won out because the source was large and could be delivered without pumps. Even with the savings, the aqueduct was expensive. The task of building the aqueduct was first entrusted to engineer Major David Bates Douglass. However, he lacked experience with large public works, especially one that required building a variety of structures: a dam, an enclosed masonry conduit, bridges and embankments, and a huge reservoir. The water commissioners replaced Douglass with John B. Jervis. A self-trained engineer with vast work experience, Jervis served as a supervising engineer on a portion of the Erie Canal (1823) and as chief engineer of the Delaware and Hudson Canal (1827).[19]

Jervis confronted the task of building what was to be the largest modern aqueduct in the world with great aplomb. Innovative design techniques were employed so that the aqueduct could remain operational across a variety of terrains and could withstand the winter cold. To maintain a uniform grade, the aqueduct ran through tunnels dug into hills and was carried by bridges constructed over ravines and streams. When the aqueduct was opened on July 4, 1842, it safely carried 75 million gallons daily. By 1860, the Croton Aqueduct was delivering its maximum, and it was pushed to provide as much as 105 million gallons daily before a new line was built.[20]

New York's need to supplement water supplied by the Croton Aqueduct became apparent in the 1870s. During droughts and in cold winter months, more water was consumed than was received, requiring the drawing of extra water from other sources in the city. Refilling the reservoirs took a great deal of time, while millions of gallons of water ran over the Croton Dam, simply unavailable for use.[21]

Notwithstanding its great overall success, the Croton system was not built without technical difficulties, input from special interests, contract irregularities, resistance from upstate citizens, and discrimination in service delivery. Construction work was sometimes haphazard, as when water mains were run through sewer lines. While aggregate water production increased dramatically, availability of supplies tended to favor the middle class over the poor.[22]

Major projects in a few other cities followed the construction of the Croton Aqueduct. Boston completed its aqueduct in 1848, another major engineering feat. Much of the Cochituate Aqueduct ran through deep trenches covered with dirt, sending water to a twenty-acre reservoir, then to two distributing reservoirs. Construction of the Washington Aqueduct, built to supply the nation's capital with water from the Great Falls of the Potomac (fourteen miles from the city), began in 1857 and was completed in December 1863.[23]

Distant sources of supply received great attention because they offered large and dependable quantities of water, but also because they provided alternatives to polluted or infected sources in the local area. Smaller communities were hard-pressed to seek distant sources, however. For many, the lack of options for dealing with polluted water supplies was the weakest link in the early systems. The transformation of protosystems into modern waterworks required methods for ensuring or improving water quality. The introduction of filtration and new techniques for water distribution held some promise for accomplishing that goal.

Complicating the search for pure water was the fact that determining what constituted a tainted supply was little understood in the Age of Miasmas. Taste and smell substituted for scientific testing in most assessments of water quality. Some physicians warned patients not to drink hard water or water with vegetable and animal matter in it, fearing that it would harm the kidneys or produce stomach and intestinal maladies. In 1873, the president of the New York Board of Health, a chemistry professor at Columbia University, advocated the consumption of lake or river water, stating that "although rivers are the great natural sewers, and receive the drainage of towns and cities, the natural process of purification, in most cases, destroys the offensive bodies derived from sewage, and renders them harmless."[24]

John Snow's research on waterborne transmission of disease inspired Dr. William Budd in his studies of typhoid fever. Like Snow, Budd determined that typhoid was spread through water supplies contaminated with human feces. Of the possible waterborne diseases that threatened American cities, typhoid fever was the worst. The typhoid bacillus could be contracted by direct contact or through contaminated food such as milk, raw fruits, vegetables fertilized with night soil, and shellfish found in polluted waters. Most often it was spread when excreta from a victim entered the water supply directly or as untreated sewage. Detecting it posed a problem because the incubation period was approximately fourteen days. Not only was the disease a threat to human life, but it could also severely damage the reputation of a city trying to attract new citizens and new business enterprises.[25]

Cities that used nearby river and lake water generally had the highest incidences of yellow fever, while those dependent on more distant sources usually fared much better. Since the organism was not discovered until 1880, statistical information prior to that time is fragmentary.[26]

By the turn of the century, various approaches to assure a pure water supply and a reduction in waterborne disease emerged from bacteriological and chemical laboratories. But the first means of water purification readily available to cities in the late nineteenth century—one that fit within the framework of the filth theory—was filtration through sand or gravel to improve water clarity, odor, and color.

Albert Stein, designer of Richmond's waterworks, was the first to attempt to filter a public water supply in the United States in 1832. Pumping water from the James River, Stein prepared a sand filter in the reservoir, but he could not get it to operate effectively. During the next forty years, several major cities, including Boston, Cincinnati, and Philadelphia, considered installing sand filters, but they were too expensive at the time.[27]

A major step forward, but not recognized immediately, was the "Report on the Filtration of River Water, for the Supply in Europe, made to the Board of Water Commissioners of the City of St. Louis" (1869) by Brooklyn engineer James P. Kirkwood. In 1865, Kirkwood had recommended that St. Louis and Cincinnati employ filters in their water systems. However, nothing came of it at the time. Later, he was hired by the city of St. Louis to survey locations for supply works along the Mississippi River. Upon recommendation of a plan that included filtration, the water commissioners instructed him to travel to Europe to obtain some firsthand knowledge of the technology. While away, opposition to Kirkwood's plan mounted, which led to a clean sweep of the commission and replacement with members unwilling to underwrite the cost of filtration. The city would not publish his report and did not filter its water until fifty years later.[28]

Kirkwood's report ultimately became a bible for those cities interested in copying the European experiments. For several years following its completion, little additional firsthand knowledge was gathered about the various European systems. By the early 1870s, a few cities began to recognize the value of filtering water. Poughkeepsie, New York, built the first American slow sand filter in 1870–72 based on the designs in Kirkwood's report.[29]

By 1880, there were only three slow sand filters in the United States and none in Canada. The Europeans, to the contrary, forged ahead with several slow sand filters, and in Buenos Aires in the 1880s experiments were conducted on the suitability of various filtering materials. Information about the experiments in filtration and other data about water supplies now were disseminated more effectively through the published proceedings of the American Water Works Association (1881) and the New England Water Works Association (1882). Other engineering societies and public health organizations also added to the rich body of data becoming available.

Along with experiments in filtration, a variety of pumping techniques and changes in pipe technology helped to transform older protosystems into modern, centralized waterworks. Aside from gravity systems, steam pumps were increasingly employed at the source and as a way of moving water to reservoirs, tanks, and standpipes.[30] Wooden pipe was adequate in low-pressure gravity systems but could not withstand the action of high-pressure pumping engines. By 1850, iron pipe came into wider use in the United States, especially in high-pressure systems. Some water utilities, however, continued to use wooden pipe until the 1930s. In the West, wood was used for large aqueducts, irrigation, hydroelectric plants, and hydraulic mining.

At first, iron pipe was not readily available through American manufacturers. The first pipes had to be imported from England. Even after domestic producers made pipe available, cost was not competitive, and some technical problems had to be overcome. But cast-iron pipe became more widespread as prices continued to drop, and the reduction in cost became a major factor in extending distribution systems after 1870.[31]

For all the improvements begun by private companies through municipal franchises, accessibility to water supply was still largely linked to class. Affluent neighborhoods and the central business district received the most water, while the working-class districts often relied on polluted wells and other potentially unhealthy local sources. As Sam Bass Warner Jr. astutely observed about a later period, "The mode of construction of water-supply and sewerage systems divides the responsibility between municipal capital on the one side and the individual installations of middle-class homeowners and home builders for the middle-class market on the other."[32] This observation applies to the pre-1880 period of private water companies as well, insofar as those outside of the middle and upper classes were unable to tap into the water supply, which at that point was available only to a limited market.[33]

While modern waterworks would flourish in the late nineteenth and early twentieth centuries, their basic form and function were established by 1880. Major cities began to devise financial plans based on enhanced revenue generation and long-term debt to plan construction and maintenance of new systems or to secure old systems from private companies. Sources of supply were no longer limited to local wells, ponds, and streams. Distribution extended over wider areas, due in part at least to the use of iron pipe and a variety of pumping techniques. And a concern for water quality led to research on filters and, in some cases, to their implementation. All of these changes occurred in the Age of Miasmas, which placed a pure and plentiful water supply squarely at the heart of environmental sanitation.

Subterranean Networks

WASTEWATER SYSTEMS AS WORKS IN PROGRESS, 1830-1880

In contrast with strides made in waterworks, the development of underground wastewater systems was meager between 1830 and 1880. Noted sanitary engineer William Paul Gerhard observed that progress in sewerage had been much slower than water supply. "This can be, in a measure, explained by the fact that taxpayers are nearly always willing to pay a small annual tax for water," he argued, "and hence the financial success of such a scheme is rarely in doubt, whereas a sewerage system does not yield an annual revenue, but, on the contrary, causes sometimes large operating expenses." "It is," he concluded, ". . . a much more difficult matter to induce communities to introduce a sewer system."[1]

Gerhard recognized what historian Sam Bass Warner Jr. has called a "first things first" philosophy, in which basic improvements in sewage disposal were being forced to wait "until the water supply problems were solved."[2] Few cities were in a position to finance two major technologies of sanitation simultaneously. Private companies had footed the bill for the early development of water-supply systems, sometimes with governmental support, but a similar path for sewerage systems was unlikely because its revenue-generating potential was limited.

Eventually, underground sewers came to be recognized as valuable in staving off epidemics, preventing flooding, and making available connections to water closets. But the initial public support for them was sluggish. Privy vaults and cesspools had been relatively effective and inexpensive disposal options until piped-in water was available. Open ditches sufficed as storm drains in communities not yet experiencing rapid expansion. In New York City, many property holders in the 1850s resisted connecting to sewers where they were available, since the law did not compel them to do so. Landlords of the poor were unwilling to make connections on their property, ready to let their tenants endure the stench of the privies and cesspools.[3]

In essence, many people failed to see the advantage of a system that evacuated something as unwanted as sewage when other methods of disposal were available. The Chadwickian notion of bringing the serpent's tail into the serpent's mouth had no more advocates in the United States than in England. With such inertia to overcome, the "pre-sewer" era carried on into the late nineteenth century. Even in places where some underground conduits and extensive surface-drainage systems existed, ordinances often prohibited the disposal of fecal matter into sewers.[4]

Expanding urban populations and piped-in water began to challenge the old methods of sewage disposal as early as the 1830s. As in England, water closets appeared in some middle-class homes soon after running water became available. Since the early market was small, the initial impact on sewage-disposal habits was minimal. For example, New York City had only about 10,000 water closets for a population of 630,000 in 1856. In 1864, Boston (180,000) had only 14,000, and in 1874 Buffalo (125,000) had only 3,000.

As running water became more common, the cesspool–privy vault systems began to fail. In 1880, approximately one-third of all urban households had water closets, and water-consumption rates were rapidly increasing. The greater volume of water used in homes, businesses, and industrial plants flooded cesspools and privy vaults, inundated yards and lots, and posed a major health hazard.[5]

The breakdown of the old methods came as a result of a clash between incompatible technical systems. Privy vaults and cesspools could not contend with a water-delivery system that increased volume so dramatically. It was the environmental implications of this clash of technologies that provided momentum for change. Flooding problems, and especially threats to health, were directly traceable to the breakdown of the pre-sewer systems. Yards inundated with wastes became new battlegrounds for programs of environmental sanitation.

Because of high costs and planning necessary to construct and maintain citywide sewerage systems, immediate change did not occur. Public health officials and engineers had been working diligently since as early as the 1850s to condition city officials for what they regarded as an essential change in the approach for dealing with wastes. They argued strenuously that although underground sewerage systems were capital intensive, the average cost over the long term would be less than the annual costs of collecting and disposing of wastes from the cesspool–privy vault system.[6]

Although Americans were not quick to accept the idea of an integrated water/wastewater system, it was widely agreed that sewers could have a major role in improving urban sanitation. The principal debate focused on combined (as opposed to separate) sewers—an issue that the English had wrestled with since the 1840s.[7] The controversy came to a head in the United States in the 1880s, but earlier skirmishes demonstrated how the commitment to environmental sanitation framed the discussion concerning alternative approaches to sewer design. Joel A. Tarr has argued persuasively, "The decision between types of sewers would appear to be one of simple engineering design. . . . Such a model, based mainly on rudimentary cost-benefit calculations, was primarily utilized in making the choice for sewerage. No such model, however, existed for choice of design."[8] Part of the reason for the lack of a model was that underground sewerage systems were new concepts at midcentury. Engineers were just beginning to work out basic functions of the system. Sanitarians struggled to determine the potential health benefits.

Combined systems came first. They handled both household waste and stormwater in a single large pipe. A newer technology, the separate system, often used two pipes—a small one for household waste and a larger one for stormwater. In some separate systems, stormwater was simply diverted into street gutters.[9]

American engineers contended with adapting systems that had been effective in Europe, but now needed to function in a different setting. For example, variations in rainfall patterns in England and the United States were important in the design of combined sewers. Rain usually was frequent but not particularly intense in England, and thus stormwater drains could be smaller than in areas of the United States where rainfall was more torrential. Lacking the natural cleansing action of a heavy downpour, English systems required more frequent manual flushing.[10]

The first planned systems in the United States (although not citywide) appeared from the late 1850s through the 1870s. Only combined systems were successfully built in the United States in the 1860s and early 1870s, largely due to cost and the lack of a successfully operating separate system.[11]

In 1857, Brooklyn began construction on the first effective planned sewerage system in the country designed to remove sanitary wastes and stormwater. James P. Kirkwood, waterworks engineer for the city at the time, hired civil engineer Julius W. Adams (known for his work on the Cochituate Aqueduct) to prepare plans. Kirkwood had previously collected information on the drainage of cities, and corresponded with several English engineers. In a report to the Board of Water Commissioners, Kirkwood recommended interconnecting the water and sewerage systems. On the question of health, he noted the long-term deleterious effects of wastes from privy vaults and cesspools percolating into the subsoil and tainting wells, which "doubtless have been long offensive to health before they have become distinctly perceptible or offensive to the taste."[12]

Armed with Kirkwood's findings, Adams decided to build a pipe system, but opted for large intercepting sewers to avoid polluting the tidewater area. The use of interceptors clearly reflected the thinking of Chadwick's rival Joseph William Bazalgette, who employed them in London.[13] The first great interceptor sewer project to follow Brooklyn began in 1876 in Boston. Although Boston lifted the ban against the discharge of excreta into any public sewer in 1833, the city continued to maintain a strict prohibition against the connection of houses to storm drains. However, clandestine connections were made in a number of instances, and all kinds of wastes flowed into the storm drains. Beyond the city limits, outlying communities chronically complained about Boston's effluent pouring into their rivers and marshes.[14]

The decision to construct the intercepting sewer was based on a report that established the basis for the future metropolitan sewerage system of Boston and adjacent communities. It called for the consolidation of more than 100 discharge sewers into one outlet in Boston Harbor, believing like "the more intelligent of our community" that "our high death-rates are connected more or less directly with the defects and evils of our sewerage-system."[15] The report also stressed the importance of adapting the new system to "the wants of a growing city." This was particularly important because the commissioners viewed the previous piecemeal efforts at drainage and sewerage not to have fulfilled that goal.[16]

In earlier years, the contours of Boston had changed due to the efforts to reclaim and fill tidal areas bordering the city limits. As the reclamation proceeded, it was necessary to extend the old sewers, whose outlets otherwise would have been cut off. But this process occurred unsystematically, and the extension lines often were laid without plan by corporations responsible for the reclamation. Constructed of wood or some inferior materials, they were prone to clogging.[17]

The interceptor was designed and partly built under the guidance of

Joseph P. Davis, who had learned about sewerage from Ellis S. Chesbrough and James P. Kirkwood and who ultimately became city engineer of Boston. But the attempt to establish a commission in 1872 to report on a comprehensive plan of sewerage was thwarted by city leaders who believed that expenses for the investigation should be shared by the neighboring cities and towns. At the urging of the City Board of Health, a commission was selected in 1876. The proposed plan attracted opposition from Boston's superintendent of sewers and several leading merchants. The superintendent did not want to support a plan that blamed his sewers for increased death rates, and he agreed with the merchants that the plan was too costly. The general public did not share the criticisms, and by 1884 the core system was completed.[18]

Chicago constructed the premiere sewerage system of the time, capturing the imagination of the public health and engineering communities in much the same way that Philadelphia had done with its water-supply system. For some time, drainage in Chicago had been poor; aqueducts served only the downtown area, and those were inadequate.[19]

Ellis Chesbrough was the chief architect of the new system. Widely known as the builder of Chicago's first water tunnel, he became equally known for his sewerage work. Born in 1813 in Maryland, his early training was in railroad engineering. In 1851, he became city engineer of Boston, and in 1855 he accepted the position of chief engineer of the Chicago Sewerage Commission. In 1856–57, he traveled to Europe for the commission to study sewerage systems. In his capacity as chief engineer of the new Board of Public Works (1861), Chesbrough recommended the building of a tunnel (1867) under the lake to an intake two miles from shore. He also planned an enlarged water system, which included the tunnel, the Chicago Avenue pumping station, and the Old Water Tower. In 1872, the year after Chicago's Great Fire, work began on a new lake tunnel and intake for the West Side. In January 1879, Chesbrough was appointed to the new office of commissioner of public works, which he held only for four months before turning to consulting.[20]

Chicago's new water and sewerage systems were impressive. The flat terrain of the city rose only slightly above the elevation of the Chicago River and Lake Michigan, and the soil and altered surfaces were so nonporous that absorption was virtually impossible and drainage confounding. The lake was a logical endpoint for runoff, but also was the greatest source of potable water.[21] The scale of the problem of sewering Chicago became evident as the city struggled to address its growing health concerns. In 1854, cholera took one out of every eighteen Chicagoans. The cholera epidemic and severe problems with dysentery led the Illinois legislature to establish the Chicago

Board of Sewerage Commissioners in the following year. Chesbrough issued a report in 1855, restating suggestions made in several previous proposals and noting health and sanitation problems facing other cities.

Chesbrough's report amounted to the first comprehensive sewerage plan for a major American city. He recommended draining wastewater into Lake Michigan via the Chicago River. He recognized that the plan created potential health hazards and obstructions to navigation by making the waterway shallower, but alternative proposals seemed to present even greater risks. Despite the numerous challenges, Chesbrough believed that he had recommended a workable plan, and so did the commission.[22] The construction of the system began in 1859, and within a decade 152 miles of sewers were completed. Chesbrough, however, did not get everything he wanted, such as two flushing conduits to ensure adequate dilution of sewage in order to protect the water supply.[23]

A unique feature of Chesbrough's plan was to "raise the city," that is, to elevate the level of the city as much as twelve feet in some areas to allow sufficient grade for proper drainage. During the construction phase, this plan gave the city "a whimsical, humpty-dumpty look to it" for about ten years. At the time, more attention was focused on the sewer outlets than the costly but successful plan to raise city buildings.[24]

Chicago employed an intercepting sewer constructed on the combined sewer model. As the lines approached the river, the streets were raised beneath the sewers, the sewers were covered with dirt, and then the new streets were paved over. Vacant lots were filled, and framed buildings were raised to the new level or torn down. Chesbrough's plan also called for dredging the Chicago River in order for it to receive the larger sewage load.[25]

Chesbrough's initial system had limits. The pollution load in the Chicago River increased rapidly, due to the sewage plan itself, population growth, and the increase in the number of packing houses, distilleries, and other businesses. Pollution in Lake Michigan also increased to the point where complaints about water quality were chronic.[26] Demands for the intake to be placed farther from shore seemed to offer the most immediate relief, but did not correct the sewage disposal problem.

From a fiscal perspective, the new system did not ensure participation of all of Chicago's citizenry, nor did it lead to rapid expansion. The sewerage board could not levy the equivalent of water rents as a permanent funding mechanism for the system, and had to rely on general funds. Under such circumstances, the sewer system competed directly with other city services for funding. Some building owners, in addition, were chary about connecting to the system because it would require spending more money for plumbing fixtures.[27]

Political and economic circumstances also affected the expansion of the sewer system in New York City.[28] The wealthiest sections of the city were not always the first to obtain new sewers, in large measure because they did not suffer the kinds of drainage problems found in slums or the neighborhoods of the poor. (This is not to suggest that sewer service favored the underclasses. It was common for sewer outfalls to terminate in poor or minority neighborhoods.)[29] Also, political machines tended to fund services to meet their own interests. For example, while the Tweed Ring made millions through kickbacks, inflated construction costs, and other ploys, it also grew wealthy on real estate deals. In areas such as suburban Harlem—where Tweed and others invested heavily—substantial networks of water mains and sewer lines could be found.[30]

At this time, disposal of sewage by dilution was the prevailing practice, that is, disposing of wastes in large bodies of running water. Sewage irrigation and chemical precipitation were well established, and intermittent sand filtration was being studied as early as 1868. But none of these methods was widely practiced in the United States before the 1880s. The Massachusetts Board of Health anticipated some of the potential pollution problems in the 1870s, and began to study relevant disposal techniques in Europe. To many engineers and others, the scale of pollution caused by the disposal of sewage was not sufficiently large in the United States to demand more than dilution.[31]

The ad hoc decisions that Chesbrough was forced to make in balancing the needs for a high-quality water supply and an effective sewerage system were leading Chicago to the development of the first major integrated sanitary system in the country. While the process was incremental, the complementary nature of the water-supply and sewerage systems was unmistakable. Few people at the time, however, recognized systems integration as a response to the needs of other cities and towns.

Eventually, a clearer connection would be made between the desire for a pure water supply and practices of waste disposal that did not create secondary pollution problems. Before 1880, sewage disposal by an underground system of pipes and complementary surface drains was only beginning to catch on. Exactly what constituted a "waste problem" was still strenuously debated.

The health principles behind the practice of environmental sanitation seemed to make clear the need to evacuate liquid wastes as quickly as possible from homes and businesses, but what to do with those wastes once they reached the end of the pipe was not always clear. And what about refuse? Was there a connection between liquid and solid wastes? To what degree was refuse a health risk? And were environmental sanitation techniques needed to eradicate cities of them if they indeed posed a threat?

On the eve of the bacteriological revolution, new technologies of sanitation had been devised. Influenced strongly by the commitment to environmental sanitation, they were still works in progress. Urbanites were becoming aware of their value, but people were still ignorant of the full potential of these systems—both as safeguards of the public health and as sources of new environmental challenges.

II

THE BACTERIOLOGICAL REVOLUTION

1880–1945

On the Cusp of the New Public Health

BACTERIOLOGY, ENVIRONMENTAL SANITATION, AND THE QUEST FOR PERMANENCE, 1880-1920

By the late nineteenth century, faith in environmental sanitation as the primary weapon against disease lost followers, especially in the medical community. Bacteriology placed more emphasis on finding cures for disease as opposed to prevention, which had been the mainstay of sanitary reform since the 1840s. Nonetheless, the commitment to develop elaborate urban infrastructure for water and wastewater services—and eventually refuse disposal—continued unabated. By 1920, many American cities could boast about plentiful sources of pure water and efficient methods of waste disposal to a greater degree than ever before.

Why did the movement for environmental sanitation persevere into the twentieth century and how, if at all, did it change? Environmental sanitation endured because: rapid urban growth increased pressure on municipal officials for pure water and adequate disposal services; Progressive reform promoted a sound physical environment to impose a "civilizing influence" on urbanites through management of city services; and more responsive city governments emerged in an era of home rule to set local objectives.

The role of environmental sanitation changed because bacteriology modified the priorities of the public health community by discrediting the community's ability to combat epidemic disease. The simple objective of preventing illness by removing waste or providing water that appeared pure was replaced by greater attention to biological pollutants. Primary responsibility for administering sanitary services shifted to—or at least was shared with—municipal engineers. Bacteriological laboratories took up the task of ferreting out biological pollutants; sanitary engineers focused on the operation of the technologies of sanitation.

In the "Era of Bacteriology," these forces produced water-supply, wastewater, and solid-waste systems that increasingly relied upon centralized organizational structures and capital-intensive technical innovations. The systems continued to provide water, evacuate wastes, and maintain municipal sanitation, but with greater attention to several pollution risks. The basic form and function of the water-supply and underground wastewater systems were well established by 1880, while comparable refuse collection and disposal systems only began to take shape after 1880. The prevailing goal in the late nineteenth and early twentieth centuries in major cities was to create more comprehensive public citywide systems that afforded permanent solutions to the delivery of sanitary services. In this way, specific project design took precedence over long-term urban planning. Between 1880 and 1920, therefore, immediate needs essentially were projected forward.

Sanitary services were part of "the technologically networked city," which emerged in the nineteenth century in Europe and in the United States. Despite variations, "most large and medium-size cities in North America and Western Europe had equipped themselves with an infrastructure of pipes, tracks, and wires" made possible by industrialism, which furnished "a range of materials and techniques not previously available."[1]

The technological transformation of American cities occurred against a backdrop of rapid increases in the number of cities and population growth. The number of urban centers increased from 939 to 2,262 between 1880 and 1910, and the number of cities with populations over 100,000 increased from nineteen to fifty in the same time period.[2] In 1880, approximately 14 million people were urbanites. By 1920, the number increased to more than 54 million, marking the first time in U.S. history that more than half of all Americans lived in cities or towns. The proliferation in urban dwellers ran well ahead of total population by an average of 18.3 percent per decade in these years. Immigration from southern and eastern Europe, as well as rural to urban migration, accounted for much of the increase.

Urban boundaries extended outward. It was not uncommon for some metropolitan areas to spread over 20 square miles or more. Improved transportation, extensive service delivery, and aggressive real estate development influenced growth. Also important was adjusting borders through annexation or consolidation. In the most celebrated consolidation of the period, Greater New York was formed in 1897 out of Manhattan, Brooklyn, Long Island City, Richmond County, most of Queens, some of Westchester, and parts of Kings County. Chicago increased its area through annexation from 36 to 170 square miles in 1899.[3]

Industrialization came to be closely identified with urban growth, especially in the Northeast and along the Great Lakes, where the most dynamic city development took place. By 1900, urban factories were responsible for 90 percent of American industrial output. Cities in a band from New York to Chicago became central places for amassing resources and labor, for developing extensive transportation and communication systems, and for opening new markets. Nine of the ten largest cities in the country were in the manufacturing belt.[4]

Coming to grips with rapid physical growth, a changing and expanding population base, and dynamic economic forces were daunting tasks for city leaders. Such a transformation inspired new political and bureaucratic responses. By the late nineteenth century, challenges to machine authority came from several sides, normally under the banner of reform.[5] "Rapid urbanization and unprecedented industrialization had produced a city of ill-fitting parts," historian Jon Teaford observes, "but many who enjoyed the comfort, security, and self-confidence associated with middle- or upper-class status believed that they could refashion the city, creating a better, less divided urban community."[6]

Urban political reformers attacked the boss system, questioning its efficiency and charging it with gross corruption. Reform groups sought to promote nonpartisanship in elections by changing the pattern of representation from wards to citywide systems, and initiating civil service reforms and noncompetitive contracts. These groups also feared the alleged political influence of immigrants and the working class. The boss system represented to them the control of the vast numbers of working-class voters and the slighting of the cities' "better sorts." However, much of what we know of the boss system grew out of reform rhetoric, and recent research reveals more complex political arrangements than the old bosses versus reformers dichotomy.[7]

Urban environmental reform in this era was broader in scope than the earlier sanitation movement, largely because Progressivism cast a bigger net. The reformers shared the view that a good society was "efficient, organized, cohesive."[8] This suggests a belief in progress through order, but also through control of human action via an improved environment. Unlike British welfare statism, sanitary reform was more appealing to Progressives because it cut across class lines and promised universal benefits.[9]

Rapid and large-scale urban growth threatened not only to promote new rounds of epidemic disease but also to further degrade the air, water, and land. Those interested in urban environmental reform openly defended city life, because they believed that cities were worth preserving. Progressives sought to eliminate physical threats to the city and its people without remaking the urban environment. Reformers were well organized and viewed pollution and health hazards as problems affecting the whole city.

Two distinctive but not totally independent groups promoted urban environmentalism in the period. One was composed of professionals and quasi-professionals with technical and scientific skills working within municipal and state bureaucracies. They were public health officials, sanitarians, efficiency experts, and most especially engineers. Their main functions were developing systems to combat disease and pollution, compiling vital statistics, and monitoring community health and sanitation. They transmitted their ideas directly to municipal decision makers and through professional organizations, but were less successful in communicating with the public.

A second group consisted of citizens with strong civic and aesthetic values who usually operated outside city government, generating influence through organized protest, community programs, and public education. Lacking the expertise or power to implement most changes themselves, they supported those in government who were like-minded. Civic environmentalists came primarily from voluntary citizens' associations, reform clubs, and civic organizations and from environmental pressure groups, such as smoke- and noise-abatement leagues and sanitation groups whose interest in city problems was more specialized. The membership of all of these civic bodies came generally from the middle and upper middle classes, but sometimes from the working classes and among minorities, and women were well represented. Civic organizations increased from fewer than 50 in early 1894 to more than 180 by the end of the year. By 1909, there were more than a hundred periodicals dealing primarily with urban affairs.

Because several Progressive reforms were rooted in industrial cities, urban environmentalists had little trouble identifying with the prevailing reform spirit of the day. In some cases, environmental reformers were urban protesters who accepted the Progressive ideology. In other cases, those who called themselves Progressives took an interest in urban environmental issues. Both shared a desire to bring order out of the chaos induced by the Industrial Revolution and an environmental determinism that led them to expect that the good in people would prevail if the evils produced by imperfect surroundings were eliminated. They also placed their faith in an expert elite to solve society's problems. Moralistic and often paternalistic, they decried poverty, injustice, corruption, and disease, but did not favor giving ethnic and racial minorities a direct role in government. Progressivism offered a hospitable framework for many favoring environmental reform and provided a set of concepts and values applicable to the fight against pollution and disease. Yet Progressives' advocacy of environmental reform was often constrained by less enthusiasm for social welfare programs that weakened their ability to direct political and social change.[10]

Central to the reformist spirit was the notion that urban problems were environmentally based. One vantage point could be found in the question, "How can we best create or foster the spirit of civic pride?" A contemporary answered, "The real solution of the problem lies in the education of the individual citizen to a higher standard of civic living and civic housekeeping."[11]

A clean, orderly environment would constrain undesirable traits, so that socially responsible people "would appear from beneath the vice of depravity." If the externalities were tolerable, problems like poverty would take care of themselves.[12] In some respects, the tenets of a burgeoning Progressivism in the cities were not far removed from Chadwick's notion of an earlier era, which linked poverty and disease. While hardly an ecological perspective in the modern sense, Progressivism promoted a broad strategy for civic revitalization that not only sought social uplift but also the preservation of urban life and the restoration of more pristine physical standards.

The urban bureaucracy underwent substantial change in the late nineteenth century, making it more responsive to developing citywide sanitary services. Well before the implementation of formal civil service laws in later years, "mayors and commissioners deferred to the judgment and expertise of professional engineers, landscape architects, educators, physicians, and fire chiefs, and a number of such figures served decade after decade in municipal posts, despite political upheavals in the executive and legislative branches."[13] The incorporation of experts into city politics made Progressivism possible.[14]

City leaders could better take advantage of a growing professional bureaucracy in a setting that shifted power away from the state capitol to city hall. Beginning in the 1870s, several cities made efforts to limit state interference in their affairs by demanding more home rule. The movement took many forms, including efforts to increase the appointive power of mayors and to gain control of various services. At first, the granting of home-rule authority was relatively selective. For example, Texas cut the number of its special acts dealing with cities from seventy-one in 1873 to ten in 1874.

Home rule proved viable in states with large cities. In some cases, the cities demonstrated political clout that they could wield at the state level. Denver was granted some home-rule powers in 1889, but this was a rubber stamp for powers the city had already accrued. In a quite different case, Louisiana granted statutory home rule to all cities in the state in 1896, except New Orleans.[15]

By the end of the nineteenth century, the reform efforts made for a conducive political setting for change. Municipal reformers made home rule a priority and gained adherents on the state level, leading to the end of many restrictive, specifically designated powers. Fiscal issues also were influential in promoting the first constitutional home-rule provisions. Deflation in the 1870s increased the burden of debt for many cities, making them unwilling or unable to assume additional financial or service responsibilities dictated by state legislators. Missouri and California were the first states to develop plans for constitutional home rule, followed by others. While the movement ebbed during World War I, it picked up momentum afterward.[16]

Greater home rule did not ensure political and financial stability for cities, but it did allow some latitude in setting local priorities. Municipal expenditures began to rise steadily as early as the 1850s—a reflection of growing demand for a range of services. Increase in the level of per capita expenditures was rapid because demand was price inelastic, that is, the cost of providing services increased while the demand remained extensive.[17]

Despite the economic turbulence of the 1870s, expenditures for many services were high. While the 1880s were a relatively prosperous time, such expenditures were comparatively low, but by the 1890s they generally began rising again. This fluctuation may be attributed to the serious debt burden on cities resulting from a 34 percent drop in the general price level in the 1870s. Assessed valuation of property—the basis for property taxes—also fell.[18]

Taken as a whole, there was a substantial increase in municipal expenditures between 1900 and 1920. Operating costs increased more than capital costs. In the early years of the twentieth century, local, non-school-related expenditures alone were greater than the budget of the federal government.[19] With greater expenditures and additional provisions for home rule, city leaders acquired more discretion in service delivery. There were some jurisdictional disputes to contend with, and there were noneconomic restraints to service choices. The reform mood of the Progressive Era "was reflected in greater checks on honesty, but also in a disposition to enforce fiscal responsibility and at the same time to allow the cities more and more financial leeway to develop their functions."[20]

The emergence of bacteriology brought a sea change in the way the world viewed health and disease. Its impact on sanitary services was subtle at first, not so much undermining the value of environmental sanitation as giving it a narrower context within a broadening health field. In some cases, advocates of the germ theory borrowed ideas and methods directly from the application of the miasmatic theory.[21] Ultimately, bacteriology challenged the very core of anticontagionist beliefs and brought into question the proper role of water-supply, wastewater, and refuse-disposal systems in the health of cities.

The germ theory—or the bacteriological theory—of disease was well established as a scientific fact by the 1880s. Only twenty years earlier, several theories competed with the emerging germ theory in an attempt to elevate contagionist views in its competition with anticontagionism. The germ theory had "an attractive simplicity in comparison with other theories" and was bolstered by information on fungal diseases in animals and humans that urged comparison with it. But progress toward an answer to disease causation was slowed by the lack of sufficient technical expertise in confronting microorganisms, especially bacteria.[22]

Few studies had been conducted based on direct examination of airborne particles. Prevailing arguments revived an older notion of unseen or invisible germs. Joseph Lister appeared to be the first to correctly identify bacteria as the cause of airborne contamination. The word "germ"—and sometimes "fungi"—remained in common use to describe bacteria, but despite the imprecision in terms, a modern contagionist theory was taking shape.[23]

The 1880s became the most fertile years for bacteriological discoveries through the work of Louis Pasteur and Robert Koch. Anthrax was the first disease proved to be caused by a microorganism, and Pasteur's earlier studies on anthrax provided the empircal base. Prior to 1885, however, little had been accomplished in the United States in applying the principles of bacteriology, and the germ theory continued to encounter strong resistance.[24]

Sanitarians and public health officials eventually embraced bacteriology as the primary tool for combating communicable disease, but the shift from anticontagionist to contagionist views was not immediate. Greater effort was made to separate sanitation/preventive medicine from hygiene by distinguishing between attention given to the environment versus a concern for the health of the individual.[25]

Before Chadwick, filth had been equated with nuisance, but the sanitary idea elevated environmental problems to the causative factor in disease. In the emerging era of bacteriology, filth lost some of its importance as a health hazard. "Smells are not dangerous," claimed engineer Ernest McCullough. "They are rarely, if ever, a cause of illness. They are simply unpleasant to our modern sense of what is right and proper." McCullough allowed that too much filth is always a "distinct menace," but "the danger to cities comes from the herding together of poor people who must live near their places of employment."[26]

Dr. Charles V. Chapin of Providence, Rhode Island, became an ardent spokesman for the "New Public Health." In a 1907 speech, he stated, "When I was elected health officer in 1884 municipal sanitation was synonymous with municipal cleanliness. . . . But the filth theory is dead, and we know dirt is very rarely the *direct* cause of sickness." He added that most sanitation programs "can be secured as well, or better, through other agencies, and the health officer should be free to devote more energy to those things which he alone can do."[27]

Over the years, sanitarians had done such an effective selling job that common sense seemed to dictate that environmental sanitation had merit as a preventive measure against disease despite the contagionist critics. And advocates of environmental sanitation continued to defend it aggressively. Sanitarian George C. Whipple stated in 1925 that "there is a tendency on the part of some health officials to belittle sanitation, to emphasize the infections in their personal relations, and to ignore the importance of environmental conditions apart from infections. From the standpoint of disease they are right; from the standpoint of health, the speaker thinks they are wrong."[28]

In the era of the New Public Health, a compromise of sorts was reached, whereby environmental sanitation assumed a complementary role with bacteriology in the maintenance of a community's health, but not the primary weapon in combating disease.[29] Support for bacteriology as the primary tool in disease eradication is best seen in the transformation of the public health profession and the health institutions themselves. Volunteer efforts were less valued as more physicians and research scientists became closely identified with the New Public Health. Physicians had always played an important role in public health, but with the advent of bacteriology their influence increased dramatically. In 1900, physicians accounted for 63 percent of all professionally trained health workers in the United States. Almost all of the early presidents and officers of the APHA were also active in the AMA. To the contrary, not all medical societies were active in public health reform.[30]

Local boards of health reflected a growing commitment to public responsibility for the prevention and eradication of disease in the late nineteenth century. In 1890, 276 of 292 cities with 10,000 or more population who reported to the U.S. Census had regular boards. While the impact of the new theory of disease was being felt at the local level, attention still was given to sanitary matters. By World War I, health departments significantly turned away from environmental sanitation, leaving it to engineering and public works departments.[31]

The effectiveness of boards of health in sanitation was further limited by distrust of the boards, especially among immigrants and laborers, and chronic problems with inequitable enforcement of the sanitary codes. New immigrants hid their sick, resisted health inspectors entering their homes, and protested forced vaccinations. Racial and ethnic prejudice also might mean overt attention to health hazards in lower-class neighborhoods, or, conversely, outright neglect of health threats, particularly in the segregated South.[32]

The bastion of the new theory was the bacteriological laboratory. The first diagnostic lab was established in Providence in 1888, tracing typhoid to the city water supply. Founded in 1886, the Lawrence Experiment Station of the Massachusetts Board of Health was intended to conduct chemical analysis of water. However, the association of drinking water with typhoid quickly led the Lawrence Station to concentrate on bacteriological problems.[33]

Throughout the period, public health remained a local and sometimes a state concern. National leadership in the field was modest until World War I. Before then, the Department of the Interior had taken some interest in the sanitary condition of schools; the Department of Agriculture had a food laboratory; the Surgeon General's office maintained a medical library; and the Census Bureau kept vital statistics. The largest single federal office devoted to public health was the Marine Hospital Service, which had assumed the responsibilities of the short-lived National Board of Health. More extensive federal involvement did not begin until the codification of some programs under the U.S. Public Health Service in 1912. The greatest test came with the influenza pandemic that spread across the globe in 1918–19.[34]

With the New Public Health, responsibility for programs in sanitation fell to municipal engineers. They had become not only the technical arm of the sanitary reform movement but also the foremost proponents of environmental sanitation. The engineering profession experienced rapid growth, and in 1900 engineering was second in number only to teaching among the professions. There were 45,000 engineers in that year and 230,000 by 1930. Engineers in increasing numbers were being employed by municipal governments.[35]

Engineers in the employ of cities were chief among the technocrat elite who built and managed the new urban infrastructure and who served alongside the newly emerging bureaucratic class of permanent city employees. David Noble has suggested that engineers tried to present themselves as "'technology' itself, the great motive force of modern civilization."[36]

Such characterizations capture the powerful faith in "progress through technology" advanced by engineers and shared by many reformers. Engineering progressivism "expected society to be saved by a technical elite."[37] Municipal engineers also acquired managerial responsibilities that made them more aware of political, fiscal, and social pressures in the administering of cities.[38] The engineers' new status was expressed in the following observation in *Scientific American* (1918): "For we live in the age of the engineer. He may be defined as the man who does things, as against the man who merely knows things."[39]

Without question, municipal engineers became central figures in the effort to manage—if not protect—the environment in the late-nineteenth- and early-twentieth-century American city. The descriptive title "sanitary engineer" came to represent what much of municipal engineering was becoming by the early twentieth century. Sanitary engineering was "one of the new social professions which is neither that of physician, nor engineer, nor educator, but smacks of all three."[40] Sanitary engineer William Paul Gerhard boldly noted that "much of the sanitary engineer's work is necessarily of a missionary character, as the public must be educated to appreciate the benefits of sanitation."[41]

The new profession was unique because it represented the only group that possessed a relatively broad knowledge of the urban ecosystem at the time. Sanitary engineers had a grasp of engineering expertise as well as current public health theory and practices. Some acquired these skills on the job. Some were trained as civil and hydraulic engineers and then later studied chemistry and biology. Others began as chemists and biologists and acquired engineering training. Before World War I, the number of sanitary engineering graduates in American colleges increased steadily.[42]

Sanitary engineering combined the generalist's perspective of environmental conditions with practical, technical know-how. As those primarily responsible for developing the new technologies of sanitation, municipal engineers were the obvious standard bearers for perpetuating the goals of environmental sanitation even in the new era of bacteriology. For the most part, they came to accept the new science of bacteriology, but they also saw less contradiction between the goals of the Old and New Public Health than many of their colleagues in the medical or scientific communities.

Engineers exhibited the greatest faith in finding technical solutions to tangible environmental problems, such as dirty water or accumulating wastes. After all, the newly developed technologies of sanitation were monuments to their efforts and their influence, and thus they would be the last to dismiss the role of sanitary services in disease prevention and pollution abatement.

Despite the engineers' fervor, it is somewhat ironic that in the years 1880–1920, modern sanitary services blossomed in the United States. It would appear that the New Public Health and the rise of bacteriology would have blunted enthusiasm for investing so heavily in various technologies of sanitation. However, the value of the early protosystems and subsequently modified systems was recognized on several levels: as providers of necessary water service, as convenient means of evacuating wastes, and as protectors of public health. The characterization of sanitary science as an instrument of community health was understood, if not always so clearly articulated.[43] Building on the momentum generated in the Age of Miasmas, many cities completed the process of establishing permanent, citywide sanitary systems in the Era of Bacteriology.

Water Supply as a Municipal Enterprise

1880-1920

B y the late nineteenth century, there was a strong feeling among municipal leaders that any respectable community needed citywide waterworks.[1] In 1870, the total number of waterworks stood at 244. By 1924, it was estimated that approximately 9,850 waterworks had been constructed in the United States. The number of waterworks in operation increased more rapidly than population growth in 1880 and 1890.

The trend in the late nineteenth and early twentieth centuries suggested a move from decentralized to centralized systems and from labor-intensive to capital-intensive methods of delivery.[2] Some were expanded versions of the older protosystems with increasingly distant surface-water supplies, more and bigger reservoirs and settling tanks, and extended distribution networks. The technology saw more efficient pumps and increasing use of electrical power. For many small towns and cities—those with a limited tax base—and for new cities, the older protosystems offered an immediate way to develop public water supplies.

Some of the most advanced systems were substantially modified, providing greater supplies and more extensive distribution networks. They also devoted greater attention to water purity through filtration and various forms of water treatment. Yet technical improvements often led to greater water demand and did not ensure freedom from epidemic diseases or pollution. Growing populations produced an almost unquenchable thirst for water, well beyond estimates. Forces that had launched the industrial age also increased the pollution load on watercourses, including those that served as sources of potable water. Ironically, water systems that had been a vital force in city expansion in the nineteenth century sometimes fell prey to the excesses of growth.

The late nineteenth century was a crucial time for expansion of older protosystems as well as a beginning point for cities that had yet to establish their first citywide systems. Cincinnati was regarded as having the most comprehensive program of waterworks improvement, with new pumping engines in 1872 and an experimental filter plant in 1907. After the fire in 1871, a new tunnel and pumping station were built for Chicago's West Side, and with extensive annexations in the 1880s, the city absorbed some smaller water systems. New systems appeared in smaller cities in the East and newer cities in the West and South.[3]

Water supply was the first important public utility in the United States and the first municipal service that demonstrated a city's commitment to growth. Officials and urban boosters promoted a variety of downtown improvements in competition with rival communities. Along with sanitarians and municipal engineers, they supported services to improve health conditions and to secure bragging rights about the cleanliness of their cities.[4]

A healthy community was central to growth. City leaders concluded that control of the sanitary quality of a city's water service would be difficult if the supply remained in private hands.[5] The push for municipal ownership, therefore, had as much to do with the desire to influence the growth of cities as to settle disputes with private companies over specific deficiencies. The "political nature" of water was important.[6]

Major cities tended to support public systems earlier than any other class of cities. In 1890, more than 70 percent of cities with populations exceeding 30,000 had public systems. In 1897, forty-one of the fifty largest cities (82 percent) had public systems. Since most of the urban population was located in the larger cities, it was not surprising that while only 43 percent of all American cities had public waterworks in 1890, 66.2 percent of the total urban population was served by public systems.[7]

The Midwest showed the greatest propensity to public systems by the end of the 1890s, which coincided with its efforts at reform of business regulation. In one major study, 73 percent of midwestern cities surveyed had public systems. The emerging industrial states along the Great Lakes showed strong support for public systems, as did the more agrarian states. For the East, the total was 42 percent, the South 38 percent, and the West 40 percent.[8]

Several factors account for the political and economic climate that favored public systems in the late nineteenth century: improved fiscal status of cities, cooperation between large cities and state legislatures in developing or expanding services, public skepticism of private companies to deliver services, and broadening regulatory power with respect to public utilities.

As noted earlier, there was a close relationship between a city's willingness to make water supply a public utility and its capacity to fund such a venture. This was made easier by improving opportunities for cities to assume debt. Between 1860 and 1922, municipal debt increased from $200 million to more than $3 billion. Legislatures were more lenient in allowing cities to float water bonds than incurring other forms of public indebtedness, since they were stable and demonstrated a good payment record. The growth of investment banking also established a national bond market, making more capital available to cities.[9]

Public systems became more widespread as criticism of private companies mounted. In some cases, a change in political environment worked against private companies, especially if reform-minded leaders criticized the franchisee as a source of corruption. In other cases, poor performance by the private company set off reconsideration of service delivery. Initially, waterworks franchises extended for long durations, offered tax exempt status for the company, and did little to regulate price. By World War I, fourteen states limited the length of a contract under provision of general law, but eighteen states still allowed perpetual franchises.[10] While as many as 850 new waterworks franchises were let during the 1880s, they were not as generous as those in the past.

The length of the franchise was usually a central point of debate, since a perpetual contract gave officials virtually no control over the waterworks company. Other key concerns included the ability to manage rates and the option to purchase the waterworks if the company did not live up to its obligations. Cities had unique leverage with respect to rates. During the 1890s, rates charged by private companies were 40 to 43 percent higher than rates charged by municipal works. Unlike private companies, cities could take a loss on operations and use taxes to make up the difference. Rate flexibility sometimes gave cities an advantage in keeping private companies from overcharging, and often made the threat to terminate the franchise quite real.[11]

As public systems became more competitive, sensitivity to liberal water contracts often resulted in a call for municipal ownership. In 1895, the New York legislature chartered Ramapo Water Company to serve customers in the New York City area. A critic claimed that Ramapo was granted "the most comprehensive powers" to do business, and the company wanted to charge $70 per million gallons supplied, when water retailed for an average of $50 per million gallons in New York City. In August 1899, the municipal assembly passed a resolution calling for municipal ownership.[12]

In some cases, a concern about local control was more persuasive than a demand for municipal ownership. Fresno, California, had considered municipal ownership since 1876, and reconsidered it after a major fire in 1882. But the city did little more than install public wells and hydrants in the 1880s, and the private Fresno Water Works continued delivery to residential areas. In 1889, a small group petitioned the city board of trustees to shift to municipal ownership, but the majority of Fresno residents favored small government and low taxes and had not lodged major complaints against the private company. When the local water and power companies went bankrupt, they were reorganized in 1902 by a local utility magnate, and were purchased by national corporations in the 1920s. Only then did the demand for municipal ownership intensify.[13]

A few cities, such as San Francisco, bucked the trend of municipal ownership. The widespread belief that the city's water supply was clear and wholesome checked the demand for a change in control. Nearby San Jose also obtained most of its water from private companies, and has continued to do so.[14] These California cases were the exceptions to the rule.

During the 1890s, a combination of factors shifted momentum decisively toward municipal ownership in most major cities. These included dissatisfaction with private companies, questions concerning the quality of the supplied water, high rates, and local interest in controlling services. By then cities had increased authorization to erect, lease, purchase, and operate waterworks, lighting plants, and, in some cases, street railways. Opportunities to issue bonds beyond previous authorized limits and tax-granting status produced additional capital.

Progressive Era reforms, such as the referendum, were employed to sanction purchases or to grant new franchises. Between 1891 and 1901, permission to own, erect, and purchase water or lighting plants had been extended to municipalities in twenty-four states. California and Kansas passed very general laws allowing for municipal ownership. Several cities had ownership clauses in their new charters.

In keeping with the regulatory trends of the period, several states established commissions with jurisdiction over waterworks. The first appeared in New York, Wisconsin, and Georgia in 1907. State regulation, however, diminished the local role in controlling public utilities, hearkening back to the days when the legislatures shaped public services without major input from the cities.[15]

The transition from private to public was not a victory of state over local control, or state and local regulation over the free market. Rapidly increasing demand for water rose beyond the capacity of most private companies to meet it. This situation provided an opportunity to promote municipal ownership as the only means to satisfy demand and to allow for city growth.[16]

The political climate had changed since the early nineteenth century. Progressive reform struck a chord with many urbanites, who recognized in its goals a strong commitment to civic rights and responsibilities. The call for an improved quality of life through efficient government, reliance on technical experts, and more equitable distribution of services added up to growing support for municipal responsibility. Public outcry against the "robber barons" and the exploitation of the trusts also cast a shadow over private water companies.[17]

In practice, municipal ownership did not automatically correct inequities in the water systems. Suburbs were often favored over central-city neighborhoods for improved service. In Detroit, uninhabited land—which had potential for new development—received priority over working-class areas of the city in the distribution and improvement of water service.[18]

In an era that clearly favored the shift toward municipal ownership, even well-reasoned criticisms were drowned out by the enthusiasm for publicly managed services. Water became a particularly favorite political issue because embedded in the water-supply service were so many concerns touching the well-being of the citizenry as well as the role of government in serving that citizenry.[19]

Metering water usage became a powerful management tool in administering the water supply in publicly run systems. Ostensibly employed as an effective way to set rates, the use of water meters was equally important as a means to check waste and to anticipate future expansions of the water-supply system. Before the mid-1870s, existing meters functioned poorly in turbid water or in water with sediment, and a substantial amount of water flowed through them without registering. As the technology improved, meters became practical devices for monitoring water flow. Meters were marketed to the public as "A Friend of the Water Consumer," "A Perpetual Invoice," or "Prevention of Wastage." Water-management officials drove home the point that meters would not only save water but also reduce rates, thus making service more equitable.[20]

Consumers were resistant to metering. They had experience with other uses, most notably for gas and electricity, and regarded it as an invasion of privacy and a way to extract more money for services they believed were already too expensive.[21] One critic noted, "No restriction should be made that would lead people to avoid bathing, or freely flushing plumbing fixtures. Anything which discourages a liberal use of water is an obstacle to social progress."[22]

Supporters of metering argued that with the growth of cities, the cost of supplying water was increasing. Metering offered one way to conserve water and thus help minimize the cost of developing new sources. Theoretically at least, funds not spent on new supplies could be utilized for other problems in the system, most notably keeping pressure high and preventing pollution. Despite the protests, some large cities and a few smaller ones began to meter their service lines in the late nineteenth century.[23]

Only modest progress was made in metering during the 1890s. Four of the 50 largest cities metered more than 50 percent of their taps. By 1920, metering had made notable strides. While only about 30 percent of the cities metered at the pump, more than 600 of 1,000 cities surveyed metered at the tap.[24]

Despite metering, new sources of supply became necessary nonetheless when old sources could no longer meet demand or became polluted. Pumped water from nearby lakes and streams proved to be no more viable for many large cities than digging new wells. These sources, as part of the local hydrologic system, often became polluted because of wastewater and industrial discharges. Local running water was the most convenient waste-disposal mechanism for many cities, and thus severely limited its value as sources for pure water.

New distant sources were one of the two best options for the largest cities, filtration and treatment being the other. Despite the success of the Croton Aqueduct, New York City faced a water crisis by the 1880s due to droughts, escalating consumption, leaks in the system, and expanding population. In 1883, the state legislature authorized the construction of the New Croton Aqueduct, which was completed ten years later. The watershed was squeezed for greater supply with the construction of several storage reservoirs and the New Croton Dam, which was the highest masonry dam in the world at its completion in 1906. Unlike the original, the New Croton Aqueduct was primarily a brick-lined tunnel for much of its length (thirty-three miles).

With the consolidation of the five boroughs, the need for an expanded water supply arose once again. The consolidation brought together 3.5 million people, consuming approximately 370 million gallons per day. The Croton system had reached nearly maximum utility, and a smaller system, taking water from wells and streams on Long Island, could not serve the new metropolitan area. In 1905, a law created the New York Board of Water Supply to oversee development strategies, and the State Water Supply Commission was established to allocate water-supply resources in the state. Between 1905 and 1914, New York City tapped the Catskill Watershed approximately 100 miles north of the city. A new aqueduct and distribution system by deep tunnel were constructed. Further development of the Catskill Watershed continued into the 1920s.[25]

On the Pacific Coast, Los Angeles is the best example of relentless penetration into the hinterland in search of water. The story is a familiar one, dramatized in the 1974 movie *Chinatown* and written about extensively in the literature of California and water resources.[26] Los Angeles was a community poised to become a major city at the turn of the century. With a population of approximately 200,000 and every physical blessing imaginable, it lacked one thing—water. A drought from 1892 to 1904 forcefully drove home that point.

The aspirations of the city's boosters were in the hands of William Mulholland, head of the local waterworks. An Irish immigrant who came to Southern California in 1877, the self-educated engineer rose to superintendent of the water company nine years later. He became head of the municipal waterworks when the city purchased the private company in 1902. Because of his confidence and apparent engineering know-how, he acquired the authority to make dramatic decisions.

Mulholland searched for new supplies in the wake of a water shortage in 1904. Former Los Angeles mayor and friend Fred Eaton told him about an extraordinary water source to the east of the Sierra Nevadas. In September, the men traveled north into the Owens Valley, where they found enough water to meet the needs of a city 100 times larger than Los Angeles. Mulholland sold the plan to city leaders, while Eaton acquired water rights in the valley. In 1905, the *Los Angeles Times* announced "Titanic Project to Give City a River." A joyous reaction spread through the city, but a reaction just short of rage greeted Eaton in the Owens Valley.

A war of sorts broke out when it was learned that water not immediately used by the city would be diverted to the San Fernando Valley for irrigation. Less than coincidentally, a syndicate of city businessmen bought 16,200 acres in the San Fernando Valley before Mulholland's plan was made public. The people of the Owens Valley realized that the land grab was also a water grab.

Owens Valley residents attempted to block Mulholland's request for a right-of-way across federal lands. President Theodore Roosevelt proposed an amendment to the bill prohibiting Los Angeles from reselling water to corporations or individuals. The bill passed as amended in 1906, but included no prohibition against the use of Owens Valley water for irrigation in the San Fernando Valley. Los Angeles scored another victory in 1908 when Chief Forester Gifford Pinchot extended the Sierra Forest Reserve to include the flatland of the Owens Valley, thus barring private entry to the lands adjacent to the water supply.

The city was poised to exploit its favored position. The aqueduct ran over a route of approximately 240 miles across mountains and desert. Mulholland constructed two hydroelectric plants in the Owens Valley, making the aqueduct the first major engineering project in the country constructed primarily by electric power. On November 5, 1913, a city starved for water was now wallowing in it. Within two years, officials annexed several communities, armed with assurances of abundant water for all who joined. The San Fernando Valley, once a semiarid grain-raising area, became an agricultural haven of truck farms and orchards. Owens Valley gained nothing.

In the years that followed, Mulholland faced the task of developing a much-needed reservoir, and in 1923 he again entered Owens Valley. This time the citizens of the area were not as complacent, and what had been a war of words became a shooting war. Ultimately, Los Angeles won its battle with Owens Valley. The forces for urban growth in Southern California were too formidable.[27]

The stories of the quest for water in Los Angeles and New York City are among the most dramatic episodes of the period, but they are not the only ones. Smaller cities and towns turned to more distant sources to meet their demands for water. Birmingham, Alabama, extended a canal six miles northeast from its pumping station. Mount Ayr, Iowa, without shallow wells in water-bearing strata, looked beyond city borders for its new source.[28]

All of these communities shared a realization that adequate water supplies were among the highest priorities for growth and the promotion of their cities. Expansion of the supply was the most significant variable in increasing consumption as well. One estimate placed total consumption in the United States at 5 billion gallons (100 gallons per capita for 50 million people) in 1916.[29] Per capita consumption figures can be misleading as an absolute measure of water use. Oftentimes commercial and industrial consumers were included in the calculations. Domestic and municipal uses of fresh water represented a relatively small portion of total water use compared to agricultural use, for example. Also, new types of plumbing fixtures such as water closets and bathtubs affected the totals, as did leaky pipes.[30]

The extension of the distribution network also made the modern water systems more complex. Technological innovations produced three types of supply systems: gravity (New York, Los Angeles, and San Francisco); direct pumping (Chicago and Detroit); and pumping from elevated storage tanks (St. Louis, Cleveland, and Kansas City, Kansas). The method of supply influenced the amount and kind of piping used in the system, and the street plan and topography determined the layout. The various demands placed on the supply also had to be taken into account. A central distribution system, for example, was particularly valuable in providing efficient service in the installation of fire hydrants.[31]

In 1890, the American Water Works Association (AWWA) established standards for all types of pipe, other materials and equipment, and some chemicals used in the field. While the distribution networks varied according to specific needs, standardization was meant to maximize efficiency of the network. Wooden pipes began to pass out of use, and lead pipes were slowly being prohibited in water systems. Cast-iron pipe was more widely used as prices continued to drop—a major factor in extending the distribution system itself.[32]

The consumer load on water mains placed stress on existing distribution systems. In 1888, the population per mile of water main averaged as little as 276 in Pacific Coast cities and as much as 830 in midatlantic cities. By 1897, the population per mile of water main exceeded 830 in several major cities.[33] Consumer load was particularly significant because although requests mounted for extension of service into the suburbs, the networks were still concentrated in the central cities. Some cities would not install lines outside of the city limits. Others might accept connections from outside the city limits only if the customers installed their own lines, as in Seattle, to increase its revenue from water without adding to the cost of construction.[34] In cities where private companies still controlled the waterworks, debate raged over how new extensions were to be added to the network. Assessing how to distribute the cost of the extensions to the customers also became a problem, especially since major extensions substantially increased the cost of operating water plants.[35]

Demand for connections beyond the inner city promoted growth and thus enhanced the city's vitality. In most cases, economic considerations favored extension of the system, with the cost falling to the new consumers. Improving the system nevertheless affected all consumers, but these costs were particularly inequitable if the improvements arose as a direct result of increased consumer load. Consumers had few alternatives but to pay the set rates. A truism was that it was easier to extend the distribution lines than improve the waterworks and the original pipes.[36]

While questions of management, consumption, and the extension of the distribution networks were crucial, the issue of water purity became an equally high priority. The revolution in public health and the efforts of science to confront water pollution not only helped to ensure a safe water supply but also modified the infrastructure of the system itself. Bacteriological laboratories, filtration plants, and treatment facilities added a new dimension to the systems.[37] Clean water also tended to increase the drinking of water—regarded as a healthful thing—and inspired greater personal cleanliness. From a civic perspective, one writer noted that "you have a right to expect as much protection from [the] water supply as from [the] police force."[38]

The earliest standards of water purity had been physical—color, turbidity, temperature, odor, and taste—and could be observed by the layperson. Complaints were common. From Massachusetts: "The odor was so bad that it would be almost impossible to take it as far as the mouth to taste it. Horses refused it at the street watering-troughs, and the dogs fled from it." From New York: "Strong, fishy odor and taste, also odor of smartweed. Popular complaint was dead fish in water-mains. Very rank." From Montana: "The water is so bad that we have had to shut off the supply each year from June to December."[39]

The senses often gave misleading evidence; to truly determine the quality of the water required further testing. Chemical standards followed physical standards, but they too could be imprecise.[40] Enthusiasm for bacteriology sometimes unduly discredited the work of chemists in evaluating water supply. Until the mid-twentieth century, greater attention would be paid to biological forms of pollution linked to epidemic disease than to industrial and other chemical pollutants.[41]

Nevertheless, the newest set of standards available in the Progressive Era was that of the bacteriologist. In 1900, sanitarian George Whipple wrote, "The microscope is no longer a toy, it is a tool; the microscopic world is no longer a world apart, it is vitally connected with our own." In 1887, the Massachusetts State Board of Health instituted the first program of sanitary chemical analysis and microscopic examination in the United States. The installation of similar facilities soon followed in most large cities. In some cases, waterworks officials and engineers exerted their authority in gaining control or influence over the laboratories.[42]

Water purity is a relative as opposed to absolute concept in the sanitation field. The term "safe" water that can be consumed without harm to humans or animals was often used instead of "pure" water, which does not exist in nature. Early attempts at filtering water to produce "clean" water—that is, water free of turbidity, odor, and smell—likewise did not ensure "pure" water. Chemists disclosed the concentrations of organic matter in water, but by 1904 few experts considered chemical analysis sufficient to determine water quality from a biological standpoint.

Development of bacteriological techniques led to greater emphasis on the bacterial quality of water. Water chemistry still proved important in improving water filtration practices and in testing for toxic agents. The routine work in the laboratories consisted of regular examination of samples from all sources of water and from the distribution system.[43]

Typhoid fever remained a scourge until the 1920s. With the discovery of the typhoid bacillus and other pathogenic microorganisms in the 1880s, public health officials and engineers began to understand the extent of water-borne diseases and sought methods for combating them. *Engineering News* reported in 1896: "The relation between typhoid fever and water supply is now recognized as being so close that continued high typhoid mortality in any city is taken as pretty conclusive evidence that the public water supply is polluted with sewage."[44]

An important precedent was set between 1890 and 1893 when the Massachusetts Board of Health laboratories devoted major attention to the transmission of typhoid fever. In the wake of an epidemic in Lowell in 1890, William T. Sedgwick, professor of biology at MIT and chief biologist of the State Board of Health, carried out a systematic investigation. For the first time, the bacteriological methods were applied to fieldwork. After a frustrating search, Sedgwick and his assistants traced the infection to an outbreak of typhoid in the nearby village of North Chelmsford. Their investigation technique soon was copied in other areas plagued by typhoid epidemics.[45]

The knowledge about waterborne disease and the subsequent widespread public discussion led to demands for workable solutions. City leaders, placing their faith in science and technology, believed that the means were now available to protect the water supply or clean up polluted sources rather than simply abandoning them. By World War I, most of the largest cities and several smaller ones invested in filtration plants and treatment facilities. The infrastructure of the modern water-supply system grew larger and more intricate. To coordinate such an elaborate system also required even greater centralization of authority and management.

While some individuals placed too much faith in filtration as the panacea for waterborne epidemics, the investment was worthwhile in improving the quality of water in almost every community that employed it. Unqualified acceptance of filtration on a national scale was not immediate, since the cost could be prohibitive, and there was debate over the best type of filtration to employ.[46]

M. N. Baker, the noted engineer, identified five eras of water treatment in America, which helps to explain the evolution of water purification.[47] The first era ended in 1870, coinciding with James P. Kirkwood's 1869 *Report on the Filtration of River Waters*. During this time, sedimentation was utilized to clarify muddy water, "natural" filtration—or the use of infiltration basins and galleries—was practiced, and several small filters strained the largest materials.[48]

A second era, from 1870 to 1890, saw the introduction of slow sand and rapid sand—or mechanical—filters. Implementation was erratic, and the principles of bacteriology were not well enough entrenched to inform the most effective use of the filters.[49] By 1880, however, there were only three of them in the United States and none in Canada.

The rapid sand filter was an American invention. Aside from the speed, the chief distinction between the English and the "American" filter was the cleaning process. The former was cleaned manually, while the latter was mechanical. The first American filters were used in paper-making and other industrial plants where very clear water was essential. Somerville, New Jersey, is generally regarded as employing the first rapid sand filter on a municipal water supply (1885).[50] By 1900, however, filtration was not widely used in the United States, and debate over the relative advantage of the rapid sand over the slow sand filter continued.[51]

The third era, 1881 to 1900, is noted for scientifically designed slow sand filters and better designed mechanical filters. The Lawrence Experiment Station in Massachusetts became the leading research center in water purification in the nation. The Lawrence Station emerged at a time when typhoid fever and cholera epidemics could devastate a community. In the fall of 1887, epidemics struck Minneapolis, Pittsburgh, Ottawa, and other North American cities. Plymouth, Pennsylvania, a city of 8,000, suffered 1,200 to 1,300 cases of typhoid two years earlier. Cohoes, New York, with a population of 22,000, had 1,000 cases in 1890–91; West Troy and Albany, on the Hudson River below Cohoes, soon had epidemics of their own.

Most instructive was a cholera outbreak in Germany in 1892. Hamburg, a city of 640,000, experienced 17,000 cases of the disease, with over 8,600 deaths. To the contrary, Altona (150,000 population), situated close to Hamburg, experienced only 500 cases and 300 deaths. Hamburg drew water from the Elbe River untreated, while Altona filtered its water. Hamburg learned a harsh lesson, but subsequently improved its waterworks, as did other German cities.[52]

Possibly the most important discovery in these years was that an improved slow sand filter could remove typhoid germs from river water, testing what only had been surmised in the Hamburg-Altona episode. Based on this work, a filtration plant was constructed in 1893 for the city of Lawrence.[53] The third of Baker's five eras, however, remained a time of uncertainty in the application of the new and relatively untested filtration technologies, especially in the South and West.[54]

The years 1901 to 1910 marked the fourth era, in which the rivalry between slow sand and mechanical filters intensified. Some decision makers were caught in the middle of the controversy, choosing sides with unpalatable results. Between 1900 and 1913, Philadelphia constructed four large slow sand filtration plants with chlorination facilities meant to reverse almost thirty years of neglect of a water-supply system once regarded as the foremost in the country. No sooner had the new plants gone on line when rapid sand filtration gained wider attention.[55]

In the first decade of the twentieth century, water treatment made important strides. Chlorination was well established, and experimentation with copper sulfate to control algae was under way. Leading sanitarians had been able to convince city leaders that typhoid fever and related diseases were preventable through a combination of filtration and treatment. Water-supply specialist Allen Hazen purportedly stated that since typhoid was preventable, for every death from the disease someone should be hanged. George A. Johnson, a consulting sanitary engineer, also forcefully asserted that "a person who knowingly takes into his mouth the excrement of another human being is certainly defective. A community having a public water supply known to be contaminated by the sewage of other cities, and using it without even attempting its purification, is certainly a victim of defective civilization."[56]

Few of its major proponents claimed that filtration was the only answer to disease prevention, but contemporary statistics indicated that it was especially effective in controlling typhoid fever. The results often were so dramatic that people lost sight of the idea that filtration's primary purpose was reducing turbidity and removing suspended matter.

Disinfection became a complement to filtration and promised to play an even larger role in removing bacteria from the water supply. Ancient civilizations disinfected what they believed to be tainted water through boiling or by using copper vessels. A professor at the Stevens Institute patented a process for the chlorination of water in 1888. But even before chlorination became popular, sewage had been chlorinated in England, France, and the United States. Bleaching powder was first used to chlorinate water in Austria in 1896. The next year, chlorine gas was applied to a test filter at Adrian, Michigan. The first continuous-use water chlorination plant was constructed in Belgium in 1902. The first use in the United States took place at Bubbly Creek Filter Plant at the Union Stock Yards in Chicago in 1908. Jersey City, New Jersey, became the first community to have its water supply chlorinated soon after the Bubbly Creek experiment. In 1909, liquid chlorine was produced, which provided a much easier method of dispersal.[57]

A dramatic decline in typhoid fever rates followed the use of chlorine in many locations. Statistics from cities utilizing hypochlorites in their water offered some striking evidence of the change (see Table 7.1). Yet despite the optimistic reports, there remained some concern among citizens about "doping the water" with chemicals. While this fear had been greater before the turn of the century, it never completely died out, and thus chlorination was not widespread by 1920. An *American City* survey for that year showed

Table 7.1. Decline in Typhoid Death Rates after Use of Hypochlorite

City	Before (1900–10)	After (1908–13)	Change in %
Baltimore	35.2	22.8	35
Cleveland	35.5	10.0	72
Des Moines, Iowa	22.7	13.4	41
Erie, Pennsylvania	38.7	13.5	65
Evanston, Illinois	26.0	14.5	44
Jersey City, New Jersey	18.7	9.3	50
Kansas City, Kansas	42.5	20.0	53
Omaha, Nebraska	22.5	11.8	47
Poughkeepsie, New York	54.0	18.5	66

Source: John W. Alvord, "Recent Progress and Tendencies in Municipal Water Supply in the United States," *JAWWA* 4 (Sept. 1917): 284.

less than half of approximately 1,000 cities using chlorine or some other disinfectant in their water supply.[58]

An object lesson was not lost on the city of Milwaukee in 1916. Because of complaints from citizens about the odor and taste of the chlorinated (but unfiltered) water from Lake Michigan, an operator shut off the chlorine one evening without permission. During the next few days, 50,000 to 60,000 cases of gastroenteritis occurred and between 400 and 500 cases of typhoid, resulting in forty to fifty deaths.[59]

Although filtration and disinfection spread less rapidly than Baker stated in his work, a movement toward purified water supplies was under way after 1910. Mechanical filters took the lead, with additional experiments in double—or multiple—filtration, a technology that came late to the United States. In Wayne, Pennsylvania, double filtration was tried in 1895–96, but then abandoned. Philadelphia completed a double filtration plant in 1911. The double filters consisted of a normal filtration system with prefilters or scrubbers designed especially for areas that were heavily polluted or burdened with fine sediment, but they were never widely used.[60]

No matter what the system, filtration rose in popularity in most large American cities and several smaller ones between 1880 and 1914. The *American City* survey of approximately 1,000 cities in 1920 showed that filtration had made good, if not spectacular, strides, especially in the East and Midwest. By that time, the broader application of filtration and the introduction of disinfection were reflected in the sharply falling typhoid fever rates in large cities.[61] Substantial decline in rates in many cities during the early years of the new century indicate the impact of filtering water, while the dramatic drop in the rates by the second decade of the twentieth century reflect the use of disinfectants, usually in combination with filtration. The

1920 typhoid death rates in the United States were more in line with those of western European countries, where water purification had been widely applied.[62]

Because purity was linked with waterborne diseases in this era, concern over pollution focused primarily on bacteria in the water (especially through sewage discharge) rather than on the impact of industrial pollutants or other toxic materials. Not until after World War I were industrial wastes viewed as a major problem affecting water purity and sewage treatment. In some cases, industrial waste was considered a germicide that, when added to water, would inhibit the putrescibility of organic material. From a much broader perspective, industrial waste was regarded by many as a necessary cost of economic growth.[63]

The U.S. Public Health Service (USPHS) was more interested in sewage pollution than in the treatment of specific industrial wastes. The USPHS turned primarily to stream studies, which led to a general theory of stream purification. A 1912 law specified stream pollution as the subject for special study, and the next year Congress established the Stream Pollution Investigation Station in Cincinnati. Water-quality experts took a closer look at the biochemical oxygen demand (BOD) characteristics of waste, natural oxidation of streams, and methods of water treatment along the Ohio River.[64]

New methods of water analysis sought to provide a precise rendering of bacterial levels in the water supply. In 1914, the USPHS issued standards, based on bacterial analysis, for water used in interstate commerce. Several states copied the standards in conjunction with methods devised by the APHA, the AWWA, and other groups.[65]

Before the establishment of standards, pollution of waterways had become a major focus of concern. The states were the centers of action for new legislation to control stream pollution in particular. Massachusetts was first to attempt to protect its inland water from pollution, in 1886. By 1905, thirty-six states enacted statutes to protect drinking-water supplies. By World War I, states were establishing boards and commissions expressly designed to regulate water pollution, and were given expanded power over industrial as well as municipal pollution. However, there were inconsistencies in the regulations, and enforcement often was lax.[66]

While the complete nature of water pollution had yet to be explored by 1920, American water-supply systems had come a long way in meeting consumption and sanitary needs of at least the major cities. There was great confidence among municipal leaders that an efficiently run public waterworks added to the city's reputation and was an effective tool for urban growth. These modern technologies of sanitation came to be regarded as one of the major achievements of city building in the Progressive Era, systems that would stand the test of time.

8

Battles at Both Ends of the Pipe

SEWERAGE SYSTEMS AND THE NEW HEALTH PARADIGM, 1880-1920

The value of public underground sewerage systems was widely acknowledged by the late nineteenth century. Between 1880 and 1920, however,
three major issues remained: Were citywide water-carriage systems finan-
cially feasible? What kind of system best suited a city's needs? How was
sewage disposal to be handled? Disposal of wastewater, due to massive increases in piped-in water, proved to be the major stimulus to underground
sewerage development. The older cesspool–privy vault methods were incapable of handling the load. Despite the fact that their functions were linked,
water-supply and sewerage systems were rarely interconnected or managed
as components in a larger sanitary system.

Private sewerage had limited appeal, especially at the cores of larger
towns and cities. A private sewer without a connection to a larger main was
of little value, although some wealthy homeowners and businesses built private sewers and let them drain into a nearby stream. Other citizens rarely
had a disposal choice, and often settled for cesspools. At the turn of the
century, the Baltimore harbor was frequently a stinking mass of decaying

garbage and other pollution in summertime, which slum dwellers close to the water were forced to endure.[1]

A company developing and operating a large sewer system could offer a useful service, but it rarely had the financial potential of water delivery. One exception was developers of new subdivisions, who provided sewerage hoping to recover the cost of construction through the purchase price of homes. But developers showed little inclination to manage the completed sewerage networks as independent systems.[2]

Resistance to private sewerage systems often was linked to health concerns, and thus the impulse to construct citywide wastewater systems arose at a time when public health was viewed as a municipal responsibility. In 1883, engineer Rudolph Hering recommended that Wilmington, Delaware, construct a modestly priced system, but the leaders balked. In 1887, the question rose again, and a group of citizens proposed forming a private company to build sewers. The council rejected the suggestion on the grounds that the city should control its own sewers. Troy, Pennsylvania, undertook to install a sewerage system with private capital in the second decade of the twentieth century, with the cost to be borne entirely by the users. Eventually, the borough began to buy up the interests of private investors and to distribute the cost of the system among all property holders.[3]

The only large city to grant a sewerage franchise was New Orleans, but it was short-lived. The city was late in building the system because of the fiscal conservatism of the region and the city's unique topography. New Orleans was built on an alluvial plain, lying between the Mississippi River and Lake Pontchartrain, with a slope back from the river to the city. This made drainage into the river difficult and costly. For its part, the New Orleans Water Works Company was to furnish water to households at a fixed price and extend its mains in order to make a greater part of the city accessible to the new sewerage system. The company failed to complete its contract, and the work was suspended. In 1898, after an attack of yellow fever the preceding year, voters agreed to levy the necessary taxes to develop a municipal system.[4]

While public sewer systems were not a natural extension of public water supplies, their value seemed more obvious when municipal ownership came in vogue and when taxes and bond revenue were being invested aggressively in the urban infrastructure. The percentage of the sewered urban population increased from 50 percent in 1870 to 87 percent in 1920. The number of communities with some form of sewer system increased from 100 to 3,000 in the same period. The highest percentages were in the Northeast and the lowest in the South and West. Cities that initiated new sewer systems, however, still represented only a fraction of communities with water-supply systems.[5] Within the largest cities, the sewer systems expanded steadily, and some very rapidly. Chicago's and Philadelphia's systems experienced a fivefold increase between 1880 and 1905; Pittsburgh's system increased by a staggering 1,522 percent.[6]

Development and expansion of sewer systems were capital intensive. Although any system adopted was meant to be permanent, it had to be justified under current budgetary conditions. The system had to demonstrate a capacity for expansion to allow costs of construction to be amortized over several years. Also, any system had to protect the public health.

Seeking the ideal system proved to be daunting, especially because of contention over combined versus separate systems. The issue came to a climax as a result of the decision to install a separate system in the city of Memphis, Tennessee. A national debate focused on the relative performance and health benefits of the approaches, on costs, and on the character of the people who embraced the technologies. Memphis faced a double crisis in the late 1870s. Among the river towns along the lower Mississippi, it had a reputation as a haven for disease. Memphis also had a corrupt government incapable of confronting its health problems. In 1873 and 1878, the city was struck by devastating yellow fever epidemics. In the summer of 1873, more than 2,000 people died; in 1878, 5,150 people died, representing one-sixth of the population. Even with stiff quarantine and sanitation regulations, 485 additional deaths were reported in 1879.[7]

The mismanaged public health program only added to Memphis's problems. In 1879, the community's business elite took the lead to repeal the city charter. Governor Albert S. Marks signed a bill creating the Taxing District of Shelby County to provide for local government in Memphis. Taxing power was transferred to the state, and local borrowing was prohibited. A commission form of government was given responsibility for daily operations, and sanitary reform had high priority.[8]

The city virtually started from scratch to build a sewer system, which now was regarded as an essential weapon against future epidemics. When yellow fever swept through the city in the 1870s, Memphis had only a few miles of private sewers in the central business district, and depended heavily on inadequately maintained privy vaults and cesspools. The poorer classes—the majority of whom were black—lived in badly constructed houses in the most unsanitary parts of town and suffered from a variety of illnesses.[9]

The crisis in Memphis was national news. The newly formed National Board of Health offered to conduct a sanitary survey of the city and a chemical inspection of its water supply. Local and state officials quickly accepted. A house-by-house inspection, completed in January 1880, revealed serious overcrowding and many inferior dwellings. The report stated that nearly three-quarters of the city's annual death rate was caused by diseases due to improper drainage and ventilation and various pollutants in the water and soil. Not surprisingly, the existing sanitary force was woefully inadequate, with several wards receiving no waste collection. Beyond the survey, it also was clear that the city's water-supply service mirrored the paltry sanitation measures. Most citizens relied on cisterns and wells. The privately owned Memphis Water Company—bankrupt in 1879—had few customers and provided meager service.[10]

The special committee made several recommendations, including the hiring of a competent sanitary officer "independent of politics," city control of the waterworks, proper ventilation for all salvageable houses and the destruction of others, cleaning the privy vaults, improved drainage, more effective garbage removal, and proper construction of streets.[11]

One member of the National Board of Health's inspection team, drainage engineer and agriculturist Colonel George E. Waring Jr., proposed the building of a unique, low-cost sewer system that promised to save the city from future epidemics. Waring's separate system was roughly based on English designs, but called for a small vitrified pipe attached to a house drain equipped with a flush tank fed by the public water supply. The sewer pipe in turn was attached to slightly larger lines and eventually spilled into even larger mains. The system, however, excluded provisions for stormwater, which accounted for the low cost. Waring argued that the main streets were sufficiently guttered to handle the runoff.[12]

Waring's plan won out, but only after a heated local battle. Some property owners believed that sewers should be privately built or were not convinced that a sewer system answered Memphis's public health plight. Among supporters for a public sewer system, some prominent engineers voiced opposition to the Waring plan. Nonetheless, the promise of an effective system at low cost won the day. By the end of 1881, most of Waring's original plan was completed.[13]

Restoring Memphis to good health was the primary justification for the system, and Waring's proposal relied heavily on the sanitary views of the day to promote his project. He was a major proponent of the miasmatic theory and strongly adhered to the notion that "sewer gas" was the primary source of infectious disease.[14] Hence, the central rationale for Waring's separate system rested on the credibility of anticontagionism.[15] Demands for economy also drove Waring's bid. "I formulated a theoretical system," he later asserted, "which had never been put into execution—which probably never would have been put into execution, but for the great needs and the great poverty of Memphis."[16] His tradeoff, for the sake of economy, was essentially ignoring Memphis's drainage needs, while focusing exclusively on a small, one-pipe system.[17]

The controversy over the Memphis system only began with the letting of the bid. While Waring claimed success, some grumbled that the system was not living up to its promise. Others criticized the idea of a separate system itself and the motives of its leading proponent. In practice, the Memphis system encountered technical problems from the start and required retrofitting, making it more expensive than the optimistic projections. The small lateral sewers were perpetually clogged, and since no manholes were built, streets had to be dug up to relieve the blockages. These extraordinary maintenance costs eventually led the city to install manholes. At the source, householders were obliged to deal with numerous obstructions in the line and malfunctioning flush tanks, requiring frequent plumbing service. Waring's system also required connection to the water system. But since it was inadequate in 1880, extension of the water pipes to supply the flushing tanks was imperative.

Although Waring envisioned sewage running into the Mississippi, the city did not authorize the building of an outlet until 1886. Instead, sewage was discharged into a nearby bayou, re-creating an old pollution problem. Waring had been overly optimistic in selling his system or, more likely, had oversold the sanitary benefits and undersold the need for effective drainage for the sake of a low bid. The city leaders were equally guilty of shortsightedness in failing to fund an outfall and for failing to take into account the extraneous costs of the system.[18]

The debate over Waring's system went deeper than technical choices and costs. It brought to a head the differences among engineers over the separate versus combined systems, and it also generated a personal attack on Waring. Before the Civil War, Waring had been a successful scientific agriculturist. In 1855, he managed Horace Greeley's farm near Chappaqua, New York, and two years later accepted a similar post on Frederick Law Olmsted's Staten Island farm. His association with Olmsted led Waring to participate as drainage engineer in the construction of Central Park. This opportunity provided him a springboard into drainage and sewerage projects across the country. The Civil War temporarily halted Waring's promising career. Entering the war as a major, he was commissioned a colonel in 1862 and served in various units throughout the conflict. Waring maintained the moniker of "colonel" throughout his life, a reminder of his attachment to military principles, decorum, and discipline.

After being mustered out of the service in 1865, Waring wandered through a series of business failures. He returned to horticulture and husbandry in 1867, managing the Ogden Farm near Newport, Rhode Island, for ten years. In the 1870s, he began accepting commissions to build drainage and sewerage systems along the East Coast and became a noted municipal engineer. His appointment to a special commission of the National Board of Health in 1879 gave him considerable recognition. While he became a leading sanitarian, he also was a self-promoter, championing various patents he held or purchased. In the 1860s, for example, he publicized the "earth closet" as a revolutionary innovation in household sanitation, even after the water closet had proved to be more practical.[19]

The system in Memphis was the result of Waring's most effective sales job. A contemporary noted that "there was so much personal magnetism in Colonel Waring that he was able to use the prestige of his sanitary achievements at Memphis to impress his views regarding small pipe sewers on a number of communities."[20] In 1881 and 1883, he patented aspects of his Memphis system, formed a company, and set out to sell the idea to various cities. Many engineers regarded this behavior as manifestly unprofessional, especially given the unpublicized but nevertheless serious shortcomings of the system. Despite criticism from his fellow engineers, Waring benefited from the widespread attention to the Memphis system. Not only did he accumulate many more sewer contracts, but he also gained vast notoriety as a leading sanitarian with his appointment as street-cleaning commissioner of New York City in the 1890s.[21]

The debate over the Waring system and its architect went beyond questions of professional ethics to the fundamentals of engineering design and economy. The separate system had challenged the combined system, a technology that until 1880 received support from the majority of American engineers. Uncomfortable about Waring's claims, the National Board of Health sent Rudolph Hering to Europe in 1880 to investigate the relative merits of the two systems. Hering, who is regarded as "the dean of sanitary engineering," was as prominent as Colonel Waring. He graduated from the German Royal Polytechnical School (1867) and served as assistant city engineer in Philadelphia from 1876 to 1880. He became interested in the failure of the old Philadelphia sewers, and in 1878 presented a paper on the subject at the American Society of Civil Engineers convention. Hering's writings drew widespread attention in the field.[22]

Hering's investigation tour in Europe lasted nearly a year, and on December 24, 1880, the National Board of Health published his report. Hering observed "a striking contrast" between the condition of sewers in European cities and American cities, which he attributed to better construction and maintenance practices in the United States. With respect to combined and separate sewers, he concluded that neither system had a greater sanitary advantage; combined systems were best suited for large, densely populated cities that were concerned with household wastewater and stormwater, and separate systems could be useful in smaller cities primarily concerned with household wastes.[23] The report added fuel to the debate over separate versus combined sewers. But from 1880 into the twentieth century, engineers became less interested in arguing over questions of superior design or relative sanitary advantage, and instead focused on local conditions and on cost.[24]

In actual practice, the implementation of separate and combined systems by the first decade of the twentieth century appeared to follow the guidelines as Hering envisioned them: larger cities favored combined systems, and smaller cities and towns favored separate systems. No large city had constructed a separate system by 1900.[25] Hering also factored in the importance of local conditions in evaluating a community's preference. "Even the customs of the people," he stated, "can materially influence designs, and certain features may be acceptable in one locality and strenuously objected to in others."[26]

Aside from the debate over methods, some cities simply battled through partisan politics and competing interests in settling on a citywide sewerage system. Sewers were expensive, and the decision to implement them was unlikely to sail through many city council meetings without some disagreement. In Boston, for example, sanitary projects absorbed about one-third of the city's total budget in the late nineteenth century.[27]

Between 1859 and 1905, Baltimore's leadership rejected proposals for a citywide system on at least four occasions. Not until the passage of a sewer bond in 1905 was the proposal successful. The lack of funds seemed to be the primary drawback, but the story was more complicated. The expansion of the water supply, the physical growth of the city, increased population, and rising real estate values made the land-intensive cesspool system inefficient. These forces had to overcome the inertia of severe political infighting. The Democratic machine wanted to turn public works projects to its own advantage, but machine members and reformers alike feared the loss of patronage to the other. In the sewer debate, cesspool interests, the oyster industry, businessmen opposed to higher taxes, wealthy neighborhoods with private sewers, and newly annexed communities objected to a municipal sewer system. By 1905, however, a more professionalized municipal government, better fiscal conditions, civic efforts to rebuild the downtown after a fire in 1904, and the desire to compete with other cities for economic development overcame the fragmentation. In the end, Baltimore got its sewer system and was able to shake its image as the city of open drains.[28]

Outward growth also exposed inequities in the delivery of services through a citywide system. In some cases, central-city businessmen had to pay additional special assessments to link to sewers. But frequently central business districts and outlying suburbs were favored in the distribution of sewer lines to the detriment of minority and working-class neighborhoods. In Detroit between 1880 and 1920, water feeders and public sewers were densest up to one mile from the city center and two miles and beyond the city center with a bias in favor of suburban land speculators and against the working class. In many cities, real estate development companies played an important role in decision making with respect to sewer construction.[29]

Concerns over local conditions helped shift attention from efficient evacuation of wastes to the pollution load on watercourses once the sewage left the pipes. A member of the American Society of Engineers stated that "no subject has engaged the attention of sanitary authorities during the past ten years more than this question of the permissible limits of pollution of water by sewage, and the best means of preventing excessive pollution."[30] This was a reflection of the rise of bacteriology and greater recognition of waterborne pollutants, but also an understanding that the new sewer systems displaced wastewater and runoff but did not eliminate them.

As these systems became more successful in capturing wastewater, they became a greater threat as a polluter. Particularly in cases where cities employed combined systems, the volume of raw sewage entering rivers and streams became increasingly burdensome. Cities and towns that relied on separate systems also contributed to the pollution load of nearby waterways with runoff and surface drainage. Nonetheless, it was becoming apparent that combined sewers might not be appropriate in cases where sewage was treated or disposed of on land, or where polluted water entered sewers.[31]

The disposal problem raised one of the most crucial issues for sanitary services, that is, the advantages of water filtration versus sewage treatment.[32] The disposal issue highlighted how sanitary services were caught in the transition from the filth to the germ theory. The first underground sewerage systems were constructed at a time when the primary health threats appeared to come from sewer gas and unattended putrefying wastes. Thus the systems had to be airtight and watertight and had to evacuate wastewater quickly. Concern over the threat of pollution to watercourses turned attention to the end of the pipe. The increased sewage load on the systems now posed a problem not accounted for in the original designs.

The threat of raw sewage, as stated earlier, was understood to be more biological than chemical. While chemical and biological concepts concerning impurities had existed side by side earlier in the nineteenth century, after 1880 solutions that incorporated biological approaches were given central attention.[33]

Between 1850 and 1900, the British scientific community began to apply principles of biology to water purification and sewage treatment. Later, American scientists followed similar lines of research. The work focused on biological organisms in the water, regarded as primary pollutants. The work on typhoid fever at the Lawrence Experiment Station in the early 1890s confirmed fears about the relationship between that disease and sewage in the waterways.[34] Sanitarians came to accept the axiom so clearly stated by George Whipple: "The danger of sewage lies chiefly in the bacteria that it contains."[35]

In the United States, support for the biological perspective on water pollution gained increasing strength as hundreds of newly constructed sewers poured their discharge into nearby waterways. While the cities that eliminated their untreated liquid wastes by pipes and drains hailed the new systems, downstream cities faced an invasion of potentially disastrous proportions.

The most common form of sewage disposal practiced in the United States was dilution. Sewage was simply discharged into natural bodies of water, sometimes with some form of preliminary treatment. Unlike in Great Britain where small rivers, urban growth, and the use of combined sewers made indiscriminate dumping of sewage impractical, American cities relied on vast water resources to serve as sinks.[36] In its earliest uses, dilution was haphazard. As one contemporary noted, "The sewers were generally discharged in the nearest bodies of water, close to the shore without proper diffusion." As a result not only were such waters grossly polluted, but local nuisances were created by sludge banks, slick, and scum.[37]

Proponents argued that disposal of sewage by dilution was a theoretically effective technique because of the ability of running water to purify itself. The extent and type of pretreatment became a major sticking point during the early twentieth century, but the method was popular through World War I.[38]

Engineers favored a systematic plan of dilution rather than simply dumping wastes into rivers and streams. To be effective and economical, care had to be taken in protecting the public health by limiting the amount of bacteria in a given watercourse and by maintaining a sufficient reserve supply of dissolved oxygen in the water. Engineers also favored purifying water supplies before use rather than treating sewage before disposal. To reduce the bacterial load on a watercourse, statistical measures were employed to know when sewage might overtax the diluting power of the watercourse and loading that a water purification plant could handle.

The first significant dilution standards proposed in the United States came from Rudolph Hering and his associates in developing a drainage canal to dilute Chicago's sewage. Although the standards were widely accepted for many years, engineers were cautious in accepting a fixed standard without taking into account local conditions. In addition to load factors, disposal techniques could be modified in order to disperse rather than concentrate sewage.[39]

Many health officials and some engineers began to question dilution in the 1890s. Increasing levels of sewage disposal, and the concern about water as a carrier of disease, stimulated interest in sewage treatment as opposed to dilution and/or filtration. George W. Fuller estimated in 1904 that more than 20 million urbanites were discharging raw sewage into inland streams or lakes, while only about 1.1 million were connected to systems employing sewage purification works. Another source estimated that in 1909, 88 percent of the wastewater of the sewered population was dumped into waterways without treatment.[40]

While debate over dilution persisted, the question of whether to treat sewage or to filter water also continued to rage.[41] The greater the ability to remove effluent from homes and businesses, the larger the problem of how to dispose of it. Viewing sewage disposal as a health problem, sanitarians embraced treatment as the best safeguard against the spread of waterborne disease. In so doing, they also were placing responsibility on those who produced the waste. If the disposal medium was a source of drinking water for the same community, then treatment made sense. If the waste flowed downstream, treatment seemed superfluous, and the assumption of responsibility by the polluter was more difficult to justify. In addition, the cost of treatment meant something different if it was to be borne by the waste producer for direct benefit to the community versus benefit to another city.

The theory of bacteriology was insufficient, in and of itself, to produce a major commitment to treatment. But concern over the bacterial contamination of sewage, in concert with its disposal in waterways, prompted litigation by downstream cities and towns. The legal issues as interpreted in most states remained grounded in common law principles. Riparian owners were entitled to use the water flowing past their property. Downstream riparian owners had the right to expect the water reaching their property to be in its natural condition, except for any "reasonable" use. The interpretation of "reasonable" varied from one court to another.

In the case of sewage discharge, the question of public versus private rights complicated the determination of reasonable use. Cities were treated as agents of the state, acting under legislative authority through a charter or special act. Discharging sewage into streams was regarded as a legal function of cities, and thus the city was without liability in damages to a lower riparian owner. However, the equity aspect of a case allowed an owner to prove that the sewage was of continuing harm to his or her property, and thus unwarranted under any statute. In this instance, the court could issue an injunction to prevent the continuation of the pollution. At the time, it was generally held that where extensive pollution of a stream had not occurred for more than twenty years—or was not authorized by statute—it could be curtailed by equity proceedings.[42]

With respect to dangers to public health, common law had to be supplemented by additional statutes because it was concerned only with property rights. In 1910, the Minnesota Supreme Court upheld a lower-court judgment in awarding damages to the widows of two men who died from drinking polluted water supplied by the city of Mankato. By the early 1920s, several courts held that a city was liable for any disease contracted from the water it distributed, and it could not dispose of untreated sewage into natural waterways if there was a threat of disease.[43]

Regulatory apparatus at the state level offered a mechanism for regularizing prohibitions against water pollution. By 1905, as many as forty-four states had laws regarding stream pollution, but only eight had workable enforcement provisions. By World War I, several states established boards and commissions expressly designed to regulate water pollution, which were given expanded power over industrial as well as municipal pollution. There were inconsistencies in the regulations, however, and enforcement was lax.[44]

Typhoid epidemics often stimulated the states to authorize regulation of stream pollution. In Pennsylvania, a state board of health was established in 1885 after a major typhoid epidemic struck the mining town of Plymouth. The state legislature did not pass stream-pollution legislation until twenty years later, again after a typhoid epidemic. The newly created Department of Health had the enforcement authority, but the law applied only to new systems or extensions of older ones.[45]

The states remained the key governmental entities in dealing with stream pollution, and after World War I more attempts were made to develop interstate agreements. But like the disputes between riparian users, all battles between upstream and downstream cities over sewage pollution could not be settled through state cooperation. St. Louis sought to obtain a federal injunction to prevent Chicago from discharging sewage from its drainage canal into the Mississippi River. City leaders claimed that the diversion of the sewage polluted the Mississippi, from which St. Louis drew its water supply. The Chicago Sanitary and Ship Canal, opened in January 1900, permanently reversed the flow of the Chicago River from Lake Michigan into the Mississippi River system. In 1907, the last two sewer outfalls emptying into the lake were shut off. The Chicagoans' rationale was that by keeping the city's sewage from draining into Lake Michigan, the greatest threat to the water supply had been averted.

Officials of the Municipal Sanitary District of Chicago had been aware for some time about St. Louis's objections to the drainage canal, but opened it without fanfare before their neighbor obtained an injunction. Ultimately, the state of Missouri petitioned the U.S. Supreme Court to prohibit the state of Illinois and the Sanitary District of Chicago from releasing sewage into the main channel of the canal on the Illinois River. The case was dismissed, but it was later discovered that the water in the Illinois was purer than the water in the Mississippi, and thus St. Louis constructed a purification plant on the latter.[46]

A variety of technologies gained attention as public health officials and others touted the value of treatment over simple dilution, and as jurisdictional battles between upstream and downstream cities heated up. Disposal of sewage on land had deep historical roots. Irrigating cultivated land with sewage—broad irrigation or sewage farming—began in England in 1858, and remained the only recognized method of sewage disposal in England until the 1870s. Large areas of sandy soil were ideal for this method. Sewage farming also found use on the European continent, in Asia, and in the United States. The method was implemented in New England in the 1870s as a way to grow crops. Pullman, Illinois, was the first municipality to use sewage farms for disposal purposes in 1881.

Sewage farming promised to convert wastes into valuable produce, but it did not catch on except in the West, where it provided some water value, especially in the arid regions. The first of the western irrigation plants was established in Cheyenne, Wyoming, in 1883. While broad irrigation flourished in California, many of the other sites were abandoned because they were poorly operated, posed a nuisance, or were replaced with water-irrigation systems.

Outside of the West, sewage farming faced criticism as inefficient and as potentially dangerous to health. Public health authorities raised concerns about the transmittal of infection to farm workers or through the crops themselves. While no conclusive evidence substantiated the claims, such criticisms undermined the widescale use of the method.[47]

Intermittent filtration, developed in England in the 1860s and refined in the United States in the 1880s, was regarded as superior to sewage farming. It was a modified form of broad irrigation, simple to construct, and required a smaller area in which to operate. Sewage was applied daily to sand beds underdrained with natural sand or tile. Intermittent application of the sewage let air enter the bed, allowing biological as well as physical actions to dispose of the waste materials. The resulting clear and almost odorless effluent was free of most suspended matter and a high proportion of bacteria. The method was first used in Massachusetts in 1887. By 1934, more than 600 such plants were in operation. Intermittent filtration was primarily confined to New England and other areas along the Atlantic Coast where natural deposits of sand were abundant.[48]

Chemical precipitation was sometimes utilized with sedimentation techniques, involving the use of coagulants (lime, alum, sulfates) to aid in the removal of solids from sewage. With origins in France and later England, it was first used in the United States at Coney Island in 1887. Because it was expensive, created large quantities of sludge, and did not purify sewage thoroughly, chemical precipitation was best suited as an auxiliary treatment, and was eventually replaced.[49]

By the late nineteenth century, the English in particular had experimented with every known method of sewage disposal in search of a last-ing solution to this knotty problem. Methods such as intermittent filtration were turning attention to biological processes validated by the growing ac-ceptance of the germ theory. The contact filter, developed in the 1890s, was among the first of the modern biological processes. "It was part of a concep-tual revolution of the principles of sewage treatment that occurred in the 1890s," states Christopher Hamlin, "the replacement of a philosophy that saw sewage purification as the prevention of decomposition with one that tried to facilitate the biological processes that destroy sewage naturally."[50]

The developer of the new process was William Joseph Dibdin, chief chemist of London's Metropolitan Board of Works and later of the London County Council from 1882 to 1897. In his official capacity, Dibdin was re-sponsible for the Thames estuary where London's sewage flowed. Public pressure to end the pollution of the estuary, but cheaply, led Dibdin to em-ploy technical means "to meet public-relations ends, convincing the public that the responsible authorities were taking responsible action." In 1887, he proposed that microorganisms in the estuary could purify sewage.[51] The revolution in sewage-treatment technology was not for Dibdin to claim ex-clusively, however. Biological treatments were developed independently in several locations in the United Kingdom and the United States.

In the early 1890s, Dibdin undertook to devise a more rapid method of sewage disposal, then a process under study at the Lawrence Experiment Sta-tion. At a site in England, he laid out a one-acre bed filled with coke, which he called a "contact system" because sewage was held in contact with the coke by closing the outlet. In 1894, he installed seven experimental beds at Sutton in Surrey. There he developed a double contact system. The contact bed—essentially a tight box holding filtering media—could use a variety of hard, smooth materials such as coke, coal, or stone. The resulting effluent was of good quality, but the beds clogged up periodically. Many such beds were placed into operation—usually in smaller plants.[52]

The trickling filter was a step beyond Dibdin's contact beds in the search for a better method of oxidizing sewage. Its design was similar to such beds, but allowed air to circulate. Sewage was sprayed intermittently over the bed, which varied in depth. When trickling filters were installed, they were utilized along with preliminary sedimentation and screening processes or other complementary technologies.[53]

Although developed at the Lawrence Station, trickling filters attracted much more attention in England. Coarse materials for use in trickling filters were in greater abundance there than natural sand. Consequently, the ac-tive refinement of the trickling filters rapidly shifted away from the United States, and thus the work of the Lawrence Station showcased a change from one-way municipal technology transfer. Until the 1890s, technological innovations had largely flowed east to west across the Atlantic, but now they also were beginning to flow west to east.

One of the earliest trickling filters was constructed at Salford, England, in 1893 and had been inspired by the Lawrence experiments. Madison, Wisconsin, built one of the earliest practical trickling filters in the United States in 1901. The first modern sprinkling filter plant for municipal use in the country was put into operation in Reading, Pennsylvania, in 1908.[54]

Among the various biological processes, the development of the so-called septic tank provided the greatest versatility. While it was not practical for the final disposition of sewage, it was most valuable as a preliminary step in treating several kinds of concentrated sewage. Its major value was the economic removal of sludge. The septic tank was essentially a sedimentation tank, which permitted the decomposition of solids or sludge in the absence of oxygen, leading to a type of fermentation. The process gained wider popularity after 1895, when Donald Cameron, the city surveyor of Exeter, England, gave it the name "septic tank."[55]

Refinements soon followed. Dr. William Owen Travis of Hampton, England, patented the Travis tank in 1903. He devised a two-story tank divided into three sections—two sedimentation chambers and one liquefying chamber, which allowed sludge to decompose for the most part separate from the flow of fresh sewage. However, only one large installation was constructed after his original plant was built in Hampton in 1904.[56] Travis was undercut by the development of a new tank designed by Dr. Karl Imhoff of Germany. Hering had been interested in Hampton's work and suggested to the sewerage engineer of the Emscher Drainage District Board in Germany that he study the design. In 1905, Hampton began to construct a Travis tank but died before its completion. His successor, Karl Imhoff, carried on and added improvements. The Imhoff (or Emscher) tank was also a two-story sedimentation/liquefying tank, but no sewage was allowed to enter the sludge chamber, and the process was faster and cheaper. By World War I, approximately seventy-five cities and several institutions employed the Imhoff tank. It soon became the most popular method for preparing sewage for secondary treatment in both the United States and Europe.[57]

Along the way, a variety of processes were devised to aid in the treatment process, such as grit chambers and screens. In Holland, the Liernur system drew sewage to a central station by means of a vacuum created by an air pump. The sewage was then deposited into barges and taken to nearby farms.[58]

In some locations, disinfection of sewage was employed to remove potential disease-producing organisms. The process began in Brewster, New York, where a chlorine plant had been installed in 1892.[59] Another innovation was the electrolytic process, by which sewage was passed through an electrolytic cell, which increased the precipitation of solids in the sewage and disinfected the sewage through the formation of hypochlorite.[60]

The most promising treatment technique—called "a modern miracle"— was the activated sludge process developed in the the second decade of the twentieth century. Aeration was used to produce a flocculent precipitate—a sludge—from sewage, which contained large numbers of bacteria and promoted the multiplication of nitrifying organisms. Seeded with nitrifying bacteria, the "activated sludge" was added to fresh sewage and then aerated briefly, resulting in the nitrification of the sewage, removal of bacteria, and the formation of a quick-settling sludge. It achieved a bacterial reduction of more than 90 percent. A small area was required for the process, but the remaining sludge was a problem not resolved in the period.[61]

Experiments with sewage aeration had been conducted for several years, but it was Dr. Gilbert J. Fowler of Manchester who discovered the activated sludge process. Fowler apparently got the idea from aeration experiments conducted at the Lawrence Experimental Station. The Manchester Disposal Works conducted aeration experiments that led to the first activated sludge process. In November 1913, Fowler and his associates publicly presented the first of his Manchester studies, in which he reported that sewage inoculated with oxidizing bacteria could be clarified and free from bacteria after six hours of aeration. Many additional experiments continued in England and the United States. The first American installation was completed in San Marcos, Texas, in 1916, soon to be followed by the widely known facility in Milwaukee, which pioneered the development of a process for the manufacture of a salable product from the sludge residue.[62]

Because of the expense and the debate over sewage treatment versus filtering water, new methods were not implemented uniformly or immediately. In 1900, approximately 1,100 of 1,500 cities and towns (with populations exceeding 3,000) had some type of sewerage system. There were only eighty-nine municipal treatment works in operation by 1905, sixty-nine of which had been built since 1894. Not until 1940 did more than one-half of the sewered urban population have treated sewage.[63]

Joel Tarr has observed that by 1890, most engineers had accepted a rational model of design choice as outlined in Hering's 1881 report on sewers. The rationale was that since neither combined nor sanitary sewer systems posed major health risks, the needs of the local community should dictate implementation. This belief rested on the assumption that large cities with combined systems could safely dispose of sewage into nearby waterways. Studies of water suggested otherwise, and the issue of treatment for sewers was given greater attention. Rather than shifting their approach to developing sewer systems to meet the new evidence, engineers relied on self-purification of streams through dilution. Such a stance pitted engineers against public health officials, with the former arguing for dilution—and increasing support for water filtration—and the latter arguing for treatment.[64]

Tarr argues that the dispute between engineers and public health officials was most intense between approximately 1905 and 1914. While this is true, it is clear that by about 1920, engineers supporting dilution had modified their view to place greater stock in some forms of pretreatment of sewage prior to disposal. Also, adherence to filtration is as much a question about responsibility for the quality of water to be utilized as it is about treatment.

It should be remembered that advocates of sewage treatment were concerned only with bacterial pollution and not chemical pollution. Furthermore, not all engineers prescribed to rigid adherence to dilution. There were many shadings of opinion.[65] What was transpiring, from the late nineteenth century through 1920 at least, was a readjustment to the changing public health paradigm and greater attention to an end-of-the-pipe pollution problem not well anticipated in the initial battle of separate versus combined systems. This latter issue was brought to the attention of both engineers and public health officials through rising concerns over water pollution. Key players included the courts, state legislatures, and, to a lesser extent, the federal government, which focused on water pollution and thus forced reconsideration of disposal options on the cities.

By 1920, a public commitment to underground sewers was widely accepted. But increased water pollution had only begun to be addressed through the debate over filtration versus treatment. In much the same way as water-supply systems spread beyond city limits, wastewater systems were developing regional import from a sanitary and legal perspective.

The Third Pillar of Sanitary Services

THE RISE OF PUBLIC REFUSE MANAGEMENT, 1880-1920

Since the advent of the sanitary idea, refuse collection and disposal had been lumped into the same category as water supply, drainage and sewerage, ventilation, and other public health problems that could benefit from the application of environmental sanitation.[1] While water-supply and waste-water systems were evolving into public functions, refuse had yet to rise above the category of "nuisance."

The "garbage problem" began to receive public notoriety in the 1880s. First, the accumulating household wastes, ashes, horse droppings, street sweepings, and rubbish became too overwhelming in the growing cities for individuals to deal with themselves. Second, sanitarians had made an effective case about the potential dangers of putrefying materials. Third, the "civic awakening" of the late nineteenth century elevated the value of a clean city as a source of pride, civility, and economic well-being. Fourth, the predisposition for publicly managed sanitary services was already established.

The quantities and varieties of wastes produced in late-nineteenth-century American cities were a testament to their growth, productivity, and consumption of goods.[2] Boston authorities estimated that in 1890, scavenging teams collected approximately 350,000 loads of garbage, ashes, street sweepings, and other discards. In Chicago, 225 street teams gathered approximately 2,000 cubic yards of refuse daily. In Manhattan at the turn of the century, scavengers collected an average of 612 tons of garbage daily. Because of seasonal variations in available fruits and vegetables, that amount increased to 1,100 tons daily in July and August. Between 1900 and 1920, each citizen of New York City produced approximately 160 pounds of garbage, 97 pounds of rubbish, and 1,231 pounds of ashes per year. Between 1903 and 1918, surveys indicated that the annual per capita production of garbage in American cities ranged from 100 to 300 pounds; rubbish, 25 to 125 pounds; and ashes, 300 to 1,500 pounds. Total per capita refuse ranged from one-half to one ton per year.[3]

Between 1880 and 1920, the United States was shifting from a producer to a consumer society. "Americans everywhere and of all classes began to eat, drink, clean with, wear, and sit on products made in factories. Toothpaste, corn flakes, chewing gum, safety razors, and cameras—things nobody had ever made at home or in small craft shops—provided the material basis for new habits and the physical expression of a genuine break from earlier times," writes historian Susan Strasser. Years before Henry Ford's famous assembly-line technique was employed to produce Model Ts, several companies used conveyor systems in meat packing, vegetable canning, and brewing of beer. In the 1880s, new machinery made flow production possible in the manufacture of consumer items, including soap, cigarettes, breakfast cereals, and canned goods. In big cities, people could shop for countless products in the new department stores; in the countryside, the Sears and Montgomery Ward catalogs brought many of those products into the home.[4]

Improvement in the standard of living was aided by a period of declining prices from the 1870s through the 1890s. However, inflation in the early twentieth century produced a boom in consumer spending. More goods available meant more refuse. However, it was easy for native-born Americans to blame new immigrants in ethnic neighborhoods or poor blacks in the ghettos, rather than affluent whites, for the piles of garbage and rubbish. A 1912 survey in Chicago correlated nationality and ethnic background with the production of refuse. It demonstrated that native-born whites produced substantially more waste than new immigrants. In essence, a fancy downtown hotel could be as responsible for producing mounds of refuse as a Little Italy or a Little Poland.[5]

Some wastes were accumulating at an alarming rate. Between 1903 and 1907, Pittsburgh's garbage increased from 47,000 to 82,498 tons (43 percent).

Other cities experienced substantial increases: Milwaukee, 24 percent; Cincinnati, 31 percent; Washington, D.C., 24 percent; and Newark, 28 percent. Population growth, greater consumption, and more efficient collection accounted for most of these increases.[6]

Franz Schneider Jr., research associate of the Sanitary Laboratory of MIT, imaginatively calculated that "if the entire year's refuse of New York City could be gathered together, the resulting mass would equal in volume a cube about one-eighth of a mile on an edge. This surprising volume is over three times that of the great pyramid of Ghizeh, and would accommodate one hundred and forty Washington monuments with ease. Looked at from another standpoint, the weight of this refuse would equal that of ninety such ships as the 'Titanic.'"[7]

By comparison, European city-dwellers produced substantially less refuse than their American counterparts. A 1905 study indicated that fourteen American cities averaged 860 pounds of mixed rubbish per capita per year, while in eight English cities the amount was 450 pounds, and in seventy-seven German cities, 319 pounds.[8]

Horses, the major source of transportation during the period, contributed a substantial share of refuse. By the mid-1880s, 100,000 horses and mules were pulling 18,000 horsecars over 3,500 miles of track nationwide. Scores of other horses served a variety of work needs in the cities. Sanitary experts calculated that the average city horse produced approximately fifteen to thirty pounds of manure daily. Another observer estimated that the 82,000 horses, mules, and cows that were maintained in Chicago produced more than 600,000 tons of manure a year. Given the approximately 3.5 million horses in American cities at the turn of the century, the problem of manure was overwhelming. Stables and manure pits also were breeding grounds for disease, and since the life span of a city horse was only about two years, carcasses were abundant. New York City scavengers removed 15,000 dead horses in 1880 alone.[9]

Collecting and disposing of refuse posed distinctive problems. A city's geographic location, climate, economic and technical factors, and local traditions were crucial. Having to contend with a wide array of wastes—garbage, rubbish, human and animal excrement, dead animals, street sweepings, and ashes—complicated the process.

Expediency often won out over health or aesthetic reasons through the 1880s. The two disposal methods that dominated were dumping refuse on land or into water, immediate solutions that merely shifted the problem from one site to another. For many cities, especially those not situated on waterways, it was common to dump refuse on vacant lots or near the "least desirable" neighborhoods. Protests from those who unfortunately lived near the stench-ridden mounds often went unheard. In the fastest growing cities, rapidly multiplying commercial and residential building and the subsequent high land values made it difficult to acquire new dumping areas.

Dumping into water was more pernicious. Cities that dumped into rivers had to contend with the wrath of downstream neighbors. In 1886, New York City dumped more than one million cartloads of refuse into the ocean, and continued to rely on this practice as a primary means of disposal for many years. It led to clogging the approaches to the harbor and polluting nearby beaches. In Chicago, the lack of "convenient and suitable" dumping grounds led the city to dispose of much of its waste into Lake Michigan.[10]

Only the use of garbage as animal feed or fertilizer demonstrated any foresight in utilizing discards as a resource. However, wastes used for agricultural purposes often were handled without proper safeguards. The transfer of garbage from city to farm sometimes required an intermediary stop at a swill yard—a breeding ground for disease—within the city limits, where the waste was unloaded for farmers to cart away.

The common practice of feeding garbage to swine also started to be questioned; studies indicated that the products of animals fed on garbage might not be fit for human consumption. Investigations in the mid-1890s in New England—where swine feeding was common—revealed that cases of trichinosis among hogs fed with garbage increased from 3 to 17 percent over a three-year period, and that the annual mortality from hog cholera increased alarmingly. The major cause of the problem was not the feeding of garbage, but the manner of handling and transporting the slop.[11]

Reform of refuse collection and disposal practices generally grew out of local demands for change, and ultimately was linked to existing programs of environmental sanitation. As long as anticontagionism held sway in a community, the potential health risks posed by solid wastes received attention among health officials. Boards of health normally did not manage public collection and disposal programs, but oversaw citizen actions and the work of scavengers.

The 1880 census revealed that at least 94 percent of the cities surveyed had a board of health, health commission, or health officer. Of these, 46 percent had some direct regulatory power over the collection and disposal of refuse, and almost all could deal with nuisances created by solid wastes. The larger American cities granted their health boards and commissions more direct influence over refuse collection and disposal than did smaller cities. The effectiveness of the boards depended on the degree to which they were adequately funded and the extent to which they were free of political interference. A large number of the boards were dominated by city officials; some boards included no physicians or sanitarians as members.[12]

Despite the constraints of local sanitary authorities, health experts tended to dominate thinking about improving refuse collection and disposal practices in the period. Noting the unsatisfactory condition of refuse collection and disposal throughout the nation, the APHA appointed the Committee on Garbage Disposal in 1887. For ten years the group gathered statistics, examined European methods, and analyzed the various approaches employed in the United States in an effort to obtain some practical answers.[13]

Civic organizations that called for improvements in sanitary conditions relied heavily upon the health argument. Concern about the refuse problem was reflected in most newspapers, popular magazines, and technical and professional journals of the day. Citizen neglect of sanitation matters was a popular theme. An 1891 *Harper's Weekly* article noted, "The average citizen, accustomed to endure nuisances as a humpback carries his deformity, saunters along sublimely indifferent to foul smells, obstructed sidewalks, etc."[14]

Ultimately, protests against inadequate refuse collection and disposal were becoming a primary function of many citizens' groups largely because garbage was a problem everyone had to confront. The Ladies' Health Protective Association (LHPA) of New York City became a leading force in the fight to bring about sanitation reform. Organized in 1884, the LHPA was the outgrowth of the efforts of women from the exclusive Beckman Hill area to force the removal of a stench-ridden manure pile from their neighborhood. The association undertook a variety of projects, including slaughterhouse and school sanitation, street-cleaning improvement, and refuse reform. While it was influenced by national trends in public health, the LHPA was a community organization without medical or technical expertise. Aesthetic considerations underscored much of the association's interests.[15]

Possibly because of the LHPA's general outlook about the nuisance quality of waste and its defilement of the city's beauty, the association put the problem in terms that citizens and political leaders could understand. Its successful efforts in lobbying for improved sanitary conditions in New York City led to the formation of similar groups in other cities.[16]

A variety of civic reform groups sought solutions to the refuse problem as part of their attempts to rid the cities of vice, corruption, and disease. The Progressive reform spirit placed great faith in the ability to alter behavior by improving the environment in which people lived and worked. Many of the important national municipal organizations during this period, such as the National Municipal League and the League of American Municipalities, embraced refuse reform as one of their goals.[17]

Middle-class women's groups assumed the leadership in civic efforts to end the refuse problem. Mildred Chadsey, commissioner of housing and sanitation in Cleveland, used the term "municipal housekeeping" to define those sanitation functions that were previously performed by individuals.

"Housekeeping is the art of making the home clean, healthy, comfortable and attractive," stated Chadsey, ". . . municipal housekeeping is the science of making the city clean, healthy, comfortable and attractive."[18]

Engineer Samuel Greeley suggested that two qualities were required for success in refuse disposal work—common sense and expert professional skills. "Women," he asserted, "are doing much to bring these qualities to bear upon the refuse disposal problems of the communities in which they live."[19] Despite the value that some placed on "municipal housekeeping," it was an acceptable reform activity for women because it was regarded as an extension of their functions as housewives and mothers. Nevertheless, the role of women in health reform had greater implications. "For the ordinary woman, the health revolution became a fundamental ingredient in women's modernization, allowing her to cope with the problems created by industrial and urban living and easing her transition into a more complex and modern world."[20] Municipal housekeeping represented more than a catchword for urban cleanliness; it was a part of the expanding mainstream of urban environmentalism.[21]

Women were aggressive promoters of refuse reform. In Boston, Chicago, Duluth, Minnesota, and other cities, women's organizations began investigations into disposal methods. The Women's Health Protective Association of Brooklyn obtained new ordinances for collection and disposal. The Louisville Women's Civic Association published and distributed 4,000 pamphlets about the garbage problem and produced a movie entitled *The Invisible Peril*. Shown to thousands of citizens, the movie depicted how a discarded hat spread disease.[22]

Historian Maureen A. Flanagan has provided solid evidence that gender roles influenced not only the type but also the manner of refuse reform. As one example, she cites the case of two city clubs in Chicago that took opposing positions on the best way to dispose of garbage. The men's club supported contracting out the service, mainly for financial reasons, because the men viewed the city "primarily as an arena to do business." The woman's club favored municipal ownership, with an eye toward improving the healthfulness of the city, because for most middle-class women, the "primary daily experience . . . was the home."[23]

The city cleanup campaign was a highly publicized form of citizen involvement in refuse reform in the late nineteenth and early twentieth centuries. At least once a year, most cities sponsored a "cleanup week" to generate interest in sanitation and other problems, such as fire prevention, fly and mosquito extermination, and city beautification. Interest in the campaigns brought together several civic organizations. In Philadelphia, the municipal government assumed the leadership of the city's campaign and transformed it into a major event. Some communities, such as Sherman, Texas, used the opportunity to campaign for better collection and disposal ordinances. Too often, cleanup campaigns were merely cosmetic exercises with few benefits other than publicity for better sanitation.[24]

Agitation for solutions to the garbage problem was rooted in a concern about the healthfulness and "civility" of the city. It also was grounded in criticism of inadequate management and random collection and disposal practices. While the issue of responsibility for water supply and sewerage had all but been resolved in favor of public management, cities were still grappling with who should provide scavenger service. Fewer than one-quarter of all cities surveyed in 1880 had a public collection system for garbage and ashes; a little less than one-third of the cities with less than 30,000 population had public collection, but an almost equal percentage were private.[25]

Figures for street cleaning demonstrate a much more significant trend toward municipal responsibility than for refuse collection. Of the cities surveyed in the 1880 census, 70 percent made public provisions for street cleaning.[26] The arteries that allowed humans, animals, and goods to move from one place to another had to be free of obstacles. The question of responsibility for street cleaning was determined more easily than for refuse because streets had no clear territorial limits. While abutters often were required to assume financial responsibility for construction and repair of streets, private responsibility for cleaning was much more difficult to monitor. Major streets, especially those in the commercial districts, were the only ones likely to be cleaned on a regular basis.[27]

Because it was increasingly impractical for individuals to collect and dispose of refuse in many cities, officials normally chose among privately arranged scavenging, city-contracted scavenging, and municipal service. The contract system was popular in large cities because it required little or no capital outlay, while still allowing for some municipal supervision. Advocates of the contract system, moreover, feared that municipally operated services bred political corruption. The contract system also was promoted as an incentive for free enterprise in the cities. Faith in the contract system was not shared by all officials and concerned citizens. As in the case of water-supply contracts, refuse collection and disposal contracts often were viewed as overly generous and without sufficient safeguards to rectify poor service.[28]

Momentum for municipal home rule and civic reform also undermined the contract approach. In 1898, the Philadelphia Municipal Association strenuously opposed the awarding of garbage contracts to the same bidders who always won them. It charged favoritism and declared that the bidding process was "a cunning scheme for robbing the city."[29]

Contract terms varied greatly, but most were of short duration, running about three to five years. Frequent renewals gave cities the opportunity to reevaluate the provisions of the contracts and review performance, but they were time-consuming and entangled in bureaucratic red tape. Contractors were unwilling to devise cleaning systems that required large capital outlays, such as constructing expensive incinerators. Attempting to hold down costs, they often employed poorly trained workers. Disaffection with short-term contracts led to increased agitation for public collection and disposal. The 1880 census indicated that 24 percent of the cities surveyed provided municipal collection service, while a 1924 survey showed 63 percent had municipal collection.[30]

The movement toward municipal responsibility was accompanied by an internal bureaucratic shift in municipal government from health department/health board supervision to management by an engineering or public works department. The acceptance of the germ theory brought into question the importance of environmental sanitation in curbing disease, and the responsibility of health officials for dealing with refuse came under fire. Some in the medical community argued that bacteriology had not advanced sufficiently to rule out the impact of decaying organic waste as an indirect cause of disease. Others argued that garbage collection and disposal had no relation to communicable disease and therefore should not be a health department function.[31] While the exact relationship between the "garbage problem" and disease was not resolved, the shift in responsibility to engineering and public works departments placed increasing emphasis on applying technical expertise to collection and disposal practices.

The appointment of Colonel George E. Waring Jr. as street-cleaning commissioner of New York City (1895) is the best example of this shift, and may have been the turning point in the development of modern refuse management. This was the same Waring who achieved such notoriety for the Memphis sewerage system. He brought to New York the enthusiasm, salesmanship, and controversy that had set off a national debate over separate versus combined sewers, now applied to the refuse problem. Waring's appointment in New York proved to be the culmination of his career. In an effort to strengthen city services, reform mayor William L. Strong had appointed Waring and Theodore Roosevelt (police commissioner), hoping that they would remake the city's corruption-ridden departments.

Waring began his duties by clearing the Street Cleaning Department of its deadwood. After years of Tammany Hall domination, the Street Cleaning Department had become heavily laden with spoilsmen who misused the department's funds and provided only marginal service. After expelling the political cronies, Waring selected young men with engineering backgrounds or military training. Even in hiring men for the most menial tasks, he sought to put "a man instead of a voter" behind every broom.[32]

To implement his extensive cleansing program, Waring devised a variety of changes. His plan represented an accumulation of the best methods attempted throughout the country. The collection of refuse posed many difficulties. Waring's first step was to initiate a system of "primary separation" at the household level. Garbage, rubbish, and ashes were to be kept in separate receptacles awaiting collection. This method allowed the Street Cleaning Department to dispose of the material using appropriate methods for each. He also initiated the building of the first municipal rubbish-sorting plant in the United States, where salvageable materials were picked out of discarded rubbish and then resold. Profits from the plant were returned to the city to offset collection costs.[33]

Street cleaning was the commissioner's most serious collection problem. The streets were crudely constructed; littering was acceptable behavior; and horse manure was everywhere. Waring enlarged the corps of sweepers, improved their competence, and raised their morale. Street cleaning was not an honored job, but the colonel gave his men an esprit de corps by increasing their pay and improving work conditions. His "White Wings" were issued white uniforms. Although such uniforms were impractical, they associated the men with cleanliness on a par with physicians and nurses. Over 2,000 strong, the White Wings performed admirably and brought unprecedented attention to the department.[34]

Waring's separation program and improvements in street cleaning were intended to raise public consciousness about the need for sanitary improvements, and to inspire active community participation. In this way he moved beyond applying new engineering techniques to the problems he addressed, as he had done with less success in Memphis.

The most dramatic example of civic involvement in New York City was the formation of the Juvenile Street Cleaning League. Initially, more than 500 youngsters participated in this program to disseminate information about proper sanitation and to inspire community participation in keeping the streets clean. Waring hoped that the children, especially from "the ignorant populations in some East Side districts," would set an example for their "less enlightened" parents. Despite the dubious class and ethnic overtones of the plan, Waring's ultimate goal was spreading knowledge about proper sanitation and developing a sense of personal commitment to city cleanliness.[35]

Waring faced his greatest test on the final disposition of refuse. Characteristically, he employed a combination of old and new techniques. Although he considered dumping wastes at sea to be the simplest method, he realized many of its pitfalls. He commissioned new types of dumping scows, and although they helped retard pollution of beaches, they were only a stopgap measure.[36]

The commissioner's treatment of garbage was based upon more innovative methods. Waring sought to retrieve resalable by-products by installing a reduction plant on Barren Island, which extracted from the garbage ammonia, glue, grease, and dry residuum for fertilizer. These salvageable materials were sold on the city's behalf. He also began an extensive land reclamation program by using waste as fill.[37]

By the end of his short tenure as street-cleaning commissioner, Waring generated considerable local and national attention, which led to similar programs in many other cities.[38] When reform mayor Strong was swept out of office in 1898, however, so, too, was Colonel Waring. Still at the peak of his influence as a sanitarian and municipal engineer, he was appointed by President William McKinley to study sanitary conditions in Cuba in the wake of the Spanish-American War. Soon after returning to New York, he died of yellow fever contracted in Havana. His tragic death cut short an active public service, although his martyrdom enhanced his reputation.[39]

Waring's accomplishments in New York City were significant. His cleansing program suggested viewing refuse as a multifaceted community problem. His comprehensive collection and disposal methods helped improve sanitary conditions in the nation's leading city, as well as making it aesthetically palatable. Waring demonstrated faith in municipal government to lead the way in providing sanitary services, aided by a technical elite freed from the influences of patronage. He also spoke the language of the Progressives, seeking to bring order to the emergence of a new industrial society.

Waring's program in New York provided a bridge between the arcane refuse practices of the nineteenth century and the newer ones of the twentieth. Municipal engineers now became the driving force behind refuse collection and disposal. While many of Waring's successors avoided his Barnum-like self-promotion, few showed the dramatic flair for rousing the public to the sanitation crusade. Municipal engineers who followed Waring placed increasing faith in technical solutions to the "garbage problem." The health official could identify a refuse problem, but it was the new professional, the sanitary engineer, who would find a solution to it.[40]

Sanitary engineers did much to lobby for municipal control of street cleaning and garbage removal and disposal. In their important book, *Collection and Disposal of Municipal Refuse* (1921), Rudolph Hering and Samuel Greeley stated flatly, "The collection of public refuse is a public utility."[41] As engineers became more deeply entrenched within the municipal bureaucracy, support for public sanitary systems became a vested interest.

Sanitary engineering was a powerful force in municipal affairs not only because of local demand but also through the network of important national engineering associations and committees. One of the earliest groups was the APHA's Committee on the Disposal of Garbage and Refuse. Its major functions included gathering statistics, inspecting local sanitation operations, and analyzing sanitation trends. The American Society for Municipal Improvements (ASMI, later the American Society of Municipal Engineers), founded in 1894, was important in refuse reform and the first national organization to unite all municipal engineers.[42]

An important function of these groups was collecting statistics about the refuse problem from North America and from abroad. The data gathered, although incomplete, provided a reservoir of information not available earlier. Sanitary engineers began to establish a consensus about many aspects of the refuse problem. The APHA's "Report of the Committee on the Disposal of Garbage and Refuse" (1897) was the first comprehensive analysis of national collection and disposal trends, and inspired other such studies and investigations.[43]

Realizing the problems inherent in interpreting data from a large number of cities, the APHA's refuse committee devised the "Standard Form for Statistics of Municipal Refuse" in 1913. Engineering groups also recommended that cities keep better records, especially in such areas as quantities of wastes produced, seasonal variations, and cost.[44]

Once sanitary engineers had a firmer grasp of the refuse problem, they turned to analyzing collection and disposal methods. They condemned the old practices, and more sophisticated criteria began to emerge. Sanitary engineers placed increasing emphasis on the need to examine local conditions before determining the proper collection and disposal methods for a community. In planning a new system, they took into account the types and quantities of wastes, the quality of local transportation, those in charge of the work, the city's physical characteristics, and the receptivity of local government and citizenry to changes in practice. Reliance on technical solutions to refuse collection and disposal was ultimately tempered by an awareness of the complexity of the problems faced. This meant incorporating few Waring-like citizen programs, however.[45]

Municipal engineers and other informed individuals came to the important, if not startling, conclusion that no single "best" method of collection existed. In practice, decisions focused on a choice between primary separation and combined refuse collection. Advocates of the former were most persuasive in cities that utilized some recovery process. In less than half of the cities with separation programs were all wastes separated. Critics of the method argued that combined collection was much easier for the householder, less complicated for collection teams, and cheaper for the city. Cities utilizing incinerators tended to favor combined collection.[46]

Collection was the costliest phase of dealing with refuse—from two to eight times as expensive as disposal. The rise in the frequency of service, promoted by reformers, was one reason for the high price tag. Servicing an expansive area demanded a large labor force and, increasingly, required additional funds for motorized trucks. Despite the mounting expenditures, service remained inequitable. Central district businesses and affluent areas received preference over working-class and ethnic neighborhoods.[47]

Disposal offered its own difficulties. Reduction and incineration were most often discussed during the period. Disposing of waste by fire was an ancient custom. "Cremation," "incineration," or "destruction" of refuse was a relatively new practice. Industrialized Great Britain led the way because of limited open space for land dumping, and constraints on sea dumping due to English maritime interests and potential conflicts with neighbors. Also, Great Britain's population was sufficiently dense to make centralized systems of disposal economical. Many engineers believed that incinerators could be an effective disposal technology, especially if they burned waste without additional fuel.

Between the late 1860s and 1910s, the so-called British destructor moved through three stages of development: low-temperature, slow-combustion furnaces; destructors that operated at higher temperatures, capable of producing steam for various purposes; and destructors capable of providing power for generating electricity or for pumping liquids.

First-generation furnaces were designed simply to burn waste. Around 1870, the earliest municipal furnace was erected in a suburb of London, but it produced disagreeable smoke and operated poorly. In 1874–75, a series of destructors were built in Manchester. The poor operation of this first-generation technology undermined confidence in cremation. Incomplete burning and the noxious smoke caused great concern. By 1885, efforts to reduce the smoke led to technical adjustments, but the cost of retrofitting destructors was expensive.

Newer destructors in the late 1880s and 1890s generated higher temperatures, reduced smoke, and produced steam for usable power. High-temperature destructors were attended by wild claims but slow progress, and public confidence sank once again. In 1905, it was widely accepted that generating electrical power through the use of destructors required additional fuel added to the refuse. Although the technical and economic feasibility of using destructors in combination with energy-producing apparatus was inconclusive, the new systems spread rapidly in Great Britain and abroad.[48]

The transfer of incineration technology to the United States brought great expectations, but many problems. Beginning in 1885, low-temperature furnaces were introduced. After the turn of the century, high-temperature destructors with adjunct boilers and power-generating capability were built. Both failed to catch on because of the same weaknesses the British experienced, and also because of the lack of understanding of the different physical, economic, and social environment into which the new technology was being introduced.

The first American incinerator was built on Governor's Island in New York City in 1885. The first municipal furnace was built in 1886 in Allegheny City, Pennsylvania, using natural gas to burn refuse. By the 1890s, cremators were popping up all across the country. In 1897, the APHA gave its provisional endorsement to incineration as an effective disposal option. However, of the 180 furnaces erected between 1885 and 1908, 102 were abandoned or dismantled by 1909.

Aside from incomplete combustion and the generation of deleterious smoke, some critics argued that British engineering expertise had not been taken into account in building the crematories in America. Others blamed the excessive use of fuels to augment the burning, which sharply increased the operating cost. Still others noted the difference between English and American refuse, speculating that the latter contained a greater water content and thus required higher temperatures to burn.

These criticisms suggest that the adoption of the British technology failed to meet local requirements of American cities and towns, which were generally less dense than their British counterparts. The availability of cheap land for dumping also provided formidable competition for incineration, as did lower fuel costs, which made hauling waste away from city centers more feasible. In some cases, American manufacturers simply produced poor-quality furnaces, taking advantage of the enthusiasm over the British designs.

About the time that the first-generation American crematories were discredited, the British were well into the second stage of development. In 1906, North American engineers made the first successful adoption of an English-style destructor in Westmount, Quebec, followed by similar efforts in Vancouver, Seattle, Milwaukee, and West New Brighton, New York. By 1910, many engineers were claiming that a new generation of burning had arrived. By 1914, approximately 300 incinerating plants were in operation in the United States and Canada, 88 of them built between 1908 and 1914.

English and German projects to convert waste into steam and electricity eventually prompted similar experiments in the United States. In 1905, New York City began a project to combine a rubbish incinerator with an electric-light plant. Energy, however, was more cheaply derived from other sources. Disposal systems that produced energy, therefore, were difficult to justify in the United States. At the point at which American engineering expertise could adapt the British high-temperature destructors to American uses, other factors intervened to frustrate the effort.[49]

In 1886, a company in Buffalo, New York, introduced the so-called Vienna or Merz process for extracting oils from city garbage. This "reduction" method was intended to provide cities with salable by-products such as grease, fertilizer, and perfume base, which would offset part of the cost.[50] Appearing in the United States about the same time as incineration, reduction underwent a similar early development: impulsive implementation, criticism, and reevaluation. The APHA's 1897 refuse committee report indicated great interest in reduction because of its promise to return revenue to the city.[51] After a period of operation, several undesirable side effects—including foul odors—led to increased criticism and protests. At the meeting of the League of American Municipalities in 1898, New Orleans councilman Quitman Kohnke complained, "We have been seduced by the glowing promises of rich rewards which the reduction process has failed to give us."[52] Plants built on the European design in the 1880s in Milwaukee, Chicago, Denver, St. Paul, and elsewhere were proving to be failures. Newer plants built in the 1890s were faring little better. By 1914, only twenty-one of the forty-five reduction plants in the country were in use.[53]

Problems with the reduction process were substantially different than those with incineration. Critics of reduction questioned the viability of the method itself, while critics of incineration questioned the design characteristics. In 1916, ASMI's Committee on Refuse Disposal and Street Cleaning recommended a compromise: reduction was fine for large cities where the revenue derived might warrant its use, but for small cities, incineration appeared to be more sanitary and less costly.[54]

The disappointing results of the two most promising disposal options in the period led to an ongoing dialogue among engineers on the most effective collection and disposal methods. Statistics for the period, however, reveal no panacea.[55] Many older practices prevailed, but criticism mounted against several of them. Dumping waste into water was the most universally condemned practice. In the early 1900s, New York City temporarily curtailed dumping refuse at sea because too much of it floated back to shore, and also because the Street Cleaning Department believed that the waste could be put to better use as fill.[56] Dumping into rivers or streams also had serious legal ramifications. Downstream cities began filing lawsuits against upstream cities that used the rivers for dumping, just as they had in sewage-disposal cases.[57]

Dumping on land also came under increasing criticism. Several cities employed land dumping because it was a practical alternative until funding for a better system could be arranged.[58] Boards of health and public works departments were continually badgered by complaints about open dumping. A 1917 report of the Cleveland Chamber of Commerce's Committee on Housing and Sanitation stated that the dumps "caused the most complaint in the collection and disposal system for refuse." According to the report, the dumps constituted "breeding places for rats and cockroaches and the homes in the vicinity are infested with them. Paper and other light articles blow over the surrounding territory and are a great source of annoyance. Fires break out on the dumps frequently not only endangering the adjoining property, but the smoke and smudge are very offensive especially when the fires often smolder for months."[59]

As land and water dumping drew increased criticism, engineers and sanitarians not only looked to reduction and incineration but also to some older methods—such as filling and burial—for alternatives. The "sanitary landfill" was the breakthrough that ultimately elevated the practice of filling to the status of a primary disposal practice. However, it did not come into substantial use until after World War II. But using fill for reclamation purposes was becoming more popular, especially as the idea of waste utilization gained support. The best-known reclamation project was begun under Colonel Waring at New York City's Riker's Island (the site of a prison) in the East River.[60]

The enthusiasm for waste utilization—reuse or recycling—fit well within the reform spirit of the day. The conservation of resources was a clear sign of a well-operating collection and disposal system. Cost factors also played a major role in determining the methods used. Like Waring, sanitary engineers sought to present their proposals for new methods with increasing emphasis upon efficiency of operation and possible return of revenue.[61]

The efforts of sanitary engineers and the zealousness of civic reformers led to substantial improvement in the refuse management system by 1920. The coming of World War I delayed some of the anticipated changes, but it also played a role in the resurgence of one of the "primitive" disposal methods—feeding garbage to swine. The wartime food conservation campaign temporarily tightened enforcement of garbage-collection ordinances and ultimately resulted in a de facto reduction in the quantity of garbage. What garbage remained, it was argued, could more profitably be fed to hogs instead of being incinerated or reduced at a time when food production had a high priority.[62]

The efforts of the Food Administration did not have a lasting impact, nor did the reduction in the amount of garbage continue to be a trend after the war. Despite the heightened recognition of the "garbage problem," attempts to apply the latest management skills, and the implementation of new technologies, questions relating to the underlying causes of waste generation were not addressed. Waste utilization efforts raised the issue of squandering resources, but few people gave much thought to the relationship between rising affluence, the consumption of goods, and the production of waste. Faith in science and technology to overcome the mounting heaps of refuse was powerful not only among engineers but also the public at large. For instance, with the appearance of the automobile as a primary mode of transportation, many people believed that the streets would finally be free of manure and other debris. The new motorcars were hailed as the best friend of the street-sweeper. No one realized the cruel irony—horse manure would be supplanted by a greater danger in the form of noxious fumes.[63]

Increased civic awareness was only one step forward in bringing refuse to the attention of the citizenry. The cleanup campaigns were ephemeral. Even efforts to obtain new ordinances or more effective methods of collection and disposal were inherently difficult. Encouraging citizens to change their habits, let alone understanding the complex problems associated with sanitation, would not be accomplished overnight. City officials also had to be convinced that improving refuse methods was a high priority among voters.

Yet it should not be lost sight of that until the 1890s, most American cities did not have systematic refuse collection and disposal systems. By 1920, that had changed. Water-supply and sewerage systems matured more quickly by the late nineteenth and early twentieth centuries, but collection and disposal of refuse had finally emerged as the third pillar of modern sanitary services. To what extent increasing reliance on public, centralized technologies of sanitation would ensure the perpetuation of "the sanitary city" remained to be seen.

The Great Depression, World War II, and Public Works

1920-1945

After World War I, neither the quality nor character of sanitary services underwent substantial change. Decision making, however, was complicated by two major disruptions: the Great Depression and World War II. Both changed the nature of city-federal relations and helped transform essentially local service delivery into systems increasingly influenced by regional and national concerns. The challenge for municipal officials, engineers, planners, and sanitarians was also to adapt sanitary services to growth characterized by metropolitization and suburbanization, as well as demand for such services in small towns and rural communities.

Water supply and wastewater were linked into regional systems or mired in jurisdictional disputes. By World War I, water-pollution issues and capturing distant sources of pure water already had demonstrated that the largest systems had impacts beyond city limits. Solid-waste collection and disposal was a major exception, remaining a local issue until after World War II. Administrative and financial considerations took priority over technical improvements, and rising concern over industrial wastes broadened the long-standing preoccupation with biological pollution.

In the 1920s, "urbanization took place on a wider front than ever before," as 51 percent of the country was urbanized.[1] Between 1920 and 1940, the urban population of the United States increased from 54.2 million to 74.4 million, and the number of urban places with populations of 2,500 or more increased from 2,722 to 3,464.[2] New commercial and service activities also reflected the transformation after World War I from a primarily industry-based to a consumer-based economy.

Sprawl and suburban expansion, accelerated by the automobile, led to regional urban networks that formed new metropolitan districts with numerous neighborhoods, factories, shopping and other commercial areas, and a variety of public institutions that were rearranged, reordered, or displaced. The number of metropolitan districts increased from 58 in 1920 to 140 in 1940, with the metropolitan population in those years rising from 35.9 to 63 million. The ratio of core-city to satellite growth tended to become smaller, resulting in declining growth rates at the centers and increasing growth rates in the outlying areas.[3]

In 1920, the growth rate of suburbs exceeded central cities for the first time. "Suburbanization had become a demographic phenomenon as important as the movement of eastern and southern Europeans to Ellis Island or the migration of American blacks to northern cities."[4] By the end of the decade, the population of suburbs was rising twice as fast as that of central cities, making the 1920s America's "first suburban decade." While the suburban boom slowed during the Great Depression in the 1930s, it did not altogether disappear.[5]

Living increasingly beyond public transportation, the new suburbanites relied on the automobile in ever greater numbers. In 1931, 435,000 people drove into downtown Los Angeles as compared to 262,000 people who arrived by public transit. Automobility undermined public transportation. In 1920, there was one vehicle for every thirteen people; in 1930, there was one for every five.[6]

The trend toward urban decentralization was promoted by political and social leaders and many professional planners. By the 1920s, most major cities had planning commissions, which drafted master plans to address the development of transportation and utility systems. The Regional Planning Association of America (1922), led by Lewis Mumford, advocated metropolitan development to merge cities, suburbs, and open spaces. While regional planning was popular in theory, narrower planning strategies utilizing zoning and traffic control were implemented, ossifying the divisions between the urban core and the suburbs.[7]

The federal government and private developers were major forces in advancing sprawl. The Federal Housing Administration (FHA) (beginning in 1934) made it cheaper for middle-class white Americans to buy a home rather than rent. FHA loans and Veterans Administration (VA) loans encouraged extensive residential building outside of the cores. While the availability of so many new mortgages led to a massive suburban housing boom, it also quickened the decay of inner-city neighborhoods. In addition, the practice of redlining minority neighborhoods, encouraging restrictive covenants that prohibited black occupancy in many suburban communities, and rejecting mortgages to nonwhite families drove a wedge between minority-occupied urban cores and lily-white suburbs.[8]

Suburbs themselves added to social, economic, and political fragmentation by resisting consolidation with central cities and looking to provide services designed specifically for their residents. Deed restrictions, restrictive covenants, and zoning laws were used in new subdivisions to prohibit rental or sale to African Americans. In western states, such prohibitions were extended to Latinos and Asians, as well as to Catholics and Jews.[9]

Steady urban growth, population deconcentration, and political and social fragmentation created new challenges in the delivery of sanitary services. The 1920s began as a promising decade as a number of cities carried out major public works projects, but soon found themselves deeply in debt. Indeed, debt payments began to assume a significant proportion of public spending in these years.[10]

The onset of the Great Depression had a multiplier effect on urban development. Accelerating demand for better services and for the repair and replacement of a deteriorating infrastructure ran well ahead of available resources. However, the need to manage debt, rising unemployment, and mounting tax defaults by property owners squelched public spending for services. In several cases, cities also defaulted on bonds that came due in the 1930s, and as a result, the price of municipal bonds plummeted.[11]

With property taxes no longer successfully undergirding city finances, with substantial debt burdens, with mounting unemployment, and with local industry in a tailspin, municipal leaders looked elsewhere for economic relief. State governments offered little support, and private financial institutions had their own solvency to consider.[12] Thus Washington quickly became the obvious source of hope.

While many cities had created problems for themselves by assuming large debts for massive public works, they also carried the burden of welfare for the unemployed virtually alone from the 1929 crash through 1933. President Herbert Hoover was reluctant to open federal coffers to the cities.[13] The Republican administration and conservative Democrats blanched at federal intervention into local affairs, especially since there had been little precedent for it. Hoover's call for voluntarism and his conviction that the

economic woes were primarily the result of worldwide economic dislocations were of little value in the mounting crisis.

As the depression deepened in 1931–32, the Hoover administration
turned reluctantly to a more activist role. The programs essentially were
designed to prop up corporations and banking institutions rather than to
provide direct relief or to fund public programs. During the Hoover years,
public works expenditures dropped dramatically.[14]

While Hoover's antidepression measures had little direct impact on
restoring public works expenditures, they began a pattern of government
involvement that was expanded upon during the New Deal. The Reconstruction Finance Corporation (RFC) in its original form financed no public
works but only loaned money to private institutions. The RFC's mission
was recast through the passage of the Emergency Relief and Construction
Act (1932). It marked the beginning of significant involvement of the federal
government in local public works, although its implementation did little to
relieve immediate economic hardships.[15]

A variety of New Deal programs had a more significant impact on the
cities and public works than those developed during the Hoover years. The
more activist and interventionist Democrats viewed the primary objective
of the programs as providing national economic relief and recovery, not developing a coherent national urban policy.[16]

During the first hundred days of the New Deal, Congress made good
on President Franklin D. Roosevelt's campaign promise to involve the federal government more directly in combating the economic crisis. In 1933,
Congress authorized the Federal Emergency Relief Administration (FERA)
and appropriated $500 million for emergency relief; the Public Works Ad-
ministration (PWA), with more than $3 billion to infuse into the economy
through a variety of public works projects; and the Civil Works Administration (CWA) for short-term relief through a nationwide work program. After
the 1934 election, the Works Progress Administration (WPA) replaced the
CWA. A massive program in work relief, the WPA spent $11 billion and
employed 8.5 million citizens. Derided as a "make-work" agency by some,
it nonetheless was involved in the construction and repair of thousands of
streets and highways, public buildings, and parks.[17]

The PWA was the central agency for establishing a basis for a long-term
local-federal relationship in public works, but not without its shortcomings.
PWA projects were continually mired in red tape. Since it was slow to spend
money and generate jobs, New Deal leaders pushed for the CWA and then
the WPA as direct employment measures.[18]

The PWA looked for projects that were "socially desirable," contributed
to coordinated local planning, were "economically desirable," and cost in

excess of $25,000. Between 1933 and 1939, the PWA spent $4.8 billion on highways, bridges and dams, airports, sewer and water systems, a variety of public buildings, and other public works projects. More than half of the monies found their way to urban areas, unlike the RFC funds. In all, there were approximately 35,000 different PWA projects—mostly small—located in every county in the nation save two.

The record of the PWA and other relief and recovery agencies in the New Deal was mixed. Compared with the programs administered under Hoover, they were a great leap forward in providing much-needed financial support for a deteriorating infrastructure and in establishing a base for new development after World War II. Many of the projects, however, were tangled in red tape and delays. In some cases, project grants were rescinded.

The presence of a new partner in municipal affairs, the federal government, introduced additional conflict as well as succor in coming to grips with local problems. Between 1932 and 1934, the percentage of receipts from property taxes for state and local governments fell from 60 percent to 45 percent, while the federal contribution rose from 1 percent to almost 20 percent. A new era in municipal affairs was at hand, an era with a profound impact on public works financing and management.[19]

While the drive for economic recovery increased federal involvement in local affairs in the 1930s, the prosecution of the war expanded federal authority and broadened its impact on cities in the 1940s. War mobilization deeply affected private lives to be sure—from the call to military service to gas rationing—but it also stimulated the urban economies through the construction of new war industries, defense housing, and the purchase of huge quantities of war matériel and other products. The placement of war industries, such as aircraft and electronics, and the development of new petroleum facilities in the South, Southwest, and along the Pacific Coast accelerated the process of decentralization already under way.[20]

Despite the scale of investment in cities throughout the war and the perpetuation of federal-city partnerships, the further development of public works was now more specifically tied to wartime needs. The Federal Works Agency (FWA) was the heir to the PWA. Established in 1939, it distributed federal grants for highways, public buildings, housing, and other community projects.[21]

Further development of sanitary services in 1920–45 was strongly influenced by the physical reality of deconcentration and the political, administrative, and fiscal changes wrought by expanding federal authority and interest in urban affairs. The Great Depression and World War II, despite the flurry of activity during the New Deal years, did retard the growth of sanitary services to some degree. Technological innovation played a less significant role in the evolution of water and wastewater systems than did financing and the nature of expansion. This was not the case in the area of solid waste, where changes in disposal technology would have profound effects in the postwar years. Nevertheless, sanitary services did not cease to be vital to community needs in these tumultuous times; they simply had to adjust to changing circumstances.

Water Supply as a National Issue

THE FEDERAL GOVERNMENT, EXPANSION OF SERVICE, AND THE THREAT OF POLLUTION, 1920-1945

D espite cycles in the economy from the 1920s to the 1940s, construction
and expansion of waterworks continued steadily. Many of the new systems were rudimentary ones in numerous small communities. In 1924,
there were approximately 9,850 waterworks in the United States, and approximately 14,500 in 1940.[1] Although the rate of growth was strongest from
the 1890s through the early 1920s, increases in the 1930s were significant
due to the infusion of New Deal funds.

Public ownership of waterworks rose only slightly between 1920 and
World War II, but the percentage of those owned by the public was relatively
high. In 1925, 78.3 percent of the communities with less than 5,000 population had public water systems.[2] Commitment to public systems reflected
the belief that municipal ownership and management had proven itself by
the quality of the service and improvements made in water purification and
treatment. Such a response also reflected the residual effects of the Great
Depression, which increased cynicism about the private sector.

In large measure, the waterworks industry appeared stable throughout the Depression years.[3] Revenues rose rapidly through the late 1920s and remained flat from the 1930s until the end of World War II. The cost of waterworks had risen steadily since World War I, but had dropped by the mid-1930s, only to rise again by the end of the war.[4]

Consumption figures indicate substantial water use in the largest American cities in the 1930s. There was some decline in overall water use, but increased metering may account for it.[5] Other variables also influenced consumption, such as new water supplies, expansion of facilities, increased population, and the type of customer base. According to a study conducted in 1926, a 20 percent increase in population resulted in a 2 percent rise in consumption. The study also indicated that a 20 percent increase in rates could decrease consumption by 13 percent; a 100 percent increase could lower demand by 40 percent.[6]

The need for greater cooperation between political units in the acquisition and delivery of water was becoming obvious, especially in response to metropolitan and suburban growth patterns. One example of government interaction was the so-called Ohio Plan, which fostered cooperation among several political bodies in the Youngstown area through a structure that resembled the New York Port Authority (a bistate arrangement between New York and New Jersey to control port activities). The Ohio Plan was given the force of law in 1919, making possible the creation of special districts for water or sewage works. By 1927, only Youngstown and a few other entities took advantage of the law.[7]

In other parts of the country, special water districts sprouted up in the 1920s, especially for the development of water resources and the delivery of adequate service. Maryland, for example, had four metropolitan sanitary districts by 1925. As a result of extensive annexations in Baltimore and Anne Arundel counties in 1918, the city acquired the rights to purchase or condemn any waterworks within the area. Other cities followed a similar path.[8]

Beginning as early as the 1880s, some cities began building regional water and sewer works. Between 1880 and 1940, Boston, Atlanta, and Oakland, California, moved to regional services, extending metropolitan authority beyond their borders.[9] Jurisdictional disputes were common, however.[10]

Another significant organizational issue focused on the potential benefits of jointly managed water and sewer facilities. By the 1920s, most major cities and many medium-size and small cities had invested in both technologies, but lack of integration had typified development of those services. The fiscal and administrative value of integration was becoming more apparent, especially in communities with public systems. But local circumstances dictated the ultimate results, and questions about what rates to charge were major points of debate. Integration would be slow.[11]

Without question, the greatest change in water-supply systems in the interwar years was the new role of the federal government. In the 1920s, municipalities averaged $119 million annually for waterworks construction; in 1933, new construction plummeted to $47 million. During the New Deal, the PWA financed between 2,400 and 2,600 water projects with a price tag of approximately $312 million—half of the total expenditures for waterworks for all levels of government. FERA, the CWA, and the WPA spent another $112 million for work relief on municipal water projects.

The greatest impact was felt in smaller communities that for the first time were able to finance public systems, treatment facilities, and distribution networks, stimulating growth and economic expansion. Many large cities also benefited from the federal largesse. A $5.5 million grant from the PWA in 1938 sparked a new filtration plant project for Chicago, and although it was delayed by World War II, the South Water Filtration Plant (the largest in the world at the time) went into operation in 1947. The U.S. Department of Agriculture also helped develop watershed programs and small water systems for rural communities and outlying suburbs. In all, thirty-five federal agencies were involved in water-resource issues in the 1930s.[12]

While the federal government became an active partner with local communities in developing water-supply systems in the 1930s, funds had arrived from Washington before then for a variety of water projects, especially irrigation, navigation, and flood control. By 1935, navigation and flood-control projects had grown significantly, while irrigation no longer dominated the federal construction allocations. The addition of municipal water supply, sewerage, and sewage treatment—and multipurpose projects—changed the very nature of the federal contributions.[13]

The first efforts to reinvigorate the economy through the Reconstruction Finance Corporation (RFC) in the Hoover years offered little for waterworks systems. Under the New Deal's PWA, grants and loans went to "federal" and "nonfederal" projects. "Federal" projects—many planned during the Hoover years—were fully funded by Washington. They constituted a little more than half of the projects, but expended less than 30 percent of the total cost of PWA construction. In "nonfederal" projects, the federal government generally shared the cost with a lower level of government. Overall, federal dollars accounted for 56 percent of the cost of nonfederal projects. Waterworks fell into the category of nonfederal projects.[14]

Between 1933 and 1937, PWA construction awards for nonfederal water and sewer systems amounted to $450 million out of $3.7 billion awarded for all projects. The WPA also played a significant role in the development of water and sewer systems, investing 9.3 percent of its labor effort between 1936 and 1940 in such activity.[15]

While federal support stimulated the development of new waterworks and provided resources for improving others, wartime priorities shifted federal funds away from local sanitary services. By the start of the war, dollars for construction of new water and sewer systems declined, due in part to reduced levels of federal support. According to a USPHS estimate in 1944, additional water-supply facilities needed after the war would cost more than $683 million. Extensions of systems in 6,455 communities would cost $502 million, and new systems in 4,863 areas would require $181 million.[16]

Some other federal water legislation incorporated features beneficial to developing local supplies. The Reclamation Project Act (1939) recognized water supplies as a major component in planning and constructing multipurpose water projects. The 1944 Flood Control Act included a provision for using federal multiple-purpose reservoirs to supplement municipal and industrial water supplies.[17]

Wartime needs held priority in getting adequate water to some residential areas. In Detroit, for example, twelve war-plant communities benefited from the rapid installation of four pumps during the summer of 1943, averting a potential water crisis. All waterworks had to abide by the regulations of the War Production Board, which required reduction in inventories and limited purchase of materials. Dislocations in personnel often occurred, since many technical specialists joined the armed services, and few sanitary engineers and sanitary chemists were being trained.[18] In some instances, internal security measures had to be established to protect vulnerable systems from sabotage, especially those servicing war industries.

Some large projects were undertaken in the 1930s. These included the 250-mile tunnel and conduit construction of the Metropolitan Water District of Southern California, which brought water from the Colorado River to Los Angeles, and the 85-mile tunnel of the New York City water supply, tapping the Delaware River and Rondout Creek. The conduit from the Colorado included some of the largest reinforced concrete pipe ever built.[19]

While many engineers admired the scale of the new projects, the political and social ramifications—including the aggressive impulses of urban growth—were sometimes lost, as in the case of the Los Angeles–Owens Valley controversy. The aqueduct bringing water from Owens Valley to Los Angeles proved inadequate in 1923. Desperate for more water, superintendent of the waterworks William Mulholland sought to invade the Owens Valley again, but this time he faced a militant response. Valley citizens organized against the onslaught, leading to a shooting war. By 1927 the city prevailed—getting its water from the Owens Valley and securing the creation of a Metropolitan Water District. By May 1933, the city had expanded its holdings to 95 percent of the farmlands and 85 percent of the town properties in Owens Valley.[20]

From a public perspective, water supplies had to be plentiful, cheap, and safe. Availability of abundant water at a reasonable cost varied from region to region, but assessment of water quality was more standardized by the late 1920s.[21] Also, the interwar years witnessed a better understanding about what constituted water pollution, and the process of purifying and treating water was significantly refined. There was more complete preliminary clarification of water prior to filtration, better mechanical filtration, better controlled use of chlorination, new procedures for reducing odors and tastes, attention to corrosive elements in water, broader use of aeration, and progress in softening water.[22] Debate, however, continued on when to filter water versus when to chlorinate it, and when sewage treatment was preferable to filtration.[23]

After World War I, hundreds of cities built water filters. In 1926, there were 635 filter plants in the United States serving approximately 24 million people. By 1938, the number of plants had more than tripled, with 37 million Americans using filtered and chlorinated water, 26 million using partially purified water, and 17 million using untreated water.[24]

Chlorination continued to receive major attention, especially because of periodic outbreaks of epidemics such as typhoid fever. Only about one-third of all waterworks employed chlorination by 1939. Some resistance to chlorination came as a result of taste and odor problems. Industrial wastes, especially phenols, reacted with chlorine to produce a bad taste in the water. Chloramine, used in Great Britain, began to gain favor in the United States in the 1930s as a replacement for chlorine. It proved to be a better bactericide, and it curbed bad odors and tastes. Municipalities tried other taste and odor inhibitors. In 1924, activated carbon was being used as a filter agent and went on to gain great popularity. By 1943, almost 1,200 plants in the United States were using activated carbon in odor control. Sometimes other chemicals were added to the water supply to protect against various threats to health. The first experimental fluoridation facilities—to improve dental health—were not installed until 1945 in Grand Rapids, Michigan, and Newburgh, New York.[25]

Lime treatment became fashionable for softening water. Laundries preferred the soft water because it took less soap to clean clothes. Boiler plants used soft water to prevent encrustation in boiler tubes. However, soft water was corrosive to iron and steel, which could adversely affect miles of pipe, and thus demand varied from location to location.[26]

While techniques of filtration and disinfection did not undergo substantial change in the interwar years, concern over water pollution emerged as a national issue. Some experts believed that public arousal over water pollution was a driving force for change. Noted sanitary engineer Abel Wolman stated,

> So fast has new construction moved under the stimulus of federal grants-in-aid, so effective after long preliminaries have been the aggressive educational efforts of health authorities and interested conservationists, and so aroused has the public consciousness become toward stream pollution abatement, that we no longer ask how many thousands, but how many millions, of people have availed themselves of treatment facilities each year.[27]

There was no accurate way to measure the level of public awareness, but a diverse array of interest groups—from conservationists to coastal oystermen and shrimpers—spoke out publicly against environmental threats to the nation's waterways.[28] Public health officials, engineers, scientists, and other experts took the issue seriously, and recognized the problem of water pollution in broader terms than their predecessors. A 1939 report of the Advisory Committee on Water Pollution of the U.S. National Resources Committee characterized the issues as contemporaries saw them: "Water pollution is a national problem. There are many sources and many types of pollution. Each type has distinctive effects on human activities. Each general type requires a special technique for abatement."[29]

Health issues still dominated the thinking of sanitarians and public health officials during the interwar years. They looked all along the water-supply system for possible contaminants. Tracking incidence of waterborne diseases—especially typhoid—proved to be a good indicator of the relative healthfulness of the public water supply, and trends in recent years had been positive. Aggregate typhoid (and paratyphoid) fever death rates per 100,000 population dropped steadily in the United States by the late 1920s. Large cities with better-developed sanitary services experienced lower typhoid rates than rural and smaller urban populations. By far, the greatest percentage of typhoid deaths was in the South, particularly the Deep South.[30]

While bacterial measures of water purity maintained a strong influence, industrial pollutants were taken more seriously as a better understanding of their composition became known and as studies pointed to the immense quantities entering the nation's watercourses. Among the problems identified was hindrance to the proper operation of water- and sewage-treatment facilities; consumption of oxygen, which reduced the dilution power of running water; fish kills; and taste and odor problems in drinking water. A 1923 report indicated that no fewer than 248 water supplies throughout the United States and Canada had been affected by industrial wastes.[31]

A common cost-free method of disposing liquid industrial wastes was discharging them into municipal sewerage systems. One study estimated that the wastes of approximately 30 million Americans "and even greater millions of population equivalents of wastes from industries" continued to be discharged without treatment into oceans, large rivers, and lakes.[32] Faith remained high in self-purification through dilution. Sufficient doubts had arisen about the limits of dilution to accept at face value absolute rates of flow necessary to prevent putrefaction of waste or to dissipate pollutants.[33]

Debate escalated on whether the disposal needs of industries granted them the right to utilize the municipal sewer system. In many cases, the industrial pollution in a city may have been equal to or have greatly exceeded the domestic pollution. Contemporaries had experienced problems with trade wastes for many decades, especially organic materials from food processing, tanning operations, textile manufacturing, and saw milling. Accelerating industrialization introduced many new wastes from iron and steel production, metal finishing, chemical production, coal mining, petroleum refining, and electrical power generation. Geographer Craig E. Colten has estimated that chemical, primary metal, and tanned leather goods manufacturers were responsible for approximately 5.7 million tons of hazardous wastes. "Among the wastes released in the mid-1930s were a variety of acids, toxic metals, carcinogenic solvents, and oils—all classified as hazardous materials by current law."[34]

A most difficult task was classifying industrial wastes, but no satisfactory system was developed prior to 1940. The array of materials was staggering, and the list kept growing dramatically as new techniques and new industries entered the market. The effects of a particular waste product on water supply—rather than the relative toxicity of the waste—tended to draw the most attention from health officials, water-quality specialists, and engineers. During the 1920s, phenol was regarded as the most serious industrial-waste problem related to water-supply purity. The major complaint was phenol's objectionable taste and odor, especially in chlorinated water. The problem was largely confined to the drinking water drawn from the Ohio River and its tributaries.[35]

Far more serious problems were in the making. Coal and petroleum distilling produced benzene, toluene, and naphtha. Lead waste from crushing and smelting operations found its way into watercourses and slag heaps, and became a serious problem of industrial hygiene in facilities where workers came into contact with incredibly high levels of lead. Arsenic, used in paints, wallpapers, and pesticides, was highly toxic and widespread. Steel mills produced between 500 and 800 million gallons of pickle liquors and other acids annually. These acids killed fish, corroded sewers, and hampered sewage treatment. Gasoline caused explosions in sewage-pumping stations and sewage-treatment plants. In Newburgh, New York, in 1929, a series of sewer explosions killed one person and injured several others. Sulfite waste from treating wood in the manufacture of wood pulp presented major difficulties in water treatment.[36]

In 1914, the Treasury Department—under which the USPHS functioned—established the first "Standards of Purity for Drinking Water Supplied to the Public by Common Carriers in Interstate Commerce." In 1925, the standards were revised to accommodate more effective means of evaluating bacteriological impurities and established maximum permissible concentrations for lead, copper, and zinc. They were revised again in 1942 and 1946, with mandatory requirements extended to cover other chemical constituents.[37]

The effort to include inorganic materials in a water standard was precedent setting, but had limited value. Health authorities continued to focus on bacteriological impurities. Earle B. Phelps, formerly of the pollution study center in Cincinnati and then an assistant professor of chemical biology at MIT, conducted a pioneering analysis of oxidation processes in New York Harbor with Colonel William M. Black of the U.S. Army Corps of Engineers. The Black-Phelps study was the first to promote the use of dissolved oxygen (DO) measures as a way of determining water quality. Phelps, like Marshall O. Leighton before him, believed that organic and inorganic industrial wastes were health hazards as well as deleterious to watercourses in general, but he proved little more successful than Leighton in generating interest in the treatment of specific industrial wastes.

The USPHS instead turned to stream studies, which led to a general theory of stream purification. Phelps and sanitary engineer H. W. Streeter developed the "oxygen sag" curve, which was the first quantitative model available to analyze changes in water quality. Indicators of the oxygen-consuming characteristics of organic waste were critical because a stream absorbed a variety of effluents, and it was necessary to know the total assimilative capacity. The Streeter-Phelps model offered a common measure, although imperfect, for determining levels of industrial pollution in watercourses.[38]

Health officials and other sanitarians retained a leadership role in the field of water purification in the interwar years. In the 1930s, training improved the quality of candidates seeking careers in environmental sanitation, but did not appreciably affect the numbers available. The Depression temporarily undercut the funding of health departments, but New Deal programs indirectly aided their revitalization through loans and grants for water and sewerage projects. Under FERA, a "health inventory" was conducted in various cities and rural communities. The demands of wartime reversed the federal funding trend of the 1930s, leaving local health departments once again strained for resources. But the war gave the USPHS a significant role in national health affairs, providing a touchstone for important health issues to be confronted after the war.[39]

There was much talk about the need to better integrate public health and engineering in these years, particularly to broaden the definition of sanitary engineering beyond water-supply and sewerage issues. Consulting engineer Edward G. Sheibley believed that engineers had failed to receive sufficient recognition for their public health work, but admitted that "since water-supply and sewage disposal are more urgent than most other municipal problems, engineers have centered their attention upon the construction details and have overlooked the opportunities for preventive work, which deals largely with environmental diseases."[40]

Training and education seemed to be the best way to broaden the function of the sanitary engineer. In 1924, Abel Wolman stated that in developing a program for training "that elusive individual whom we have been in the habit of calling the 'sanitary engineer,'" it was necessary to "avoid the Scylla of the structural engineer and the Charybdis of the laboratory devotee." The end product would be "sanitarians of environment."[41] Sanitary engineering and related topics, however, were not deeply entrenched in the technical curricula of many universities at the time. Furthermore, in 1923 there were no practicing sanitary engineers among the forty-five elected counselors of the APHA's Section of Sanitary Engineering. In 1925, there was only one.[42]

The Depression severely undercut the employment of engineers in municipalities. By April 1933, 44 percent of the engineers employed by cities in 1927 had lost their jobs. In 1941, the APHA went on record stating that there was a notable lack of public health engineers working on municipal environmental sanitation. The paucity of work on the local level was offset somewhat by the employment of engineers on federal projects. Optimism exhibited in the 1920s about the broadening role of the profession of sanitary engineer was significantly stalled in the interwar years.[43]

By 1945, sanitary responsibility in the United States still remained divided among local, state, and federal entities. Locally, the design of water-supply and sanitary systems and the supervision of construction were the responsibility of internal engineering staffs in large cities and sanitary districts, and the responsibility of private consulting firms in smaller cities and towns. Municipal ordinances did not effectively confront the broad problems associated with water pollution, but the establishment of regulatory bodies in some cities worked to decrease pollution.[44]

The states were the centers of action for new legislation to control stream pollution. In the early and mid-twentieth century, this issue remained the major concern over the disposal of municipal and industrial wastes. The states generally opposed any extension of federal regulatory authority in water-pollution control, preferring to confine federal involvement to investigation and research.[45]

Public health and sanitation laws increased dramatically at the state level by the end of the nineteenth century, particularly so in the Northeast and slower in the South. The first state legislation to control stream pollution, written in 1878 in Massachusetts, gave the State Board of Health the power to control river pollution caused by industrial wastes. By World War I, states were establishing boards and commissions expressly designed to regulate water pollution. By 1927, all but four states had boards of health with divisions of sanitary engineering.[46]

The results of checking water pollution were often disappointing. There were inconsistencies in the regulations, and enforcement was lax. A survey conducted by the American Water Works Association in 1921 found that only five states granted ample authority to the state's pollution agencies, and in nearly all cases the lack of appropriations hampered enforcement. Several laws failed to provide penalties for infringements, and most laws exempted specific industries, specific wastes, and certain streams.[47]

In theory at least, the individual or company responsible for a nuisance could be liable for violating a law prohibiting pollution by industrial wastes. In general, state boards preferred cooperation to placing themselves in an adversarial relationship with industry. Trade associations such as the American Petroleum Institute and the Pulp and Paper Association soon became involved in joint projects for pollution control or waste utilization, although their motives were sometimes questioned and questionable. State agencies often justified cooperation on the grounds that drastic control by court action would hinder economic growth and might result in incomplete investigations of actual stream requirements and of the applicability of treatment processes.[48]

Before national legislation was enacted, interstate compacts were the primary institutional means, other than the courts, to deal with water-pollution issues among the states. Chronic interstate rivalry could spill over into other areas; thus one possible solution was interstate agreements to control or abate pollution. The compacts could be formal agreements ratified by state legislatures and approved by Congress, or they could be informal arrangements. The first interstate river compacts were drafted in 1922 for the Colorado and La Plata rivers. Both compacts were concerned primarily with water-rights issues, but others involved flood control, irrigation, drainage, and conservation.[49]

In 1925, New York, New Jersey, and Pennsylvania signed the Tri-State Compact regarding the sanitary protection of the Delaware River, and forbade the discharge of untreated sewage and industrial wastes into the Delaware or its tributaries. In 1931, New York, New Jersey, and Connecticut created the Tri-State Treaty Commission to study and make recommendations about pollution abatement in New York Harbor. The Tri-State Compact became effective in 1936. By the late 1930s, several states had developed compacts of their own, especially along the East Coast and the Great Lakes.[50]

The interstate cooperative approach was an incomplete tool. The interstate compacts did not serve as regional plans to abate municipal and industrial pollution, since they were drawn more narrowly to deal with the level of discharge into water. They also did little to further the definition of what constituted environmental liability, and could do little to establish national standards of pollution control.

To a limited degree, federal regulation in the period sought to accomplish what state and court actions could not. There was, however, no overriding national vision for dealing with pollution. In 1912, the federal government assisted the states in evaluating water pollution through the USPHS's Stream Investigation Station in Cincinnati. In 1938, a loan and grant program for the states was established through the USPHS's Division of Water Pollution Control. Before the end of World War II, the USPHS conducted medical research and provided some medical services, but its powers were limited.

The beginnings of concern about hazardous substances was noted in the passage of the Pure Food and Drug Act of 1906, the Insecticide Act of 1910, and the more substantial Food, Drug and Cosmetic Act of 1938. In addition, industrial safety and hygiene began to receive a national hearing by midcentury. Not until the passage of the Federal Water Pollution Control Act of 1948 did the federal government enact comprehensive national water-pollution control legislation.[51]

Two pieces of legislation became important precedents for future action in dealing with water pollution and industrial waste—the so-called Refuse Act of 1899 and the Oil Pollution Control Act of 1924. Section 13 of the Rivers and Harbors Act of 1899, commonly called the Refuse Act, prohibited the discharge of wastes—other than sewer liquids—into navigable waters without a permit from the U.S. Army Corps of Engineers. It also suggested a strict prohibition against dumping that seemed to go beyond the primary goal of the law. For several years, the Refuse Act functioned as a minor statute to protect navigation. By the 1960s, the act was used, as one commentator noted, as a "cause célèbre for the environmental movement." Or as another suggested, "a piece of legislation that was aimed at keeping carcasses of cows and other floating debris from obstructing the smooth flow of commerce seems to have been turned into a useful bit of antipollution legislation by some enterprising conservationists and politicians concerned with the environment."[52]

The Oil Pollution Control Act of 1924 prohibited the dumping of oil into navigable waters, except in emergency or due to unavoidable accident. Because pollution was not regarded as an impediment to the economic fortunes of the oil industry, it did not receive much attention from oilmen in the 1920s. Nonetheless, groups most directly affected by the pollution began to protest. Oil-pollution problems were linked to water contamination, due essentially to tanker discharges and seepage problems on land. The former attracted the most attention, since the polluting of waterways and coastal areas directly affected commercial fishermen and resort owners. Conservationists also decried the discharges because of the impact on fish, waterfowl, and estuaries and bays.

Secretary of Commerce Herbert Hoover was the leading government proponent of oil conservation and antipollution at the time. As an engineer he opposed waste, and in his capacity as secretary of commerce he felt compelled to protect American fisheries, despite his conflicting responsibility to commercial shippers. The resulting government action, although precedent setting, fell short of Hoover's goals.

Within the oil industry, the call to end polluting practices was met with apprehension. The American Petroleum Institute (API) was defensive, but it soon realized that industry studies could control the flow of information on pollution. The API did not accomplish anything substantial at first, using its data simply to reduce criticism of the industry. In Congress, a bill to control oil pollution met stiff resistance, and a much weaker bill than Hoover wanted was sent to President Calvin Coolidge. The Oil Pollution Control Act of 1924, the first federal pollution control act since 1899, had inadequate enforcement provisions and dealt only with dumping fuel at sea by oil-burning vessels.[53]

A start was made in dealing with the complex issue of water pollution in the interwar years. The preoccupation with biological pollutants and epidemic disease was broadened significantly to confront a variety of toxic chemicals. Between 1920 and 1945, not only had water pollution become a national issue but the development and maintenance of water supplies in general were receiving countrywide attention as well. New Deal programs stimulated the further expansion of waterworks, especially into smaller communities. The larger point was that decision making concerning water supplies ceased to be a purely local function. For large cities, in particular, this meant a new orientation in planning for the future. The extension of metropolitan boundaries required new approaches for delivering water beyond the old urban core. Changes to come would have to take into account both the scale of growth and shifting priorities, especially concerning water quality.

Sewerage, Treatment, and the "Broadening Viewpoint"

1920-1945

Sewerage systems changed in scale more than in kind after 1920, as attention focused on approaches that could keep pace with urban growth. The debate over separate versus combined systems failed to rise to the previous level of intensity, but performance of those technologies had decision makers questioning their choices. In addition, the independent development and maintenance of water and sewerage systems were reconsidered, as treatment facilities were strained by rising volumes of wastewater related to increased water use.

Laying sewer line had long since become an essential feature of American urban infrastructure. In 1870, 50 percent of city dwellers lived in communities with sewers, although that represented only 11.7 percent of the total U.S. population (4.5 million of 38.6 million).[1] By 1920, 87 percent of the urban population lived in sewered communities, which represented about 45 percent of the total U.S. population. These figures were impressive because the number of communities with sewers had increased from approximately 100 in 1870 to 3,000 in 1920. Sewering of the urban population was almost universal by the end of World War II, at least in the largest cities, and extended to 8,917 communities.[2]

Although sewering American cities in the interwar years resulted in extensive coverage of urban communities, construction of new sewers and extensions of existing lines were susceptible to the same vagaries of the Great Depression and the war as water-supply systems. Construction volume dropped significantly through the early 1930s, rose sharply during the New Deal years, sagged during the early years of the war, was stimulated briefly by wartime industrial demand, sagged late in the war, and began to rise again in 1946.

Amid depression and war, many cities still faced population increases and outward growth, pressuring officials to supply basic services. In the case of sewerage, cities had to rely on public resources, although some explored privately operated utilities. In 1945, 8,154 out of 8,824 sewerage systems were publicly owned.[3] In some cases, cities simply were outgrowing their sewerage systems. As areas experienced the building of more paved streets and structures, effluent and runoff increased. The need for greater capacity became a common complaint.[4]

Separate systems came under great scrutiny, especially those that had neglected to develop adequate storm sewerage. In 1945, there were approximately 6,844 separate systems in place, compared with 1,470 combined systems; 373 communities utilized both.[5] The case of Houston, Texas, is noteworthy. About 1900, the system was a hybrid of separate and combined pipes with insufficient storm sewerage. By World War I, the city had a rudimentary separate system on the Waring model, emphasizing sanitary considerations over drainage. However, sewerage was so inadequate that in the early 1920s, private parties were constructing three times the mileage installed by the city. In 1937, only 175 of the 792 miles of sewers in Houston were storm sewers. Although they were larger in circumference than the tinier separate pipes, they were inadequate to contend with the increasing runoff and an average rainfall of forty-two to forty-six inches annually.[6]

Outward growth itself posed a significant problem for wastewater service in several communities. Between 1920 and 1926, sewage flow into Los Angeles's main outfall increased from 33 million to 78 million gallons per day, roughly matching population growth. Metropolitan growth in several directions made finding a solution to the increased volume more difficult. San Diego experienced similar problems in 1940 when its population increased from 200,000 to nearly 300,000 in a single year.[7]

Contending with overloading and the impact of growth—not to mention demands for sewage treatment—required a significant financial commitment. A city's bonded indebtedness and potential revenue sources had to be taken into account in setting priorities. Also, the onset of the Depression derailed many efforts to confront the sewerage problem. More than water supply, such items as sewage treatment seemed to drop rapidly down the list of municipal priorities.[8]

New Deal recovery programs offered some respite from the downward spiral of funding sewerage and sewage-treatment projects. Beginning with the RFC, and then the CWA, PWA, and WPA, businesses with staffs of engineers and chemists looked eagerly to federally funded projects to offset the hapless industrial market of the early Depression years. Funds for sewerage projects were slow in coming, however.

Ultimately, grants and loans for sewer-system projects were let. Through the PWA, they were on a scale exceeding water-system projects in allocations if not in number. The PWA supported 1,850 sewer projects with approximately $494 million in loans and grants, compared to 2,582 water-system projects for about $315 million. Between 1933 and 1939, the PWA constructed approximately 65 percent of the country's sewage plants. WPA resources also contributed to many sewerage projects.[9]

Sewage projects could be found across the country. Cities in New York State improved their planning apparatuses to attract greater federal support. The PWA expenditures for waterworks and sewerage improvements increased by a factor of four between 1933 and 1935. The development of activated sludge plants took place not only in New York City but also in Chicago and Columbus, Ohio. Stimulation in the construction of new sewage-treatment facilities was particularly noteworthy in parts of the country that had lagged behind in such development. In the South, cities such as Atlanta; Memphis, Tennessee; and Greensboro, North Carolina, made significant improvements. In the far western states, modern sewage-treatment plants essentially date from the advent of the PWA.[10]

Federal involvement in local affairs did not come without a price. On a few occasions, the PWA used its financial leverage to influence the construction procedures and fiscal practices of municipalities. With respect to sewerage, the best-known example was a project to complete disposal plants for the Sanitary District of Chicago, which were meant to arrest the dumping of raw sewage into the Chicago River. The district had begun the project prior to PWA involvement, but it had been curtailed because of the Depression and other problems.

Before providing funds to restart the project, the PWA insisted that a sewer rental charge be imposed on meat packers, believing that they utilized the sewers disproportionately. The Sanitary District considered an appropriate means of allocating costs, but was barred from inspecting the packing plants by the courts. The PWA then moved to examine the disposal practices of the packers. It also insisted that legislation be enacted to permit the levying of a sewage-treatment tax on any and all "extraordinary users" of the new plants. Under pressure from the PWA, legislation was written to allow the Sanitary District to recover monies secured in an earlier judgment against the city of Chicago. Since the Sanitary District had no alternative funding option, the PWA's leverage was powerful.[11]

Federal-local partnerships raised many questions of jurisdiction and authority. Federal financial support, however, did not resolve chronic problems of maintenance and operation or capital costs of the growing public works infrastructure. In addition, the federal government was not always responsible for precipitating political action over infrastructure funding. In the South, the sudden availability of federal dollars for public works programs mobilized local authorities in several communities. In Atlanta, aggressive contacts with federal administrators brought half of Georgia's entire WPA allocation for construction to the city. The highest priority for the expenditure of funds was a metropolitan sewer system.

In New Orleans, federal construction projects became deeply ensnared in state politics. In 1933, city officials requested PWA support for a project to be supervised by the local sewerage and water board. The application was delayed because Secretary of Interior Harold Ickes, in charge of the PWA, did not trust the Huey Long–dominated state government, which claimed control of all state contract work. In 1935, PWA funds were authorized for the project nevertheless, but as expected, Long and local authorities squared off for disposal of the funds. As soon as the federal funds were released, Louisiana governor Oscar K. Allen obtained a court order preventing the city from beginning the project, and urged the legislature to create a new sewerage and water board for the Crescent City. In response to the maneuverings, Ickes froze the funds. After Long was assassinated in September 1935, the freeze was lifted. The new Democratic machine in Baton Rouge wanted to bring federal dollars to the state, and thus New Orleans went on to complete the sewer project with federal funds.[12]

In the area of finance, the cost per capita of sewers and sewage disposal compared favorably with the cost of other sanitary services.[13] Repairs, extensions, and new construction were constant expenses, and the cost of treating wastes accelerated as water use increased and more connections to the system were added. As late as the 1940s, no clear pattern had emerged for developing a uniform rate structure for sewage disposal. In 1938, more than 600 municipalities in thirty-five states were using revenue bonds and income from sewer rents and sewer-disposal charges to finance new projects. The initiation of PWA grants provided much of the impetus for municipal use of revenue bonds.[14]

Cities attempted several approaches to distributing the cost of sewage services. One method was to levy a tax based on the assessed value of property without regard to the service rendered. Another approach was to base the charge for sewerage on water use.[15] Financial difficulties of the Depression stimulated use of service charges for sewage disposal and also refuse collection. Smaller cities found rental charges attractive because of the cities' limited ability to raise taxes or to seek revenue bonds. A 1945 survey indicated that only 184 cities of over 10,000 population were actually using sewage rental charges to increase revenue.[16]

Efforts to resolve fiscal problems, to cope with demand arising from growth, and to maintain service in the face of depression and war influenced the management of urban sewer systems in much the same way as water-supply systems. Attempts to relate sewerage costs to water use stimulated efforts to promote joint management of the services. Efficiency and economy were bringing water and sewage systems together, especially in smaller communities that could not afford elaborate bureaucracies and were not caught up in the kind of political tug-of-war that made such mergers more difficult in larger cities.[17]

In some cases, incipient planning efforts grew out of the necessity to fund costly sewerage projects. Some small and medium-size communities pooled their services into joint water and sewage facilities, or simply connected several communities to a single system. In Essex and Union counties in New Jersey, eleven communities signed a contract in 1927 to finance and construct the joint Outlet Sanitary Sewer. Seven communities on the east side of San Francisco authorized a regional sewage-disposal survey in 1940. However, city rivalries sometimes led to lawsuits rather than cooperative agreements.[18]

Calls for planning of sanitary services were more frequent after World War I, but what kind of planning? Other than joint management and cooperative measures attempted by smaller communities, the advancement of sanitary districts offered a means to rationalize development and funding of sewerage systems over multicity areas. Such districts organized for sewer-disposal purposes were slow in developing. Another governmental tool was the "municipal authority," but it tended to be more geographically restrictive than sanitary districts.[19]

Regional planning had yet to expand much beyond cooperative programs, except in the area of sewage disposal. Conflict between upstream and downstream cities over the dumping of sewage and industrial waste had been fought in the courts and addressed through interstate sanitation compacts. Advocates began to call for regional planning as a means to anticipate problems rather than trying to deal with the results. Pennsylvania set out to classify streams within its borders as preserved in nearly natural condition or for sewage disposal. In Oregon, Portland and sixty-five other communities were developing common plans for sewage disposal. In the Seattle area, efforts were being made to establish a Metropolitan Sewer District.[20]

The debate over treatment of sewage versus filtration of drinking water dissipated but did not altogether disappear after 1920. Although dilution was still widely practiced, sewage treatment as a necessary technical process and urban service was firmly established in these years, with discussion shifting to "how much?" and "what kind?"[21] In the 1920s, the basic principles of sewage treatment did not change, but the methods were modified significantly. George B. Gascoigne, consulting sanitary engineer from Cleveland, noted confidently in 1930 that "during the 21-year period sewage treatment has developed to the extent that today, with proper operation, it is possible to treat sewage continuously by several well-tried processes, to any desired degree of purification, and without offense." The Imhoff tank remained popular, and the activated sludge process emerged "from the toy of 1915 to a full-fledged industry in the large plant at Milwaukee." In the area of oxidation treatment, the trickling filter was the standard. Septic tanks and Imhoff tanks were considered primary treatment methods; chemical precipitation, activated sludge, trickling filters, intermittent sand filters, and land application were considered intermediate or secondary methods.[22]

Research trends in sewage treatment slowly turned toward a broader assessment of water pollution and how to address it. Dr. Willem Rudolfs, chief of the New Jersey Sewage Experiment Station, anticipated a new direction for sewage treatment when he stated in 1927, "The trend in sewage disposal is now in the direction of bio-physico-chemical treatment and studies on the biological treatment of sewage must take greater account of the physico-chemical factors." He realized, however, that practically all contemporary treatment was still along biological lines.[23]

Older methods used to attack waterborne diseases were refined to adjust to greater volumes of waste or to greater demands placed on treatment facilities, and to redefine what constituted a pollutant.[24] According to one sanitary engineering professor, "Industrial wastes, as an element in sewage and as a source of river pollution, are now beginning to receive the deferential attention which they have long deserved but have hitherto failed to receive."[25] Yet advances in the control of industrial wastes were not so encouraging.[26]

For the first time, in 1930 the increase in population living in communities having sewage treatment other than dilution caught up with the overall growth of the population in sewered communities. Between 1920 and 1945, the population served by sewage-treatment plants increased from 9.5 million to 46.9 million (494 percent). In 1945, 62.7 percent of people living in sewered communities also had treated sewage, while only 37.3 percent disposed of raw sewage. In 1945, there were approximately 5,800 municipal treatment plants in the United States.[27]

Chemical precipitation made a return largely because of changes in the 1930s. As the price of chemicals dropped, and since biological methods required too much land area in densely populated communities, chemical precipitation was adopted in several locations. Also, increasing recognition of trade, or industrial, wastes as pollutants turned attention to the value of chemical treatment, which was more reliable than other methods. By 1938, about 100 plants used some form of chemical precipitation.[28] Ambivalence about the long-term value of chemical treatment persisted nevertheless. Chemical treatment came to be viewed as just one step in sewer treatment. In many locations, chemicals were only used for seasonal operations, otherwise the cost of the chemicals became prohibitive. Combined with subsequent filtration of the effluent, the process was regarded as comparable to biological treatments. This suggests the continuing experimental nature of sewage treatment itself during this period.[29]

Researchers and engineers professed confidence about the technical progress being made in sewage treatment, but they did not claim to have identified solutions. "Dual disposal" techniques—the treatment of mixtures of sewage and ground garbage—were being tested in Lansing, Michigan, and Gary, Indiana. Also, a variety of experiments were being conducted on gas-sludge utilization, incineration of sewage sludge, and sludge-disposal processes. The handling and disposal of sludge, a "problem child" of sewage treatment, received attention on a par with sewage effluent.[30]

For activated sludge, the most highly touted biological sewage-treatment process, facilities were expanding, and a variety of techniques were being tested. However, the theory of the process was still under investigation, while patent suits over the process added to the confusion. The Milwaukee activated sludge plant—the largest at the time—was placed into operation in 1925. Important research and testing also were conducted in Chicago, Houston, Indianapolis, and elsewhere. Few plants went on line in the 1920s, but by 1938 hundreds were in operation. In 1939, Chicago claimed the world's largest activated sludge plant.[31]

Looking back on trends in sewer treatment, Paul Hansen of Greeley and Hansen Consulting Engineers (Chicago) said that the two "pressure periods"—the Great Depression and World War II—had a "marked effect" on the design of water purification and sewage treatment works overall. He added that sewage treatment developed "more marked trends" than water purification in the interwar years. Water purification had secured public approval earlier, but sewage treatment achieved greater acceptability due in large measure to federal financial stimulation. Sewage treatment also was a more complex process than water purification, given sewage characteristics, industrial wastes, and the disposal requirements imposed by the bodies of water receiving the waste.[32]

Harold W. Streeter, senior sanitary engineer of the USPHS in Cincinnati, referred to advances in sewage treatment and water purification as a "general broadening viewpoint" on stream pollution and control. Streeter credited "the growth of public interest in stream pollution as a nation-wide problem" as the stimulus for this broadening view.[33] More likely, the deepening experience of research scientists, engineers, public health experts, and sanitarians was responsible for the budding environmental insight.

In the 1930s, about 75 percent of the sewage produced originated in the drainage basins of the Northeast, North Atlantic, Ohio River, the Great Lakes, and the Upper Mississippi. The area also saw a great concentration of industrial pollution. In a 1939 government study, this urban population corridor was referred to as "the municipal pollution belt of the United States." It was the primary focus for those interested in decreasing water pollution.[34]

The "broadening viewpoint" was an indicator that the old debate over environmental sanitation versus individual health was beginning to be put aside. The bio-physico-chemical perspective on treatment reflected the emergence of a more sophisticated outlook. A full-blown ecological perspective had yet to emerge, but the further development of sewage-treatment and water-purification techniques suggested movement toward a new environmental paradigm that would influence sanitary services after World War II.

Streeter noted that the general broadening viewpoint on stream pollution and control had "brought with it a marked change in the nature and complexity of the engineering problems to be solved" but "not so much as regards the actual treatment of wastes, as regards the formulation of workable plans for pollution control." Such plans, he concluded, had to be designed "to restore and maintain complete river systems in proper condition for various water uses, and must avoid, as far as possible, the wastefulness of over-correction in some areas and the ineffectiveness of under-correction in other areas." He saw progress in what he described as a "water-shed consciousness" in the surveys of the Ohio River valley beginning in the early 1920s.[35]

Streeter's view was closely akin to the "wise use" conservation perspective applied to natural resources in the Progressive Era.[36] Yet it also depended on a systems perspective about protecting watercourses, which was a far cry from older notions of nuisance and the spread of epidemic disease. Attention to industrial wastes also expanded the notion of water pollution beyond a threat from disease and as a function of domestic sewage disposal. To some extent, Streeter's comments moved beyond the strict notion of sanitary services as primarily an engineering, or a technical, problem.

The interest in water pollution by professional groups, such as the ASCE, was manifest in their conference proceedings and transactions. Formed in 1920, the Conference of State Sanitary Engineers included the chief sanitary engineers from the various state boards of health. In 1928, the Federation of Sewage Works Associations was established to coordinate the activities of engineers and technicians in the field. Led by such well-known figures as George W. Fuller and Harrison P. Eddy, the federation collected and distributed a wide array of technical and scientific information on sewage and industrial waste treatment. The dividing lines between sanitary engineers and public health officials over corrective measures to deal with sewage and water quality blurred in the interwar years.[37]

Skepticism prevailed as to whether effective efforts were under way to eliminate stream pollution in these years. Harvard engineering professor George C. Whipple declared in the early 1920s, "There is now greater indifference to stream pollution, a greater laxity in enforcing laws, than was the case before the World War."[38] In a slightly whimsical tone, one writer characterized those seeking sewage treatment as favoring "purity at any price," "purity if we can afford it," or "let the fisherman go elsewhere."[39] If we accept this characterization, most engineers and sanitation experts would probably fall into the middle category. Some groups, like the environmentalist Izaak Walton League, would probably fit in the first group, as would oyster growers or fishermen. That citizen environmental groups were beginning to take up a concern for sewage treatment and water purification suggests the emergence of serious public dialogue on the issue of water pollution. The efforts might not have gone as far as "the growth in public interest," as Streeter suggested, but it was a start.

In the late 1920s, the Izaak Walton League conducted its own national study of the pollution of streams, with attention to the impact on aquatic life. In Illinois in the 1930s, the league promoted the slogan, "Clean Streams for Health and Happiness," and was credited with arousing public sentiment in favor of scientific treatment of sewage. In the 1940s, Kenneth A. Reid, executive director of the Izaak Walton League of America, proclaimed, "What more logical public works program could be found for the benefit of all of the people of the United States than a frontal attack on water pollution as the No. 1 postwar public works project?"[40]

Concern over stream pollution and sewage disposal helped to accelerate regulatory responses to water pollution at the state level, but federal action lagged considerably. Locally, attention focused on municipal liability for water pollution, and the extent to which industry could pollute watercourses through the use of public sewers. By the 1920s, it was established by law that a city could not, as one lawyer noted, "plead its governmental capacity as a shield against liability for negligently furnishing an unwholesome [water] supply." More problematic was the notion that, irrespective of negligence, a city implicitly could warrant that its drinking water was safe. Some court cases found cities to be negligent in not properly inspecting the water to keep it free from contaminants. In several cases, the courts were not willing to find that a public or private waterworks was liable on a theory of implied warranty of wholesomeness, as distinguished from a negligence test. This meant that if untreated wastes were dumped into water, the city was liable for any illness caused by the pollution.[41]

This was a relatively narrow definition of pollution, which did not take into account chemical as well as biological contaminants. Courts did affirm the liability of municipalities when discharges from sewage-disposal plants threatened a plaintiff's land or water, and seemed to be more lenient with municipalities situated adjacent to tidal waters as opposed to fresh streams. Those municipalities along tidal waters were granted more leeway in disposing wastes, as long as they did not create a public nuisance. The courts looked favorably on statutes of limitation where municipalities or states reduced the chance of liability by protective laws and charter provisions limiting the time in which an individual could sue the city.[42]

A debate grew in the 1930s with respect to industrial waste as to whether the disposal needs of industry justified granting it the right to utilize the municipal sewer system. A USPHS survey concluded that "the organic pollution contributed by industry is about equal to that contributed by the entire population. In many cases, the industrial pollution in a city may have been equal to or greater than the domestic pollution."[43]

Some forms of industrial pollution had impacts that cities would have preferred to avoid. Acids killed fish, corroded sewers, and hampered sewage treatment. Inflammable wastes caused explosions in sewage-pumping stations and sewage-treatment plants. Sulfite pulp waste presented major difficulties in water treatment.[44] The regulation of industrial practices followed no uniform path during the interwar years, and left cities with little recourse in many instances.

As with general water-pollution legislation and regulation, the most significant advances in regulating sewage disposal and treatment in these years occurred at the state level. In the 1920s, several states passed laws making it mandatory for cities to provide treatment works for sewage. The state health departments often assumed some responsibility for enforcement of these laws.[45] In too many cases, responsibility for stream pollution was divided among several state departments, with no clear authority granted to anyone.

Interstate rivalries over sewage pollution either had to be resolved through compacts or in the courts. Probably the most famous court action of the period involved the Chicago Sanitary District. From 1900 to 1906, the city became involved in litigation over its drainage canal. Officials in St. Louis objected to Chicago diverting sewage from Lake Michigan into the Mississippi River via the Illinois River. The courts ordered the Sanitary District to cease disposing of untreated sewage into the canal. While carrying an appeal to the U.S. Supreme Court, the Sanitary District began efforts in 1909 to pursue waste treatment. By 1922, it had planned and was constructing sewage-treatment plants. After many years of wrangling, the Supreme Court issued a decree that limited the amount of lake water the Sanitary District could divert. As economic historian Louis P. Cain stated, "This decree set a precedent for court-ordered limitation rather than governmental regulation." In essence, the decision validated the idea of sewage treatment and undercut the notion that raw sewage or industrial waste could be satisfactorily disposed of by simple dumping. The idea of dilution was crippled by the decision, but not wholly discredited.[46]

While sewerage and sewage-treatment techniques did not undergo major transformations in the interwar years, the importance of waste collection and disposal was strongly felt. There was a contrast between the limited commitment of resources to maintaining and upgrading systems during the Depression and war, and the assertion of the "broadening viewpoint" about the nature and extent of water pollution. For their part, engineers loosened their commitment to dilution, and raised questions about biological threats as the primary focus of antipollution efforts. Rethinking the problems associated with delivering pure water and effective disposal of sewage was shifting the environmental dialogue to a higher plane.

The "Orphan Child of Sanitary Engineering"

REFUSE COLLECTION AND DISPOSAL, 1920-1945

In a 1925 article in *American Journal of Public Health,* George W. Fuller referred to garbage disposal as "an orphan child of sanitary engineering." He went on to suggest that "engineers have had only random contact with [refuse collection and disposal] and their authority and opportunities for needed research have been inadequate to ascertain what the problem really is in different cities and how works can be best built and operated."[1] Samuel A. Greeley was not as charitable: "Garbage disposal is a phase of sanitary engineering perhaps less closely related to the public health and in which less progress may perhaps be recorded than in the fields of water-supply and sewage treatment."[2]

While Fuller and Greeley underestimated the contributions of engineers such as Rudolph Hering and George Waring to the advancement of refuse collection and disposal, they nonetheless accurately characterized the place of refuse vis-à-vis water supply and sewerage. By comparison, refuse collection and disposal did not receive the attention in technical and sanitation communities that water and sewerage received. With the end of the filth theory, refuse disposal declined in significance as a way to avert severe health hazards.

Regarded as an engineering problem more than as a health issue, refuse disposal assumed second-class status among the sanitary services. By the 1920s, several of the advocates of refuse reform had died. Many of the new generation of engineers tended to be university trained and more interested in applying their skills to specific problems than attacking broader environmental, aesthetic, or social concerns.

Refuse collection and disposal remained a local issue in the interwar years. Health and other environmental risks associated with solid wastes did not receive the attention given to typhoid or to industrial effluent. Massive quantities of waste, nevertheless, continued to mount. An ill-conceived and jarring demobilization put the United States in an economic tailspin following World War I. Eventually, a surging recovery led to the boom mentality of the 1920s and unmatched prosperity. The 1920s saw the growth of several consumer-oriented industries and the expansion of a white-collar workforce, built upon war-generated capital, industrial planning, technical innovation, and peacetime demand. Productivity rose twice as fast as the population and the middle class were expanding, but the few still controlled much of the nation's wealth. Businesses that catered to the consumer market were the most successful of the era. Mass-production techniques, an emphasis on high-pressure advertising, and easy consumer credit tended to produce democratized materialism, especially for the middle class.

The attractiveness of American consumer goods derived from variety and price. The chemical industry produced an array of new fabrics, kitchen utensils, floor coverings, and cosmetics. With access to confiscated German dye patents and expertise from innovative chemists, DuPont introduced rayon and cellophane. Other synthetics—such as plastics—were produced. Of all the consumer goods, the automobile was king, making private transportation widespread. By 1930, nine of the twenty leading corporations in the country specialized in consumer goods as compared with one of twenty in 1919.

Advertising entered the ranks of big business in the 1920s with major campaigns directed at the larger urban markets. Easy financing made consumer goods very attractive. Even religion was merchandised. Madison Avenue ad man Bruce Barton wrote *The Man Nobody Knows*, a bestseller that depicted Jesus Christ as the greatest salesman of all time.[3]

The devastation of the Great Depression obliterated the escalating economic growth of the 1920s. However, the Depression only temporarily derailed the consumer trends and habits begun earlier. Generation of solid wastes rose steadily between 1920 and 1940, from 2.7 to 3.1 pounds per capita per day.[4] In large cities during the 1920s, the average person contributed between 150 and 250 pounds of garbage per year and four-tenths of a cubic yard of rubbish and ashes.[5]

Despite its low status as a sanitary problem, refuse was hardly inconsequential for city governments. Strides had been made in gathering statistics on refuse, in developing more systematic techniques, and in extending municipal responsibility for collection and disposal. Collection of refuse, especially in the wake of expanding city limits, gained substantial notice because of the extraordinary costs involved. Attention focused on new administrative arrangements and possible advantages of mechanization in transporting waste to its final destination. A reevaluation of techniques was in the offing for disposal, although the focus remained on finding a single, technical panacea rather than developing an integrated system.

Like water supply and sewerage, the refuse problem was aggravated by the Depression and the priorities of the war. Unlike those services, refuse did not appear to pose the same environmental danger. Ironically, given their inferior status as sanitary services, refuse collection and disposal and street cleaning placed a large financial burden on major cities in the 1920s and 1930s. While they required lower construction costs, they utilized a much larger workforce than water supply and sewerage.[6]

The debate over municipal or private collection and disposal of refuse, which had appeared to be settled by the early twentieth century, resurfaced in the 1920s and 1930s. A central issue was who should bear the costs of collection. In several cities with municipal solid-waste functions, the idea of a service charge gained a following. Proposals abounded for linking the cost of collection to the weight and volume of discards, the hauling distance, the type of residence or business, and even the number of rooms in a home or the size of water bills. In some cases, special taxes could be levied to underwrite the cost of the service. No single approach was completely satisfactory.[7]

The erosion of revenue during the Depression made cities hard-pressed to maintain adequate services, let alone invest in capital projects. The Depression also meant that markets for salvageable materials diminished. New Deal programs provided some support, but were meager compared to programs for water supply and sewerage. As of March 1, 1939, only forty-one nonfederal refuse projects (as compared to 1,527 sewerage projects) received federal loans and grants. CWA workers picked up rubbish as part of fire-protection programs, but relief funds were not made available for regular collection and disposal services. Federal loans and grants, however, could be used to build incinerators, improve dump sites, and conduct cost studies.[8]

Despite the Depression, refuse collection and disposal practices underwent noteworthy technical modifications and changes by World War II. Some engineers were not convinced that progress was being made in the early 1920s. "The practice of refuse collection and disposal in American municipalities," Greeley said, "has been characterized by some well-informed observers as a mess of mistakes. They see a wide variety of disposal methods in use, and in some cities apparently abrupt changes from one method to another, sometimes involving the abandonment of seemingly useful and expensive going plants. They see large and costly disposal works built and operated for a few years and then abandoned to gradual disintegration."[9] In some respects, Greeley was responding to one of the immediate impacts of the search for permanence. Reliance on new technology to solve problems of disposal exposed a certain rigidity of thinking and a lack of adaptability of those technical fixes to changing urban circumstances. Also, Greeley was reacting to the impulsive implementation of disposal technologies foisted on cities by private vendors.

It had long been apparent that collection and transportation of refuse were the most costly aspects of refuse service, and ones that posed several problems.[10] Collection was the phase of refuse service most likely to remain a private responsibility or to be contracted out. Along the Pacific Coast, in particular, private waste scavengers and other forms of private collection dominated. Cities with populations of more than 100,000 across the country usually employed municipal service.[11]

Aggregate statistics tell a more complete story. A 1929 survey of 667 cities (with populations over 4,500) showed that only 247 (37 percent) had some type of municipal collection. A 1939 survey of 190 cities in the United States and Canada found that 149 (78 percent) had some form of municipal service. This survey, however, failed to indicate how many cities did not respond or had no systematic service, and thus the figures for municipal service may have been artificially high.[12]

To better determine the most efficient collection service, some cities initiated time studies to evaluate the location and standardization of receptacles, the size and types of wagons required for service, the length of the haul, and the requirements for secondary transportation.[13] Separating wastes at the source remained the preferred method in theory in order to facilitate a range of disposal options. But practices varied widely.[14] The greatest changes in collection in the interwar years were the increased use of motorized vehicles, the addition of transfer stations, and the use of secondary transportation. By World War II, the motorized truck replaced the horse and cart as the standard collection vehicle. Transfer stations were set up to centralize wastes for more economic hauling to the final disposal destination. Secondary transportation, such as large trucks, railroad cars, or barges, was used to increase the volume of hauls to disposal locations.[15]

No standard disposal method was employed across the country during the interwar years. Despite criticism going back to the late nineteenth century, the practice of dumping refuse on land was more frequently used than other methods. It was estimated that 90 percent of cities and towns with populations of less than 4,000 relied on open dumps. Dumping was convenient and simple, but it was notoriously unsanitary, attracting vermin, giving off offensive odors, threatening groundwater supplies, and posing a fire hazard. In 1929, a special APHA committee recommended that refuse dumps not be sited along the banks of streams because of leaching and the washing off of materials into the water after storms.[16] By the late 1930s, dumps were disappearing from the outskirts of some cities because of ordinances that banned their use.[17]

Ocean or sea dumping was no more agreeable than its terrestrial counterpart. Dumping wastes into the oceans and seas, however, was going out of fashion. In 1933, New Jersey coastal cities went to court to force New York City to terminate ocean dumping. That same year the U.S. Supreme Court sustained the lower court's decision to terminate the practice. Also in the 1930s, the state of California passed a law prohibiting the discharge of garbage into navigable waters or the ocean within twenty miles of shore. Other coastal cities, such as Bellingham and Port Angeles, Washington, and Vancouver, British Columbia, continued dumping refuse at sea.[18]

The most promising technology emerging in the 1920s was the sanitary landfill. Ashes and rubbish had been used for fill for years, but the use of organic wastes to fill ravines or to level roads was highly objectionable. The sanitary landfill combined filling and open dumping. The basic principle was to dispose of all forms of waste simultaneously and to eliminate putrefaction of organic materials. Typical sanitary fills were layered: garbage was covered with ashes, street sweepings, rubbish, or dirt; then another layer of garbage; and so forth. Chemicals were sometimes sprayed on the fill to retard decaying.

To some authorities, this method was nothing more than glorified open dumping, and they opposed it. The idea of a "sanitary" fill was intriguing because of its promise to deal with a wide array of refuse.[19] After World War II, the sanitary landfill became the first universally accepted disposal option in the country. It did not catch on immediately because it was expensive and labor intensive. Early attempts were tried at Seattle, New Orleans, and Davenport, Iowa, in the 1910s.[20] The modern practice began in Great Britain in the 1920s as "controlled tipping." The American equivalents appeared in the 1930s in New York City, San Francisco, and Fresno, California.[21]

Jean Vincenz was most responsible for developing, implementing, and disseminating the sanitary landfill. Commissioner of public works in Fresno, California, from 1931 to 1941, Vincenz was the first to use the "trench," or "cut and cover," method. Prior to developing his sanitary fill in Fresno, Vincenz studied British tipping techniques, visited several California cities, and consulted with a New York City engineer active in developing that city's fill. The sanitary landfill in Fresno opened in October 1934 at the city's sewer farm. Vincenz began a second one in that year, and soon the city was filling 4.3 acres per year with approximately 24,000 tons of mixed refuse.[22]

Sanitary landfills in San Francisco and New York City received more immediate attention than Vincenz's fill in relatively obscure Fresno. San Francisco began its operations in 1932, initially as an emergency measure, and not until 1936 did it operate effectively as a primary disposal option. Unlike the Fresno fill, San Francisco's was constructed along tidal flats on the bay, and was utilized for reclaiming land. Such modifications of the shoreline and leaching problems from the fill eventually raised major concerns.[23]

The New York City landfill began in 1936, and was similar in design to the Fresno enterprise, only larger. It was located at Riker's Island, the site of a city prison. Pleased with the project, city officials authorized other sites in the 1930s with the expectation of reclaiming additional land. Debate broke out on the degree to which the sites were indeed "sanitary," and political battles arose over the conduct of the Department of Sanitation in carrying out its disposal policy.[24]

Momentum slowly shifted to building sanitary landfills in the 1930s and early 1940s. During World War II, the U.S. Army Corps of Engineers experimented with them, and in 1941 Vincenz accepted a post as assistant chief of the Repairs and Utilities Division in the corps. Although he was skeptical about the army's extensive adoption of sanitary fills without sufficient supervision and adequate equipment, he followed his orders to implement them. By 1944, 111 posts were using sanitary landfills, and by the end of 1945 almost 100 American cities had adopted them.[25]

Waste utilization, including scrap-metal salvage, recovery of other salable items, and recycling/reuse of food wastes, had long had a place among disposal options. Swine feeding and reduction continued to be practiced after World War I, although they showed increasingly less promise. Out of favor in the late 1890s, swine feeding saw a brief resurgence when the country was impelled to increase its food supply during World War I but declined afterward. The Depression and World War II helped reinvigorate the practice, since droughts in 1934 and 1936 led to severely reduced corn crops and higher pork prices.[26]

In the 1930s, scientific studies demonstrated that the use of raw garbage as feed was an important factor in the infection of pigs with *Trichinella spiralis*, which could be transmitted to humans in undercooked pork. Although the morbidity rate for trichinosis was low and mortality was rare, the link between the disease and feeding practices turned several cities toward other methods. However, the practice of feeding garbage to swine continued in some areas until the 1950s.[27]

Reduction competed with swine feeding for garbage supplies. The reduction method, as stated earlier, utilized chemical and mechanical processes to extract salable grease and tankage from food wastes. Because the plants were so expensive, they were largely confined to use in large cities. The plants slightly increased in number after World War I, but as grease prices dropped the numbers declined again. Unfortunately, not a great deal is known about the reduction plants because about half were run by private contractors. Aside from the high cost of construction, the demand for them was limited because the plants were placed at a distance from the source of waste, due to the horrendous odors emanating from them. The increased hauling charges, as well as contaminants entering local water supplies, added to the method's limitations.[28]

One writer bemoaned the demise of reduction plants, which he believed eliminated garbage efficiently as well as recovering useful material. "Oh, for a Moses of the Garbage Can!" he proclaimed. "The tragedy is not so much in the money loss involved as in the fact that the state of the art will be set back a good forty years. There is no other method of garbage disposal which has been widely accepted as satisfactory."[29] Not even an improved reduction facility could live up to his expectations, however. In 1939, only fourteen reduction plants remained in operation.[30]

A few other methods adopted the principles of utilization, but with only modest success. Composting processes employed either to produce humus or to dispose of refuse were studied primarily outside of the United States. George Waring's rubbish-sorting facilities, opened in New York City in 1898, were designed to return some funds to the public coffers. But city officials never sustained Waring's commitment to utilization, and the plants were shut down by 1918. Newer rubbish-sorting programs played down utilization, regarding it simply as a useful way to reduce waste hauled to the dumps. Ready markets for cans, bottles, scrap metal, rags, rubber, and paper were never sufficiently reliable to attract more attention to rubbish sorting. Private philanthropic organizations had some success with recovery efforts, but their goals were more social than environmental.[31]

Wartime salvage efforts proved much more successful than peacetime counterparts because public motivation was heightened, and the need for recyclables was great. In World War I, the Waste Reclamation Service of the War Industries Board was modeled after the National Salvage Council in Great Britain. In World War II, the United States again borrowed ideas from other countries in establishing a salvage program. Many of the programs of the War Production Board's Salvage Division were based on British practices. The National Salvage Effort relied on cooperation from approximately 1,600 local authorities, which directed the work of volunteer groups, including women's organizations and the Boy Scouts of America. Material from scrap metal to rubber tires was collected for the war cause. The paper recycling rate by the end of the war was 35 percent, but began to drop when peace was declared.[32]

Utilization methods were successful in targeting some wastes for recovery and reuse, but none had a universal following. Swine feeding and, to a lesser extent, reduction indicated some commitment to utilization. In both cases, only organic wastes were disposed of in this manner. Sanitary landfills were quite new, and open dumping continued despite its shortcomings.

A strong showing for incineration demonstrated its versatility, by handling inorganic materials or by burning mixed refuse. Incineration had a checkered history, however. As expectations about disposal changed, the role of incineration also changed. Prior to 1920, incineration failed to maintain a prominent place among disposal options. In the early 1920s, incinerators were most likely to replace open dumps in suburbs and smaller communities where plant locations were available, where occasional incomplete combustion was not considered objectionable, and where the furnace was not utilized to generate power. A 1924 report indicated that 29 percent of the cities surveyed burned or incinerated their refuse. But interest in using incinerators to convert waste to energy was tepid in an era of cheap energy.[33]

The limits to using incinerators in the 1920s did not result from serious environmental concerns. It was widely believed among engineers that incinerators were an acceptable disposal option from a sanitary standpoint. If properly installed, they did not pose a smoke and odor nuisance since most American furnaces at the time burned only garbage and possibly rubbish.[34] After a decline in interest in burning wastes in the early part of the century, incinerators became a relatively popular disposal alternative by the end of the 1930s. The total number of incinerators probably exceeded 700 in the decade, but incineration was responsible for disposing of only about 5 to 10 percent of refuse produced.[35]

The units developed by the standards of forty years later were, as one expert noted, "hopelessly in violation of air quality standards," but were not subjected to rigid regulation at the time.[36] In the 1930s, a growing interest in the relationship between air pollution and burning waste was developing. But the view that incinerators were relatively sanitary was not successfully challenged in this period.[37]

Part of the problem with the failures of incinerators was purchasing them from the stock shelf without an engineering study of local factors. Manufacturers were unable or unwilling to supply them because of competitive costs. Incinerators were gaining an advantage in some urban areas because open dumps were too unpopular to maintain in central cities. Yet incinerators still faced problems of NIMBYism. One proponent noted: "Uninformed and misguided public opinion frequently delays construction of incinerators. Each section of a community wishes the incinerator on another part of town, under the mistaken belief that the production of odors and smoke will be detrimental to health. What is needed is public education to the certainty of odor elimination in the high temperature furnaces of modern design."[38]

In a reversal from the 1920s, incineration in the 1940s began to find favor in larger cities that could afford the technology for disposing of their garbage and combustible rubbish, and simultaneously assigning ashes and non-combustibles to landfills. Of the major cities, only New York City turned to landfilling in this period, but still relied on incinerators for disposing of some waste. In cases where incinerators were in competition with landfills, debates focused on cost and convenience more than environmental risk. Incineration expanded in appeal in the 1940s also due to extensive use at army camps and other military installations. Yet high construction and operating costs remained the greatest obstacle to its widespread use in many average-size communities.[39]

The invention of the in-home garbage disposer offered a different approach to refuse disposal. The idea can be traced to grinders and shredders utilized at sewage plants in the 1920s and 1930s to convert large solids into fine pulp. The first attempt to grind garbage and dispose of it into sewers took place in Lebanon, Pennsylvania, in 1923.[40] In 1935, General Electric adapted the idea of the municipal grinders into the design and manufacture of the "Disposall" for use in the kitchens of private homes.[41] Soon the devices were being publicized in science magazines like *Scientific American*, and also in women's magazines such as *House and Garden*. General Electric and other manufacturers directed their marketing effort at homebuilders, kitchen remodelers, and municipal officials to capture a larger consumer base. Major installations awaited the end of World War II.[42]

In 1930, Nathan B. Jacobs, a consulting engineer from Pittsburgh, queried whether refuse disposal had made any progress and what the future held in store. He thought refuse disposal deserved more attention, and concluded that an appeal to taxpayers had to stir civic pride as well as take into account the pocketbook. He argued that centralized control and management of services, such as collection and disposal of refuse, would be essential as metropolitan growth continued.[43]

The Committee on Refuse Collection and Disposal of the APWA also recognized the broader context of the refuse problem in the 1941 edition of *Refuse Collection Practice*. The committee applauded the "remarkable improvement in sanitary service in Europe" due in large part to the exchange of ideas among the officials in various countries and the coordinated effort to find solutions. It also recognized that translating those successes into an American context would not be easy. First, the per capita amount of refuse produced in Europe was far less than in the United States, which the committee attributed to the "more frugal habits" of Europeans and "less abundant supply" of various materials. Second, the committee argued that the "individual habits of cleanliness, orderliness, and thriftiness" in Europe were manifest in a willingness of the people to "submit to rigid enforcement" of sanitary ordinances and regulations. Third, the committee believed that the "respect accorded the public service generally and the refuse collection service in particular" was greater in Europe. And fourth, all refuse was collected together in Europe for efficient processing, and the range of disposal methods was less varied.[44]

While the committee likely exaggerated the efficiency and effectiveness of European collection and disposal services, and undervalued the American counterpart, it raised a similar issue expressed by Nathan Jacobs in 1930. The idea that confronting the refuse problem required more than a technical fix and was an interactive issue between the consumers of goods and those responsible for disposal seemed to have been lost. After World War II, refuse collection and disposal would rise in status. But not until later in the century did the fear of an impending "garbage crisis" confer on it serious national attention.

III

THE NEW ECOLOGY

1945–2000s

14

The Challenge of Suburban Sprawl and the "Urban Crisis" in the Age of Ecology

1945–1970

Relentless peripheral growth and central-city deterioration characterized post–World War II urban conditions and affected technologies of sanitation. Sprawl placed stiff demands on providers of water supply, sewerage, and refuse collection and disposal. For its part, suburbanization proved to be a financial drag on service delivery. Social scientist Dennis R. Judd noted that a study of metropolitan areas "found that the cost of central city services was explained more by the suburban population level than by any other factor."[1]

For the first time, concern over a decaying infrastructure raised important questions about the permanence of the sanitary systems devised and implemented earlier. However, an array of mounting social ills—characterized as an "urban crisis"—increasingly shifted attention away from infrastructure problems. Sanitary services not only were situated within a new social and political context after 1945 but within a changing environmental context as well. The emergence of the "New Ecology" and the modern environmental movement produced a paradigm in which sanitary services were viewed with different eyes, even among engineers, public health officials, and other sanitarians influential in establishing the original systems.

After 1945, an increasingly larger share of the urban population could be found in metropolitan areas. Between 1920 and 1970, the number of Americans living in cities rose from 50 percent to 73.5 percent. From 1940 to 1950, metropolitan growth swelled by 22.2 percent, while the total population of the United States only increased by 14.5 percent. In the 1950s, the metropolitan population grew nearly five times the nonmetropolitan rate. In 1960, 63 percent of Americans (almost 113 million) lived in metropolitan areas.[2]

Territorial growth was equally impressive, especially in metropolitan areas where population was on the rise. Between 1930 and 1970, striking expansion took place in cities such as Houston (from 72 to 453 square miles), Dallas (42 to 280), San Diego (94 to 307), Phoenix (10 to 247), Indianapolis (54 to 379), San Jose (8 to 117), and Jacksonville (26 to 827).[3]

Annexation became a particularly attractive means of growth for the newer Sun Belt cities, especially in locations with limited competition from other urban areas and favorable state laws. Annexation limited governmental fragmentation, circumscribed independent suburban development, and helped to fill the tax coffers. In 1945, 152 cities with populations of 5,000 or more completed annexations, which greatly surpassed prewar levels. By the late 1940s, almost 300 municipalities annexed additional territory, and by 1967, 787 municipalities had done so.[4]

The growth patterns for core cities and suburbs were essentially mirror images between the years 1940 and 1970. By 1970, more than half of all metropolitan residents lived in suburbs. The percentage total Standard Metropolitan Statistical Area (SMSA) growth for core cities dropped from 40.7 percent to 4.4 percent between 1940 and 1970, and rose from 59.3 percent to 95.6 percent for suburbs. Between 1950 and 1970, Chicago and New York City both lost population. In Detroit, core population dropped by 20 percent, while the immediate suburbs increased by more than 200 percent. Even in Los Angeles—the only large city with an increase in core population—peripheral growth outstripped central-city growth by 141 percent to 43 percent.[5]

The Sun Belt migration after World War II was essentially a suburban movement, particularly in states such as California, Florida, and Texas.[6] While urban population concentration for SMSAs was greatest in the Northeast, the largest increases took place in the South and West. Thus the gap in growth between core cities and suburbs was a phenomenon primarily focused in the East and Midwest, especially until the 1960s.

Trends in housing most graphically demonstrated the acceleration in outward growth. Between 1946 and 1960, builders constructed approximately

14 million single-family homes, and by 1970 almost two of three American families owned their own homes. Billions of dollars of FHA mortgage insurance and low-interest VA loans proved essential in fueling the postwar housing boom. The availability of federal mortgage guarantees promoted the unprecedented building of new subdivisions in suburban America.[7] Kenneth Jackson has stated that "the American suburb was transformed from an affluent preserve into the normal expectation of the middle class."[8]

Suburbs had a push-pull effect on the urban middle class, especially whites. Pull factors included the attractive FHA and VA mortgages in new, pristine housing developments. Push factors included inner-city decline and racial prejudice, which encouraged white flight. In addition, middle-class urbanites vacated central cities to avoid being taxed for escalating expenditures there. In this case, both whites and blacks pulled up stakes for the suburbs.

As early as the 1930s, the number of blacks living in the suburbs of America's largest cities increased. By contrast with the white suburban population, the numbers were quite low, as might be expected given the efforts to exclude minorities, the poor, and the elderly from suburban communities. Black suburbanites, and Latino suburbanites in the Southwest, often found themselves segregated in the new surroundings. A ghettoization of suburbs followed the pattern of central-city segregation. By the 1970s, segmented residential patterns along racial lines continued but over a larger area. Active exclusion of minorities in the suburbs did much to dampen the enthusiasm of African Americans, Latinos, and the poor for suburban life.[9]

Jobs and retailing followed the outward migration of Americans. In the 1960s, the number of central-city jobs declined in the fifteen largest metropolitan areas, while the number of suburban jobs rose. New factories were being built in suburban communities at a much faster rate than in central cities. The shopping mall, designed for the automobile culture, soon replaced the downtown department store as the primary American retailing icon. By 1956, there were 1,600 shopping centers in suburban communities, with 2,500 more being planned or under construction.[10]

The postwar suburbs were at least as dependent on the automobile as the older suburbs had been on the electric streetcar. While the U.S. population increased by 35 percent between 1945 and 1965, automobile registration increased by 180 percent, or from 26 million to 72 million vehicles. By the early 1960s, approximately 80 percent of American families owned cars.[11] Federal funds swelled for highway construction, with approximately half of the estimated cost of the U.S. highway system under the 1956 Interstate Highway Act utilized for 5,500 miles of freeways running through urban areas.[12]

The outward push into the urban periphery did not ensure escape from central-city problems. As early as the mid-1950s, citizens in the inner belt of the most populous suburbs demanded more schools as well as new sewers, pure water supplies, and adequate refuse-disposal facilities. Resources were spread thin in some suburban communities, resulting in efforts to attract new businesses and industry and to raise taxes. Political antagonisms between suburbs and central cities accelerated, exposing the consequences of separation, especially in terms of racial segregation and the deepening financial problems of the central cities.[13]

After World War II, the central cities became the most beleaguered urban areas. The inner-city housing market began to collapse as potential purchasers looked to the new construction along the cities' peripheries.[14] Eventually, as business followed the outmigration, commuting from the suburbs to the central business districts dropped off. Retail sales slumped in most major cities, and hotel revenues deteriorated, while suburban shopping centers and motels experienced brisk trade.[15]

An early response to the growing divisions between core and periphery was promotion of downtown revitalization and urban renewal, which addressed some economic and physical problems but hardly resolved many of the fundamental underlying ills. The renewal programs sacrificed low-income neighborhoods for highways or private and governmental office buildings, reflecting changing views concerning the use of the newly revitalized central cities. Slum clearance, despite the impact on local residents, became the prime objective of revitalization projects and destroyed more homes than were built.[16]

By the mid-1960s, throngs of experts were bemoaning an "urban crisis," characterized by declining central-city economies, deteriorating health and environmental conditions, rising violent crime rates, race riots, and community-wide despair. This dismal portrayal of inner-city life was matched by a growing apathy toward city problems expressed by those along the periphery.[17]

The provisions for new sanitary services and the need to improve existing systems were caught between the momentum of outward growth and the mire of the core in the nation's metropolitan areas. Development and improvement of sanitary services were vying for priority status among a wide array of social, economic, and political concerns. In addition, the needs and wants of the core cities and suburbs within metropolitan America diverged sharply, making any uniform policy difficult, if not impossible, to implement.

City administration and finance underwent major changes, which strongly influenced sanitary services. The presence of the federal government in urban affairs diminished somewhat in the immediate postwar years. Peace and rising affluence diffused the emergency atmosphere of the

Great Depression and World War II, and weakened support for New Deal policies, which appeared to have little impact on the problems surrounding metropolitan growth. Republican presidential victories in the 1950s, less dependent on the core cities than during the Democratic years, shifted priorities away from urban issues. To many, the need for a broad national urban policy and the continuation of extensive federal aid were out of pace in the new era.[18]

Increased local control was on the rise partly because of metropolitan fragmentation. Suburban development fostered a variety of new governmental units, which reinforced economic, social, and racial divisions between core cities and suburbs, and among the suburbs themselves.[19] In 1967, there were 20,703 local governing bodies in 227 SMSAs, including 4,977 municipalities, 3,255 towns and townships, 404 counties, 7,049 special districts, and 5,018 school districts. By the 1990s, the number of local governments swelled to 80,000.[20]

Federal-city partnerships did not disappear during the 1950s, although the Eisenhower administration wanted more private interests involved in urban affairs. Direct federal assistance to large cities rose between 1952 and 1960, but it represented only 1.2 percent of total municipal receipts. While cities did not face a fiscal crisis in the 1950s as they had in the 1930s, budgets remained tight. Local governments reduced debt loads substantially during the war, but faced a backlog of public works. As cities expanded after 1945, the demand for services and an inflationary economy pushed municipal expenditures upward again. Municipal borrowing in the late 1940s made up the difference between revenues and expenditures.[21]

With the return to Democratic leadership in the White House in 1961, urban affairs achieved priority status. In 1960, only 44 federal grant programs provided funds for big cities; by 1969, the number of programs exceeded 500. The belief that urban problems could be managed locally faded in the 1960s for several reasons. Demands for services continued to mount as metropolitan communities expanded. Inner-city decay far exceeded the ability of local governments to deal with it. Revenues failed to keep pace with expenditures. And local leaders realized that the Democratically controlled White House and Congress owed their political success to urban majorities. As a result, new funds were allocated to several existing programs for housing, urban renewal, transportation, and pollution control. With Lyndon Johnson's Great Society, federal resources were directed to a whole new group of urban programs. Several sought to reinvigorate neighborhood participation, reduce municipal fragmentation, and revive regional planning.[22]

Despite growing federal support, American cities faced serious financial problems that strongly influenced decisions over which services received priority. Revenue-generating mechanisms of local authorities were constrained by federal and state governments preempting some forms of taxation, such as income and sales taxes; the inequitable distribution of needs versus resources in the various jurisdictions; and the possibility that changes in seeking new revenue streams could drive out industry and commerce or people of color, the elderly, and those prone to white flight.

Both core cities and suburbs faced financial stress—the former were losing their tax base, the latter were facing increased demands for services. In both cases, expenditures were rising faster than local revenue. At the same time, the nation's tax burden remained unchanged between 1957 and 1967. The competition for revenue sources was great among different levels of government and the cluster of local units. The general property tax remained the most important source of local income. Before 1932, it provided almost 75 percent of local governments' general revenue, but by the end of the 1960s property levies financed only 34.5 percent of the budget of local governments. Rising costs and greater demands for services made dependence on property taxes less feasible and politically volatile after World War II. An increasingly popular strategy was to have a larger portion of services financed at the state and/or federal level. Another approach was to pursue new local sources of revenue, particularly municipal income taxes and local sales taxes. Both sources allowed central cities in particular to capture revenue from suburban commuters.[23]

Service charges became valuable income generators, especially for sanitary services and water-pollution control. Water, in particular, remained one of the greatest net revenue producers in relation to operating expenses, along with electricity, gas, and port facilities.[24] In some respects, service charges accounted for a relatively low percentage distribution of expenditures for sanitary services in large SMSAs. Local government expenditures in 1967 on capital outlay for sewerage alone was 8.4 percent of the total; only education, streets and roads, and local utilities captured larger percentages. Growing capital requirements also meant that city indebtedness increased substantially.[25]

After World War II, complexities of urban fiscal problems, the patterns of urban growth, fragmentation of local government, the changing role of the federal government, and deterioration of older infrastructure posed major challenges to city leaders. Significant in future strategies for new sanitary services, and the repair and replacement of the old, was a paradigm shift in environmental perspective. The traditional emphasis on public health impacts of a pure water supply, adequate sewerage, and the disposal of refuse was being replaced by a focus on biological and chemical assessments of water purity, the nature of waste, and the various effects of pollution. Constructed upon a more holistic theory than the health views of the past, the modern ecological viewpoint broadened the assault on pollution by giving greater regard to chemical standards and a wider array of physical threats to health. It minimized the value of environmental sanitation and gave increasing attention to individual, or private, health matters as opposed to traditional public health concerns.

At the heart of the paradigm shift was the emergence of the New Ecology. The basic concept of ecology revolves around "the relationship between the environment and living organisms," particularly the reciprocal relationship between the two.[26] The emergence of ecology as a science coincided with the industrial era beginning in the early twentieth century.[27] The New Ecology in practice was not an all-embracing environmental concept as much as shifting emphasis toward different priorities. By the 1960s, ecology changed from a scientific concept to a popular one as the questioning of traditional notions of progress and economic growth became more intense.[28] Rachel Carson's *Silent Spring* (1962), a grim warning of the dangers of pesticides, seemed to best capture the new spirit. Career ecologists were beginning to make it clear that "respect for the biosphere, like respect for justice, must continuously have a place in law and government."[29]

Ecology was a helpful blueprint for constructing national environmental policy. In its simplest form, the New Ecology could guide the country from utilitarian conservationism of the past to an era emphasizing environmental quality and personal health and well-being. Also, "from the late 1960s on, environmental issues were propounded in terms of consumerist property rights in a manner potentially contradictory to capitalist accumulation."[30]

At the forefront of the environmental movement in postwar America were citizen and public-interest groups as well as a variety of experts. Between 1901 and 1960, an average of three new public-interest conservation groups appeared annually; from 1961 to 1980, eighteen per year. While the 1970s witnessed the most dramatic rise in the modern environmental movement, the 1960s helped to build momentum through older preservationist groups such as the Sierra Club (1892) and the National Audubon Society (1905), and newer organizations such as the Conservation Foundation (1948), Resources for the Future (1952), and the Environmental Defense Fund (1967).[31]

Although environmental groups had yet to identify a common agenda, the tone and spirit of environmentalism were changing. Quality-of-life issues, pollution control, wariness toward nuclear power, a critique of consumerism, and insistence on preservation of natural places indicated a giant step away from "wise use" of resources and a challenge to traditional faith in economic growth and progress.[32]

Despite the persistent battles fought in the public arena over preservation versus development, and economic growth versus environmental quality, between 1945 and 1970 there were major strides in national environmental legislation, such as the landmark Wilderness Act (1964). Yet legislation alone did not guarantee improved conditions. The demand for expertise was reminiscent of the spirit of the Progressive Era, when the conservation movement had first emerged. What connected the older movement with the new one was the hope that science and technology could find the way to solve a potential environmental crisis. The almost blind faith in science and technology exhibited in years past was replaced by a more complex, and sometimes schizophrenic, relationship. The nation's triumphs in science and technology were sometimes blamed for the excesses of the new consumer culture. At the same time, the advice of scientists and technical experts was sought to help eradicate pollution and restore a more amenable quality of life.[33]

Abel Wolman observed that "only in comparatively recent years have people been complaining that most of the sins of modern society are attributable to the engineer's works." But Wolman and many other engineers strongly defended the profession's impact on the environment, suggesting that "without technology, modern society would literally collapse." "The task of optimizing the use of the world to the benefit of man is inescapable," he added.[34]

While engineering practice seemed to reflect an interest in controlling the environment, sanitary engineers were beginning to consider the need for a clearer understanding of their function and their professional status in light of changing environmental views. There was a growing concern that sanitary engineering, in particular, did not have a single professional home, although it held primary ties to civil engineering.[35] The notion that engineers would have to keep pace with the rapidly changing world around them was not lost on them. They began to realize that they needed to address many of the pressing problems of modern life. Wolman had pointed out in 1946 that "a scope of activity has been defined for the sanitary engineer of the future which can be no longer delimited by the purely technological." Assistant Surgeon General Mark Hollis shrewdly observed that "speaking of our rapidly changing technology, it is probably the rapidity of change rather than the change itself that will exert the greatest impact on the practice of sanitary engineering."[36] Engineers, indeed, were facing a major transition in how they were perceived and how they perceived their function in the postwar world.

Change also was afoot in the public health field. The trend away from preventive medicine and greater attention to social and behavioral concerns were evident after World War II. This was ironic given the heightened sensitivity about an array of pollution problems emphasized by the modern environmental movement. On the other hand, the curbing of many communicable diseases and the relegation of sanitation functions to technical entities sharpened the focus on personal health.[37] Although there was a clear disparity in life expectancy based on race, between 1940 and 1970 expectation of life at birth increased for the whole U.S. population from 62.9 to 70.8 years.[38] Between 1920 and 1956, there were an average of twenty-five waterborne-disease outbreaks per year, but between 1950 and 1956 no deaths were reported.[39]

Limited interest in environmental sanitation was not related to any reduction in governmental public health units. Quite the contrary; the broadening of public health activity in the social and behavioral areas—even with less attention to preventive medicine—was reflected in the expansion of public health institutions. However, while 116.5 million Americans in those 1,434 health units had full-time health services, another 43 million in 1,594 counties did not.[40]

Increasingly, health issues were being addressed nationally through agencies in the Departments of Agriculture, Commerce, Housing and Urban Development (HUD), Interior, and Labor, as well as in the Civil Defense Administration. The Department of Health, Education, and Welfare (HEW), which ultimately managed the USPHS, was the principal federal agency concerned with health. The Bureau of State Services administered federal-state and interstate health programs. Also important was the World Health Organization, founded in 1948.[41]

In the years 1945 to 1970, the maintenance and development of water-supply, wastewater, and refuse systems would be challenged by the changing dynamic of urban growth and by the transition to a new environmental paradigm. The ability of sanitary services to protect the public health and to function efficiently would depend less on technical merits and more on an array of external forces from within and without city government.

A Time of Unease

THE "WATER CRISIS" IN AN EFFLUENT SOCIETY, 1945-1970

A *Fortune* article on infrastructure in December 1958 stated flatly that water supply and sewerage "remain a signal failure in public works." "These vital deficiencies," it added, "are being attacked haphazardly, reluctantly, and locally, instead of on an area-wide basis, which is the only effective approach. And not only are water and sewerage facilities woefully deficient, their potential as a powerful tool for shaping communities is being almost totally overlooked."[1]

This assessment was harsh, but it is clear that many sanitary systems built in the late nineteenth and early twentieth centuries were in decline by the mid-1940s. The Committee on Public Information for the AWWA reported in 1960 that of the approximately 18,000 functioning water facilities in the United States, one in five had an inadequate supply, two in five had poor transmission capacity, one in three had defective pumping, and two in five had flaws in their treating capacity. The committee estimated that improvements in distribution were required in 57 percent of the systems and that many waterworks needed to upgrade their administrative and accounting procedures.[2] A 1961 survey of 6,370 communities reported that 58 percent believed their systems were adequate, but 30 percent rated them inadequate, and another 8 percent noted impending deficiencies. Problems included low peak pressure, insufficient storage, poor water quality, and difficulty in meeting the needs of new residential areas and industry.[3]

Decisions about improving water-supply systems had to be made within a context of rapid urban growth and increasing water usage, especially with the availability of new appliances such as automatic dishwashers, washing machines, and air conditioners. The automatic dishwater, for example, increased per capita consumption of water by as much as thirty-eight gallons per day.[4]

New waterworks continued to come on line, especially in the metropolitan periphery and in smaller cities and towns no longer able to depend on private wells or rudimentary water systems. In 1945, there were approximately 15,400 waterworks in the United States supplying about 12 billion gallons per day to 94 million people. By 1965, there were more than 20,000 waterworks supplying 20 billion gallons per day to approximately 160 million people.[5] The overwhelming majority of these new waterworks served smaller communities rather than increasing capacity in major cities, but the heaviest workloads fell to the fewer than 400 waterworks serving the major metropolises.[6]

Urban expansion strongly influenced the growth of water systems in the postwar years, accompanied by increasing expenditures. By the mid-1960s, 83.4 percent of water-supply facilities (in cities with 25,000 or more population) were publicly owned.[7] The annual value of the water placed water works within the nation's top ten largest industries.[8]

With substantial growth and higher operating costs, expenditures for water utilities on a national scale outstripped revenues beginning in 1950 and continued into the 1970s. Compared to electrical and gas utilities, investment in water systems was high, but not commensurate with expanded service.[9] Despite growing inadequacies of existing waterworks, demand for water rose. In 1965, the national average in the continental United States was about 157 gallons per capita per day (gcd) for public municipal systems as compared to approximately 137 gcd in 1955. Residential use accounted for 46.5 percent of total withdrawals. The majority of the water used in the home went for flushing toilets (41 percent) and washing and bathing (37 percent). In suburban communities, watering lawns and gardens was a large percentage of total domestic usage. The greatest withdrawals of water came from the North Atlantic, Great Lakes, and California areas, which accounted for 55 percent in 1965. In the West, irrigation in particular outstripped all other uses. In fact, national water-consumption figures (municipal and agricultural) suggest that the scale of consumption by irrigation and industrial uses greatly exceeded municipal uses.[10]

Postwar water-supply problems were exacerbated by several causes: uneven distribution of facilities for storing, pumping, and transporting water from available supplies to areas of greatest demand; chronic shortages in places such as the arid West; and increased pollution of traditional sources of potable water. Some cities had long depended on distant supplies of surface water. San Francisco's primary reservoirs were as much as 150 miles from the city; Los Angeles and San Diego drew water from as far away as 550 miles. In some cases, supplies were tapped beyond state boundaries, leading to intense battles for control. For example, Southern California struggled with Arizona interests over the use of the Colorado River.[11] Of the 100 largest cities in 1962, 66 relied on surface supplies compared with 20 that relied on groundwater, and another 14 that utilized both. A 1965 survey of cities with populations of 25,000 or more showed that 865 of 1,514 (57.1 percent) of water-supply facilities depended on groundwater, and another 126 utilized both groundwater and surface water.[12]

Distribution problems resulted from the location of water facilities in central cities, which often serviced the larger metropolitan area or outlying suburban communities. In the 1960s, the central plant in Chicago supplied water on a contract basis to approximately sixty suburban communities.[13] The number of special districts and other administrative arrangements were increasing in response to the need for water. In 1962, there were approximately 1,500 special districts—many of which had developed out of a consolidation of smaller service systems.[14]

It was often in the interest of the central city to extend water lines to the suburbs, and for suburbs, growth was impossible without adequate services. In some cases central cities were reluctant to extend distribution lines outward, especially into unincorporated areas, if there was no guarantee of future annexation. In San Antonio, for example, the older, central city got more public utility connections than the newer, thinly settled areas.[15] Some cities, such as Los Angeles and Milwaukee, used water service as leverage to impose annexations on outlying areas. Real estate developers or alternative public entities constructed pipelines beyond the existing city limits, ultimately making outlying suburbs attractive to future annexation. In Houston, which benefited from a potent 1963 state annexation law, developers were expected to supply water and sewage services.[16]

It was common for central cities to offer water to fringe areas on a contract basis. To minimize the risk of extending water lines to the periphery, out-of-city water charges were often levied. Cost of extensions was quite high because of the larger lot sizes in suburban communities and the low population densities along sections of proposed mains. A twist on this theme occurred in the early 1970s. HUD secretary George Romney faced several legal and political challenges with respect to segregated residential patterns in numerous cities. One solution was offering water and sewer grants to communities willing to accept subsidized housing. However, suburbs showed little interest in the program because the grants offered such disagreeable conditions.[17]

From the vantage point of the total water system, the cost of distribution represented as much as two-thirds of a utility's investment.[18] Difficulties of extending new distribution lines were exacerbated by undersized pipes, overextension of mains, and problems in design. Internal corrosion of distribution lines reduced carrying capacity, produced discolored water, and caused bad tastes and odors.[19] In some cases, the inadequacy of the distribution system reflected lack of sufficient planning; in others, it suggested an inherent inequality in service provided. For example, in *Hawkins v. Shaw*, a notable service equalization lawsuit, the issue was not simply the lack of water and sewer service for black versus white families but also variations in the quality of the service. Because of smaller pipes in African American neighborhoods, water entered homes more slowly, and water to fight fires was not readily available. Thus fire insurance was higher in those areas. The age of the poorer neighborhoods where many black families lived was a major contributor to the size of the mains.[20]

Filtration also faced serious scrutiny. Mathematical models were employed in the 1960s to predict the performance of filters and to determine filtration patterns.[21] Pretreatment, in particular, was studied because it aided the performance of rapid sand filters. While there had been little improvement in filter design for fifty years, new pretreatment equipment was becoming available. Filtration in general remained an expensive component in water treatment, if not the most expensive. Ironically, as pretreatment operations became more efficient, filters added less to improved water quality. A new concern for engineers was that the cost of filters seemed to be out of proportion to their contributions.[22]

Even the widely adopted practice of chlorination generated doubts. In 1966, it was employed by 99 percent of all municipalities that chemically disinfected their supplies. Concern arose over the ability of chlorination to keep up with increasing water demand, and tests showed that chlorine was not effective against all microorganisms, at least in the concentrations used in existing waterworks systems.[23]

A new water additive, fluoride, was meant to offer additional health benefits via the water supply. Some water sources contained natural fluoride, but for those that did not, it could be added. Communal fluoridation began in Grand Rapids, Michigan; Newburgh, New York; Evanston, Illinois; and Brantford, Ontario, in 1945.[24] Fluoridation was used to reduce dental caries (tooth decay), a chronic disease especially prevalent in children. Proponents of fluoridation, including the USPHS and the American Dental Association, believed that it could reduce cavities by 60 percent. In 1951, two councils of the American Medical Association and a special committee of the National Research Council issued statements declaring that there was no evidence of toxicity in adding fluoride to drinking water.[25]

After 1951, when the benefits of fluoridation were first announced, public acceptance was on the rise, but organized opposition to the use of fluoride in drinking water also intensified. Opposition was similar to that raised against chlorination, that is, skepticism about any compound added to a natural water supply. With the advent of the nuclear age, fear of radioactive contamination added to concern about the purity of national water supplies and the possible introduction of foreign substances into the water.[26]

The fight over fluoridation ultimately became the most controversial water-treatment issue in the postwar years. Local lawsuits were posted, and national bills to ban fluoridation were introduced in Congress. Some cities dropped the practice after an initial trial. Between 1953 and 1963 fluoridation was halted at sixty water-supply systems, although twenty-six eventually reinstated it.[27] Some criticized city-sponsored fluoridation as an assault on personal freedom. Seattle radiologist Dr. Frederick B. Exner said the practice "violates the most sacred rights of God and man." The most outlandish claim was that fluoridation was a Communist scheme to kill or weaken Americans. In Stanley Kubrick's 1964 black comedy, *Dr. Strangelove*, the deranged General Jack Ripper—who had independently unleashed a bomber wing with nuclear warheads on the Soviet Union—tells his executive officer that fluoridation was a monstrous Communist plot to sap "our precious bodily fluids."

Setting aside extreme claims, the antifluoridation groups' less strident criticisms and half-truths about fluoridation kept the controversy alive. Sodium fluoride was rat poison. Fluoridation was "unnatural" forced medication. And there were questions: What were "safe" levels of fluoride? Wasn't fluoridation wasteful, since only a small amount of the water supply is ingested?[28]

In the long run, support from key public health and scientific groups and the positive results on dental health led to increasing use of fluoridation. In 1951, more than 360 communities had adopted the process; by 1968, more than 4,000 had done so.[29] After ten years of use in Newburgh, New York, there was 41 percent less tooth decay among sixteen-year-olds than in Kingston, where no fluoridation took place. Among a six-to-nine-year-old group, the results were more dramatic, with 58 percent less tooth decay.[30]

While problems of water distribution and treatment were serious, public attention was directed toward dramatic claims of water shortages and drought. Unease over long-term effects of drought was exaggerated for much of the country with the exception of the West and Southwest. A more serious problem was the pernicious impact of source depletions. In California, demand for water was on the rise as drought lingered across the state. Agricultural interests wrung their hands, and the Public Utilities Commission ordered a brownout on all of California north of the Tehachapi Mountains to conserve hydroelectric power. While San Francisco and Oakland avoided water problems, smaller towns with no connections to the larger systems were suffering. Southern California was hardest hit, with below-normal rainfall and supplies in reservoirs falling. In several Illinois communities, cities and towns depending on deep wells were using more water than could be replenished. Heightened awareness of a brewing water shortage drew attention to droughts and depletions across the country. In 1953, all but three states (Idaho, Rhode Island, and Mississippi) reported water shortages.[31]

Drought was only one in a growing set of problems over several years, which aggravated matters by drawing down many local water supplies. Reservoirs were often poorly protected against siltation; limited attention had been given to water reuse and other conservation measures; stream pollution ruined billions of gallons annually; and experts had done poorly in identifying available groundwater. In an observation that would become more common as the New Ecology blossomed, the head of the U.S. Conservation Service concluded, "Taken together, our existing and prospective water difficulties amount to a very serious national water sickness, contracted as part of our rapid national development and hidden for years by the lingering memory of our original abundance."[32]

Contemporaries may have overgeneralized about a "national problem" of water depletion. A dialogue about excessive water use and possible remedies grew out of these public revelations, but there was no consensus about the nature of the problem or the best solutions. Some believed that a water shortage was at hand, while others claimed that there was plenty of water. Shortages were essentially local in nature, some said, and could be prevented by reducing useless squandering. Still others believed that growing population and industrial expansion were putting a "fearful strain" on municipal water supplies. Some even contrasted the shortages with periods of flooding. One hydrologist who recognized the problem as a distribution issue coined the phrase "The People-Water Crisis."[33]

Possible solutions for the water problem drew wide responses. Some proposed solutions were stopgap, such as restrictions on water use. Other approaches focused on making systems more efficient and capturing new supplies. In Dallas, construction began in the late 1950s on a reservoir on the Sabine River (fifty miles southeast of the city) meant to supply water until the 1980s. Further planning was under way to address the community's needs to the year 2000.[34]

Dependence on groundwater was leading several communities to consider potential surface supplies. By the late 1960s, assessments of water needs were hampered by the lack of good data. The last major study of water in the United States had been conducted in 1954. Other ways to augment supplies included artificially recharging groundwater sources and desalination. Both were expensive and not widespread.[35]

The federal government responded to the crisis by addressing specific problems, establishing commissions, and creating dam-building programs. Such projects seemed contrary to growing sensitivity about resource conservation and other environmental concerns. Projects such as Hetch Hetchy in the Sierra Nevada area; Echo Park on the Green River; Glen Canyon, Marble Canyon, and Bridge Canyon on the Colorado River; and Rampart along the Yukon faced stiff resistance in the 1960s. Environmentalist Wallace Stegner stated that the problem of the dams was that "water, once paramount, has become secondary." The cost of irrigation and dams was paid for through the sale of hydroelectric power. "The questionable dams are never simple water holes. What dictates the damsite is as often as not the power head: efficient generation of power calls for a higher dam, and hence a bigger lake, than a simple waterhole does."[36]

Investigations by federal water councils and commissions helped focus attention on water issues. Yet they did not always provide accurate projections or suggest clear planning strategies for future water supplies. For example, the president's Materials Policy Commission (Paley Commission) projected 1952 water-use estimates forward to 1975, but they substantially underestimated the rate of growth.[37]

Questions of quantity of water were only part of the postwar water problem. Water quality was becoming increasingly significant, and attention to water pollution broadened considerably from the focus on biological contaminants to a variety of chemical contaminants. One reason was the substantial decline in a number of waterborne diseases, and confidence in available methods and technologies in combating those that remained. One report noted that between 1945 and 1960, outbreaks of typhoid were only 31.4 percent of the total of the previous twenty-five years. For waterborne diseases in general, the total number of outbreaks dropped dramatically during 1946–60. Another report indicated that from 1946 to 1960, there were 228 outbreaks of waterborne disease or poisoning, with 25,984 cases attributed to drinking water. Of that total number of outbreaks, only 70 occurred in public systems. The largest percentage of outbreaks were a result of untreated groundwater or inadequate control of treatment.[38]

Municipalities and industries alike were blamed for the poor state of available water supplies, as was sedimentation from soil erosion. Ground-water pollution, caused primarily by industrial wastes, was of particular concern because it was much more difficult to abate than contamination of surface water.[39] Aside from the known health hazards, there were unknown or little-known dangers from thousands of new organic chemicals in such products as detergents and pesticides, from a variety of industrial processes, and from nuclear technology.[40]

Criticism of water quality led some to portray water pollution as a "national disgrace" and to decry the United States as an "Effluent Society." As urban growth reduced the physical distance between the water intake of one city and the sewage outfall of another, the USPHS stated in 1960 that "pollution of our streams is occurring at a faster rate than treatment of water for reuse."[41] The scene of a "burning" Cuyahoga River (running through Cleveland and Akron) and charges that Lake Erie was "dying" were graphic examples of inland water pollution. Such occurrences were not limited to the industrial East; worsening conditions for American waterways could be traced nationwide. Fish kills were one sign of the increasing pollution load affecting more than 2,200 miles of rivers and 5,600 acres of lakes.[42]

The newer Drinking Water Standards gave increasing attention to toxic chemicals and soluble minerals that made their way into the water supply. Following the revisions in 1946, the AWWA voluntarily adopted the standards for all public water supplies. In the 1960s, the federal standards were supplemented by the International Standards for Drinking Water (1963) sponsored by the World Health Organization and AWWA's Quality Goals for Potable Water (1968).[43] Concern remained, nevertheless, that the standards had not gone far enough in dealing with toxic or potentially toxic substances. Some experts also claimed that the existing bacteriological techniques widely in use were not sufficiently exact to determine the sanitary quality of water.[44] Reservations faulted on the side of caution. The encouraging news was that several new analytical techniques were utilized as early as the 1940s to detect and monitor some trace elements and toxics, such as DDT. Yet by 1970, the ability to evaluate hazards from microchemical and microphysical substances was far from comprehensive.[45]

It was becoming clearer that available treatment facilities were woefully inadequate in number and in technical proficiency. More effective planning was necessary, and more money was needed for research and construction of facilities. By the end of World War II, construction of new sewage and industrial-waste treatment plants had literally ceased. One estimate in 1961 suggested that to meet adequate fresh water needs for the immediate future, municipalities would need to spend $4.6 billion; another $575 to $600 million per year was needed just to retire the backlog of industrial-waste treatment facilities.[46]

Local authorities were ill prepared to identify sufficient resources to satisfy the treatment requirements, let alone to curb further pollution. The predictions about the inadequacy of facilities nonetheless spurred efforts to organize new water and sewer authorities and sanitary districts. State-level activity proved somewhat more comprehensive. In 1946, only four states— Alabama, Missouri, South Carolina, and Utah—had no specific laws prohibiting the discharge of sewage and other wastes into waterways. By the 1950s, more than twenty states still did not have a truly comprehensive water-pollution control apparatus.[47]

State health boards played a prominent role in water-pollution control, especially by the 1960s. In 1946, twenty-eight states gave boards of health regulatory authority for water-pollution abatement. In some cases, agencies allied with health boards or special commissions were given that authority. Yet state enforcement of water-pollution laws varied widely.[48] Furthermore, state action and interstate compacts did not result in uniform national water-pollution control. The realities of having to deal with industries vital to the states' economies and with other interests (sports and recreation groups, for example) meant designing policies and laws to balance or compromise state objectives.

The role of the federal government in water-pollution matters, except in major interstate rivalries, had been minimal. New Deal financing of water and sewerage systems contributed to mitigating some of the worst water-pollution problems. Since 1897, about 100 bills relevant to stream-pollution control had been introduced in Congress, but none had been enacted.[49] In 1948, the Eightieth Congress passed and President Truman signed the first major federal water-pollution control legislation. The Water Pollution Control Act empowered the federal government to participate in the abatement of interstate water pollution. It also authorized financial assistance to municipalities, interstate agencies, and the states for construction of facilities that would reduce water pollution. In July 1949, Congress made the first appropriation to implement the act. The USPHS also began to collect data about water pollution. The act, however, limited federal enforcement to interstate problems with the consent of the participating states.[50]

In practice, the 1948 act was unwieldy and difficult to enforce. Regarded by many as experimental, it initially received a five-year life. At the end of that trial period, it was extended an additional three years until June 30, 1956. The 1956 amendments—although cumbersome and often ineffectual—maintained state dominance, but represented the first permanent federal water-pollution control effort. Although federal investment in new construction was not great, the financial incentives produced positive results. The rate of treatment-plant construction more than doubled in the four years after the enactment of the legislation, and the act produced the highest local-to-federal ratio of any national aid program. More than 1,300 projects were started or planned.[51]

The 1948 and 1956 legislation by no means ended the controversy over water-pollution control. Between 1945 and the mid-1960s, debate in Congress revolved around federal enforcement powers vis-à-vis water pollution and the level of federal financial assistance deemed necessary to assist in the construction of waste-treatment facilities. Little attention was paid to the generation of waste and how to curb it.

The Federal Water Pollution Control Act Amendments of 1961 continued the cooperative approach between the federal government and the states. However, federal jurisdiction expanded beyond interstate waters to include navigable waters, or essentially all major waterways in the United States, but the government's abatement power proved difficult to enforce.[52]

Construction of water and similar projects expanded significantly. Contract awards in 1961 increased by an average of 25 percent compared to the first four years of the original program; in 1962, there was a 51 percent increase. Since the program had started, 4,437 projects had been approved for grants, and in 1963 a backlog of 1,657 applications for grants remained. Despite the apparent stimulation in sewage-treatment projects, only 23 percent of the grant funds in 1961 went to cities of 50,000 or more in population. Like New Deal water and sewer programs, the water-pollution acts were a greater boon to smaller cities and towns, rather than a complete answer to the water-pollution problem of many watersheds. However, financing of water and sewer projects came from other federal sources as well.[53] Research funding also proved limited in the 1950s and early 1960s. In fact, after World War II, research in private laboratories and universities, in certification programs, and in state health-training programs "never got back into high gear."[54]

Possibly the most important, and certainly the most controversial, water-pollution bill between 1945 and 1970 was the Water Quality Act of 1965. The federal government became more deeply involved in water-quality management, which had been a local and state prerogative. Debates over federal versus state authority raged again, as did the battle over what constituted excessive pollution levels and adequately clean water. The momentum of the new environmental movement and numerous stories about the threat of water pollution brought considerable public attention to the issue.[55]

In Congress, bills to amend the federal Water Pollution Control Act were introduced in 1963. Industries directly threatened by a new water-pollution bill—pulp and paper, chemical, and petroleum companies—strongly opposed provisions that would increase pressure to abate their pollution of interstate waters. Many state and interstate water-pollution control agencies objected to provisions that threatened their authority. Additional opposition came from farm groups and from some professional engineering societies. Not surprisingly, environmental groups such as the Izaak Walton League, the National Audubon Society, and the National Wildlife Federation supported the new measures, but expressed reservations about whether the legislation would be tough enough. They were joined by a variety of groups, including the League of Women Voters and the National Council of Mayors.

The most controversial part of the key Senate bill, sponsored by Senator Edmund Muskie (D-Maine), involved the provisions concerning water-quality standards for interstate waters. From 1963 to 1965, Muskie and his supporters attempted to maintain standards called for in the bill despite opposition in the Senate and in a diluted standards provision in the House bill. The compromise provisions accepted in September 1965 offered several concessions. The bill did maintain the role of the secretary of HEW in promulgating and enforcing standards. President Johnson signed the bill in October 1965, although he predicted that additional legislation would be required.[56]

Advocates hoped that the new law would change the strategy underlying the country's water-pollution programs from "one of containment to one of prevention."[57] The 1965 act substantially modified the bureaucracy of water-pollution control to carry out its objectives, particularly the creation of the Federal Water Pollution Control Administration (FWPCA). The FWPCA was the most visible symbol of the new federal role as activist agent in formulating and enforcing water-quality standards. Almost immediately it became the focal point of criticism from state agencies. The USPHS also disapproved of the rival upstart agency, viewing the transfer of functions as a blow to its prestige.[58]

Under the 1965 act, states were required to produce water-pollution criteria by June 30, 1967, along with standards to "enhance the quality of water," which would in turn become federal standards. If the states failed to address the issue, the FWPCA would then step in. Implementation of the provisions proved to be difficult. Anxiety also heightened over ambiguities in the enforcement procedures.[59] Finding sufficient funds to address water-pollution control also posed a problem for municipalities and state agencies. The loss in momentum of many Great Society programs and the rising costs of the war in Vietnam made that financial goal more difficult to achieve.[60]

Having confronted the water-pollution control issue in a serious way, the 1965 act placed the federal government squarely in the middle of setting and enforcing of standards. It also raised questions about what actions were practicable in protecting water quality, given divergent interests. While no one was completely satisfied with most standards, the time had come when the technical and economic limits of confronting water-pollution control on a national scale would be severely tested. The years 1945 to 1970 represented a period of unease in confronting mounting problems with the nation's water supply. The droughts and local shortages raised concerns about a pending

water crisis. Shifting focus to chemical pollutants in almost every important inland watercourse marked a new phase in the controversy over the
extent and severity of water pollution. In the early years of the New Ecology,
confidence in the ability to deliver pure and plentiful water was questioned
seriously for the first time since the nineteenth century.

Beyond Their Limits

DECAYING SEWERS, OVERFLOWS, AND FOAMING PLANTS, 1945–1970

Surging metropolitan growth after World War II presented a challenge to city officials and engineers in developing new sewerage systems or augmenting older ones. Statistics suggest that expansion of sewer systems kept pace with growth. Between 1945 and 1965, the number of sewered communities rose from 8,900 to more than 13,000—an increase from about 75 million to 133 million people served by combined and/or separate systems.[1] Unsewered areas were located primarily in rural communities or in outer suburbs.[2] Between 1942 and 1957, population clearly outstripped the expansion in sewerage; from 1957 to 1962, sewerage grew slightly faster.[3]

Some contemporaries argued that investment in public sewers fell behind that of public water supplies because of the challenge in establishing sources of revenue to meet bond obligations and the public's difficulty in accepting sewerage as a utility on a par with delivery of water. Citizens regularly turned down sewer bond issues, and the states were often indifferent to local wastewater needs. Sewerage and sewage-treatment plant construction, it is clear, could not keep pace with metropolitan growth.[4]

To pay for operations and to maintain existing systems, increasing numbers of cities turned to sewer rentals. Like metering water, some believed that sewer-service charges encouraged conservation. In 1950, at least 273 cities of more than 10,000 in population had dropped expenditures for sewage disposal from the general tax ledgers in exchange for service charges.[5] In 1951, led by the ASCE and the American Bar Association, a committee of eight organizations produced a report and manual on water and sewer rates. The salient principle was: "The needed total annual revenue requirement of . . . sewage works shall be contributed by users and non-users . . . approximately in proportion to the costs of providing the use and benefits of the works." In the postwar years, legal limitations and restrictions made it necessary to finance new sewer construction with revenue bonds payable from the proceeds of sewer-service charges.[6] The problem of funding sewerage projects persisted despite the fact that Americans often used their existing sewers to or beyond their capacities.[7]

Developers frequently extended sewer lines to new outlying communities, while public efforts to rebuild sewerage systems in the core cities received much less attention. The connection of existing water and sewer systems—or the construction of new ones—in suburban areas contributed to further metropolitan decentralization, while not necessarily improving the quality of existing systems.[8] Even in suburban communities, there was no guarantee that adequate sewer lines would be developed to meet demand. "The evidence to date," said sanitary engineer Joseph A. Salvato, "suggests that it is in the suburbs where the battle for sound liquid waste disposal systems is being lost."[9]

Septic tanks were more economical for smaller, slowly developing subdivisions, especially those in predominantly rural areas. In 1945, approximately 17 million people were served by individual home sewage-disposal systems, especially septic tanks, compared to 75 million served by public sewers. Between 1946 and 1960, over half of new residents of metropolitan areas used septic tanks. Most of the growth was in communities of 50,000 or less.[10]

As population increased, septic tanks often proved inadequate to meet community needs. One contemporary account noted: "In an attempt to make quick profits, many real estate operators erected immense housing subdivisions, often equipping each home with a septic tank as a substitute for a costlier community waste-disposal system. The closely arrayed tanks soon began to exceed the carrying capacity of the soil, resulting in the widespread seepage of toilet and kitchen wastes." Suburban areas thus began to suffer from extensive groundwater contamination.[11]

In 1946, the USPHS had begun conducting extensive studies of household sewage-disposal systems. By 1957, changes in design criteria were being offered. As one engineer stated, "At best, septic tanks and subsurface sewage disposal systems are poor substitutes for the treatment of domestic sewage before the advent of household washing machines, synthetic detergents, and garbage grinders."[12]

The heavy investment in a central-city sewer system made connecting suburban lines attractive. For suburbs, often characterized by small political entities ill equipped to take on major engineering projects, the chance to connect to an existing system was likewise advantageous. Inducements could come in the form of low rates, but more commonly suburban communities were asked to pay higher rates to offset the cost of the connections and the added wear and tear on the existing sewerage plants.[13]

Expansion of sanitary services into suburban areas boosted the regional approach to service delivery. Cost and efficiency were central factors in the development of regional—or at least metropolitan-wide—sewer systems. A variety of organizational and jurisdictional arrangements were attempted. Multicity consolidations were popular in some areas. Smaller communities, such as Ewing and Lawrence townships in Mercer County, New Jersey, phased out their cesspools and septic tanks in the mid-1950s and connected to a consolidated system representing approximately 75 percent of the population of the townships. By 1960, almost every municipality in the Philadelphia area had entered into an agreement to let the central city receive and treat its sewage. An alternative approach was a county-owned but city-managed scheme, which accelerated the process of integrating and consolidating the sewerage systems.[14]

In several cases, control of the sewerage service by the central-city government posed real difficulties for outlying communities, who believed they could exert little influence over the arrangement. The development of special sewerage districts offered an alternative method for bridging city and county boundaries and possibly mitigating jurisdictional battles. Many of the special sewerage districts had elected officials serving on their governing boards and acquired the power to raise revenue.[15] An alternative were sewerage authorities, particularly popular in Pennsylvania, where state officials had been aggressive in favoring sewage-pollution laws, but where construction of purification plants through general taxation plans had been difficult. They had the advantage of being free of assessment restrictions and constitutional limits on borrowing. Being divorced from politics proved an advantage in some cases, but also could be a shortcoming by reducing leverage with the public. Some communities balked at either special districts or authorities because they increased the layers of government.[16]

Urban growth was not the only challenge. Confidence in the sewer systems' functioning and permanence was shaken by overloading and deterioration. In Manhattan, the brick sewers built in the 1860s and 1870s were "approaching failure." Major collapses in large trunk sewers running east and west on the island already had occurred. In Brooklyn and Queens, cement sewers—considered innovative in the 1890s—began to fail by the 1960s.[17] In Lewiston, Maine, Hurricane Edna poured seven inches of rain within a few hours during one day in March 1955. The rushing floodwaters blew a hole about fifty feet in diameter where the ground had collapsed above an eighty-year-old sewer, resulting in the need to add a new length of concrete pipe. Locals were amazed that the damage was not worse.[18]

The need to reevaluate sewer-pipe technology was a lively issue in the 1950s and 1960s as engineers became increasingly concerned with corrosion, leaking joints, and greater amounts of infiltration water overloading the sewers.[19] An additional threat to the existing systems was design, with many sewers stressed beyond their capacities. Problems with overflows introduced a whole new round of debate over the virtues of separate versus combined sewers. Inadequate storm-sewer complements were regarded as the prime culprit for existing systems to fail.

In absolute numbers, there were more sanitary sewer systems than combined systems in the United States. A 1958 USPHS study found that 8,632 of 11,131 sewered communities employed sanitary sewers (77.5 percent), compared to 1,451 with combined systems (13 percent) and 428 with both (4 percent). However, many of the sanitary systems served small and medium-size communities, while combined systems were more prevalent in the densely populated Northeast, Great Lakes, and Ohio River regions.[20]

By 1964, it was well established that combined sewer overflows (CSOs) represented a substantial pollution source nationwide.[21] At the Federal Water Quality Administration's Symposium on Storm and Combined Sewer Overflows held in Chicago in June 1970, one participant noted that "the water pollution problems [including overflows] which have become the target of public opinion and public official concern are the sins of the past being imposed on the present."[22]

Discharge of excess water from combined sewers or discharges of stormwater from separate storm sewers carried an array of pollutants into watercourses. Approximately one-third of cities surveyed in one study indicated that infiltration exceeded their sewers' specifications. Sanitary engineers and public health officials originally believed that overflows were sufficiently diluted to pose no serious water-pollution problems. As urban growth caused a major increase in runoff volumes and rates, the overflows increased in frequency and duration.

Combined sewer overflows (approximately three-quarters of all overflow sources) and other overflow problems often occurred in places where the land was used for industrial purposes and where discharges primarily emptied into streams. A large amount of industrial waste was discharged into the sewer systems, representing effluent equal to a 69 percent increase in population in surveyed areas. Stormwater discharges from separate storm sewers occurred frequently and often for longer durations than from combined sewers, and even "clean" stormwater was polluted. Nevertheless, discharges from combined sewers, which intermingled stormwater and a variety of other wastes, was regarded as the more serious pollution threat.[23]

The solution to overflows most widely advocated in the 1960s was conversion to sanitary sewers. Some states implemented regulations prohibiting the construction of new combined systems or additions. But a financial commitment to convert to sanitary sewers was regarded as too steep for many cities. In addition, some experts argued that separation was not the panacea for all of the problems associated with combined sewers, particularly if it was carried out on only a portion of the entire sewerage system.[24] Short of separation projects, some cities searched for ways to protect watercourses from pollution. In the 1960s, Chicago officials promoted a storage concept where stormwater could be held temporarily. Until that time, relatively few storage tanks had been installed in the United States, although they were popular for many years in Great Britain, Germany, and Canada.[25]

Although overflows raised serious concerns, sewage treatment remained a major focus of antipollution efforts. One observer commented that "sewage treatment is unique in that it is the only municipal function that is performed more for the benefit of others than for the municipality itself."[26] The number of urbanites served by treatment plants remained relatively static immediately after World War II, but began to increase in the 1950s and 1960s. In 1945, 62.7 percent of the sewered population received waste treatment in 5,480 communities; by 1957, waste treatment had increased to 77.7 percent in 8,066 communities.[27]

A vigorous building campaign was under way after 1945, temporarily easing the wartime backlog of treatment facilities. Between 1946 and 1949, 646 new municipal sewage-treatment plants were built. By 1950, more than 1,100 treatment projects were in some stage of completion, a level never before attained. A large portion of the projects included repairs, enlargements, modernization, and additions to existing plants. By the end of the decade, however, estimates showed approximately 15 percent of all treatment plants still needed to be replaced, another 10.4 percent required enlargement, and 11.4 percent required the addition of other treatment units. A 1958 study indicated that approximately one-third of the sewage plants surveyed only provided primary treatment, although two-thirds offered secondary treatment.[28]

In larger cities and in areas with compact, dense populations, activated sludge techniques continued to dominate. In 1951, the Hyperion Activated Sludge Plant in Los Angeles began operating. It was the largest and possibly the most modern plant of its type at the time.[29] In addition, various modifications in the activated sludge process occurred in the 1950s.[30]

Treatment technology had not advanced sufficiently to reduce concern over the impact of sewage flows on water pollution. J. K. Hoskins of the USPHS estimated that in 1946, U.S. inland waterways received waste from domestic sewage equal to the raw sewage of 47 million people. The same waterways received industrial waste equivalent to the raw sewage of 55 to 60 million people.[31] A 1948 issue of *Business Week* stated that untreated sewage and industrial waste dumping in the Ohio River and its tributaries was so bad that "when the water is low, one quart out of every gallon of it is from a sewer. Yet, more than 1.5 million people get their drinking water from the Ohio River."[32] A 1951 estimate of the total pollution load carried by water in the United States was conservatively estimated to exceed raw, untreated sewage from a population of 150 million.[33]

In 1964, some experts claimed that the increase in population, in water withdrawals, and in the types and amounts of waste "soon rendered the dilution and self-purifying properties of the receiving waters incapable of restoring our water supply. In fact, many waterways became open sewers." They added that the primary response to the pollution—constructing waste-treatment plants—barely kept up with current needs.[34]

Sewage discharges after World War II were complicated by changes in the constituent materials entering the sewer lines and treatment plants, including petrochemicals, pesticides, and potential radioactive materials.[35] One of the most controversial effluents was synthetic detergents. Their growing use created a problem for sewerage operations because of foaming at plant sites and because sewage containing detergents had to be treated before disposal. In 1950, it was estimated that 500 million pounds of synthetic detergents were being used, and the amount was expected to increase to one billion pounds. The sale of synthetic detergents was responsible for a 200 percent rise in the production of polyphosphates.[36]

As soon as the issue of synthetic detergents was raised, some interested parties openly disagreed on the polluting characteristics of the new products. Procter and Gamble Company, a major producer of synthetic detergents, presented experimental data that discounted the negative impacts of their product. Some experts claimed that available data was circumstantial. They argued that foaming was a minor problem for sewage-treatment plants, and that detergents might only interfere with operations if present in abnormally large quantities.[37]

Treatment-plant operators continued to be concerned with foaming and other problems attributed to synthetic detergents. The first recorded case of detergents creating foaming problems at a sewage plant took place in 1947 in a small town in Pennsylvania. Some saw foaming as only the most obvious problem for sewage-treatment plants. A main difference between conventional soaps and synthetic detergents was the bactericidal properties of the latter. They claimed that detergents also interfered with the operation of settling. Also, phosphorous salts entered lakes and ponds and encouraged the rapid multiplication of unwanted agents.[38]

The debate over synthetic detergents was a microcosm of a more general questioning of the impact of industrial pollutants on sewerage systems, treatment plants, and water pollution. Biological threats faded in the postwar years as the focus turned to chemical wastes. However, federal officials in 1960 optimistically announced that pollution had been reduced on 21,000 miles of interstate streams, and some industry officials suspected that federal pollution estimates even exaggerated the seriousness of the pollution problem. Many plants simply ignored warnings of increased pollution loads and resisted demands to improve conditions. Industry groups opposed new regulations, while some municipalities and regional bodies expressed concern about the loss of local control in dealing with pollution.[39]

No fixed standards of quality for sewage entering watercourses existed immediately after World War II, although the USPHS and other agencies devoted some effort to assessing existing water-pollution problems. In 1947, six measures calling for federal control of stream pollution were under consideration in Congress. Despite the hesitancy to advocate more federal authority, local governments did not want to continue a cycle of building and rebuilding sewage-treatment plants only to fail to keep pace with pollution abatement.[40]

The Water Pollution Control Act of 1948 created the federal authority to participate in water-pollution abatement and established the policy of financing construction of abatement facilities, but the 1956 act was more significant for sewage treatment. Appropriations from the 1956 act resulted in a 62 percent increase in the construction of waste-treatment works. The act was a welcome boost to construction, but it did not go far enough. American cities would have to build sewage-treatment plants at a rate of $533 million per year for ten years to meet the nation's water-pollution abatement requirements. Actual construction costs were about half that figure. The maximum dollar limit was too low to attract large projects and did not encourage cooperative programs among neighboring communities.[41] In the late 1950s, when the spirit of the New Deal was fading, some feared that even this modest federal program in water-pollution abatement would inspire controversy. "Those opposed to it will scold that it is another example of 'creeping socialism,'" one story noted.[42]

The reintroduction of federal grants for water-pollution abatement was not the only reason for a conservative backlash against the act. Giving the federal government greater enforcement power over antipollution measures was threatening to opponents. But the federal role in water-pollution abatement did not die in the 1950s. In 1961, Congress amended the 1956 act by increasing the annual appropriations for treatment facilities and raising the ceiling on individual grants. It also required the reallocation of unused state allotments to other states needing more funding. Even with additional federal aid several states did not provide adequate supervision of existing facilities by the end of the 1960s.[43]

In the 1960s, grant programs for treatment facilities continued, and funds became available for comprehensive pollution-abatement planning, training personnel, and research into water-quality criteria. The question of establishing some form of sewage-disposal standards was taken up on the national level through an emphasis on water-quality criteria. The Water Quality Act of 1965 required that the states develop water-quality criteria for all interstate waters and that the discharge of liquid wastes meet these standards.[44]

Yet the gap between aspirations for cleaner water and more carefully monitored sewage discharge and actual federal support was not closed by the end of the 1960s. The Johnson administration only requested a modest amount for sewage-treatment facilities in 1969, faced with continually rising demands from Great Society programs and the war in Vietnam. With pressure from home, Congress was considering a much larger package as authorized by the 1966 Clean Waters Restoration Act. But the promises of federal funds had been long delayed and left many cities with little ability to plan for the future. The funding gap promised to stall the progress made since the late 1950s.

Between 1957 and 1969, $5.4 billion was spent on building sewage-treatment facilities, with $1.2 billion coming from federal grants. Most of these funds were expended in communities with less than 25,000 population. Despite the acceleration in funding for sewage-treatment facilities, many of the largest cities continued to battle with unsympathetic state legislatures, federal funds trickled down to smaller cities at a much slower rate than demand, and the water-pollution problem continued to outstrip construction.[45]

As in the case of water-supply systems, the postwar economic boom and the dynamic expansion of suburban America masked the chronic deterioration of the infrastructure and the inability of cities to keep pace with sanitary needs. The broadening perspective on water pollution also presented a challenge to cities that went beyond the battles against epidemic disease. While often not as dramatic as a yellow fever epidemic, modern water-pollution problems were more insidious and possibly more difficult to resolve. The water and wastewater systems of the early twentieth century—systems that had inspired pride in the "can do" spirit—were failing to live up to expectations and foretold a soon-to-be unsettling fear of a new era of crisis.

Solid Waste as "Third Pollution"

1945-1970

By the 1960s, solid waste had become a national environmental issue. In his 1970 book, William E. Small stated, "Today, there is a general recognition that solid wastes are a cancer growing on the land, awful in themselves and awful in the way they further foul the already polluted air and waters near them—a third pollution inextricably interlocked with the two that have been long considered as unacceptable environmental hazards."[1]

Land pollution joined air and water pollution as a triad of blights deserving federal action. The editors of *Sourcebook on the Environment* proclaimed that the solid-waste problem "took the nation by surprise in the 1960s."[2] The waste problem, it appeared, was now being viewed with different eyes—not as a nuisance or a health hazard but as a ubiquitous pollutant. In 1972, Environmental Protection Agency (EPA) administrator William D. Ruckelshaus asserted that solid-waste management was "a fundamental ecological issue. It illustrates, perhaps more clearly than any other environmental problem, that we must change many of our traditional attitudes and habits. It shows us very directly and concretely that we must work to adjust our institutions, both public and private, to the problems and opportunities posed by our traditional disregard for the pollution effects of disposal, and particularly for our misuse of natural resources."[3]

The prosperity that produced the "throwaway" culture in the post–World War II era was stigmatized by its role in generating increasing volumes of waste. For example, a two-person household in New Haven, Connecticut, earning $6,000 per year in the late 1960s and early 1970s generated 800 pounds of waste annually. A four-person household earning $12,000 per year produced 4,000 pounds annually. The doubling of family size, in this case, was less significant than the increased income in the creation of waste.[4]

The volume of municipal solid waste (MSW) rose to staggering proportions by the mid-twentieth century. Daily production of residential and commercial waste increased from two pounds per day per person in 1940 to four pounds per day per person in 1968. While the per capita amounts leveled off at about four pounds per day, population boosts resulted in a steady rise in aggregate waste disposal. In 1969, 256 million tons of residential, commercial, and other municipal waste were generated, of which only 190 million tons were collected.[5]

The problem of quantity was exacerbated by the types of wastes, including relatively new materials such as plastics, other synthetic products, and toxic chemicals. As more was learned about the waste stream, older collection and disposal methods came into question, but were rarely abandoned.

The nature of the waste stream suggested important front-end problems. Of all the discarded items, paper, plastics, and aluminum grew steadily as a percentage of total MSW.[6] The unprecedented growth of the packaging industry was an important component in the availability of products with short useful lives. Packaging took on special importance in the late 1940s because of the rise of self-service merchandising through supermarkets and other consumer outlets. This new direction in marketing required packages that would help sell the product and reduce theft or damage.

The popularity of these packaging materials resulted in new uses for paper and a substantial increase in packaging waste. The widespread consumption of discardable goods also seriously aggravated the litter problem in urban as well as rural communities. Use of paper stock rose from 7.3 million tons in 1946 to 10.2 million tons in 1966. For convenience and less cost, nonreturnable bottles replaced returnables. In 1966, paper and paperboard accounted for 55 percent of packaging materials consumed; glass, 18 percent; metals, 16 percent; wood, 9 percent; and plastics, 2 percent. In 1966, packaging cost the American public 3.4 percent of the gross national product (GNP), not including the expense of collection and disposal. Packaging material in that year amounted to 52 million tons of waste.[7]

Paper represented the largest percentage of municipal waste, while yard and food waste steadily declined. Plastics were a relatively small but rapidly increasing portion of the total. Inorganic materials were scant in terms of percentage, but included many household chemicals with potentially serious environmental implications.[8] Among the materials collected nationwide in 1969 were 30 million tons of paper and paper products, 4 million tons of plastic, 100 million tires, 30 billion bottles, 60 billion cans, and millions of tons of grass and tree trimmings, demolition debris, food waste, and sewage sludge, not to mention millions of discarded automobiles and major appliances.[9]

The cost of solid-waste management continued to be an important factor in addressing collection and disposal problems. In the late 1960s, local governments were spending approximately $1.5 billion per year on collection and disposal. By the early 1970s, the forty-eight largest cities in the United States were spending nearly 50 percent of their environmental budgets on solid-waste management. Several cities began to investigate whether municipal services were an economical way to deal with refuse. Total collection and disposal costs, including municipal service, private service for industrial and other wastes, and self-disposal, exceeded $4 billion by the end of the 1960s. Pickup frequency and location were important in determining the cost of collection. A California study in the 1960s stated that about 47 percent more waste was collected when twice-per-week collection replaced once-per-week collection, and the labor required increased approximately 128 percent.[10]

By 1950, more than 300 cities had instituted service charges for collection and disposal services. This trend followed the user-fee approach commonly employed to underwrite sewer maintenance and treatment-plant operations.[11] As a part of total solid-waste management costs, collection remained the largest percentage. A 1968 survey indicated that the typical community of 400,000 spent $2.5 million for collection each year, but only $900,000 a year for disposal. Refuse collection was characterized more by economies of density than economies of scale, thus population deconcentration, typical of the modern metropolis, was the enemy of efficient and economical collection.[12]

The major technical problems associated with collection revolved around familiar themes—the manner and kind of pickup service provided. Source separation was largely curtailed because of cost, although some cities required yard wastes to be separated from garbage and other household wastes. Multiple collections also were expensive, and often cut back.[13] The introduction of compaction vehicles helped take advantage of economies of density, and use of transfer stations eased the problem of distant collections. Reliance on vehicles to collect refuse, however, had environmental repercussions. One study noted, "The emissions from the massive fleet of garbage trucks, the traffic congestion aggravated by the trucks, the littering caused by trash blown off the trucks, and the disposal processes themselves all contribute to environmental disturbances which would not have occurred if there were no wastes."[14]

While some solid-waste managers hoped that mechanization would decrease reliance on hand labor—and thus reduce accident claims and forestall messy labor disputes—collection remained labor-intensive. Labor plagued pickup services in many cities, especially in cases where race was intertwined with an assortment of worker grievances. The workforce was primarily unskilled or semiskilled, and the work was highly unattractive.[15]

Home-disposal technology had some impact on collection practices. The in-sink grinder became popular in the 1950s. Although it had yet to become a standard home appliance, the grinders made great in-roads in new middle-class subdivisions. The device did not single-handedly solve the collection problem, disposing of about one-tenth of the total volume of refuse collected. The grinders also faced stiff opposition from people in wastewater treatment because they could clog sewer lines and overload sewage-disposal facilities. In 1949, opponents succeeded in banning electric disposers in New York, New Haven, Philadelphia, Miami, and much of New Jersey. In Detroit, Denver, and Columbus, Ohio, legislation sought to increase the use of the device, but usually in new construction only. While the public health benefits of home grinding became more widely accepted, the belief that the device could make cities garbage-free was unrealistic.[16]

City officials usually focused more attention on traditional collection and disposal practices than on new home technologies to solve their refuse problems. Initially, when municipal service expanded in large cities, private firms lost residential accounts and concentrated on commercial and industrial waste collection. The trend toward municipal collection began to shift back to private companies by the early 1960s.[17] APWA surveys show that cities utilizing only contracting for collection remained steady as a percentage of the total between 1939 and 1964, but exclusive municipal collection dropped by 10 percent in those years. Household collection showed the strongest attachment to municipal collection.[18] A 1968 survey demonstrated that public carters conducted 56 percent of the collection of household waste, while public collection handled only 25 percent of commercial waste and 13 percent of industrial waste.[19]

As cities annexed greater numbers of suburban communities, they often contracted with existing private firms rather than expand municipal service. Also, the shift to a more competitive system was accelerated by the change from separate to mixed refuse collection, and by demands for greater efficiency. Increased volume of collectible wastes led several cities to contract part of the service, and dissatisfaction with municipal service encouraged private collectors to claim that they could compete favorably with their public counterparts.[20]

Debate over privatization of collection became the battleground for confronting a range of problems. Major issues included: who had control, and who made decisions about the disposition of the wastes? During the 1960s, a few large hauling firms absorbed smaller companies in some cities, beginning a trend toward consolidation. Three companies organized in the 1960s came to dominate the industry in the 1970s: Browning-Ferris Industries (Houston), Waste Management Inc. (Chicago), and SCA Services (Boston). Although they handled less than 15 percent of the waste at the time, these conglomerates and others soon acquired residential and industrial customers and also offered an array of services. By 1974, the large firms held contracts in over 300 communities. Between 1964 and 1973, 65 percent of cities surveyed operated some sort of collection service, but the proportion of cities with exclusively municipal collection decline from 45 to 39 percent.[21]

It was not uncommon to find consolidations accompanied by criticisms of unscrupulous business practices. At least since the 1950s, the garbage-collection business was alleged to be controlled by the underworld. In response, in 1956 the New York Department of Consumer Affairs was given the responsibility of licensing carters and setting maximum rates. Repeated investigation revealed that racketeers continued to be involved in an activity where the market was not competitive, and customers were often defrauded. The number of firms in the collection market declined dramatically since 1956, when regulation was initiated. Merger as opposed to company failures was common, with no significant new entries. In 1957, as a result of a protracted state senate investigation into labor racketeering, it was discovered that members of New York's underworld had acquired control of Teamster Local 813 and were using it to pressure carters who did not accept their customer allocation scheme. Between 1956 and 1984, New York conducted no less than fourteen investigations into the solid-waste industry.[22]

Finding effective disposal methods continued to be a serious challenge. Land disposal remained the most popular form of disposal. By 1970, there were 15,000 authorized land-disposal sites in the United States, and possibly as many as ten times that number of unauthorized sites. Many of these areas could not be regarded as sanitary landfills.[23] Yet since the 1950s, sanitary fills had significantly increased in number and were competing effectively against other disposal options.[24]

In the 1950s, the Sanitary Engineering Division of the ASCE prepared a manual on sanitary landfilling that became a standard guide. It defined sanitary landfilling as "a method of disposing of refuse on land without creating nuisances or hazards to public health or safety by utilizing the principles of engineering to confine the refuse to the smallest practical area, to reduce it to the smallest practical volume, and to cover it with a layer of earth at the conclusion of each day's operation, or at such more frequent intervals as may be necessary."[25]

During the 1950s and 1960s, the prevailing wisdom was that sanitary landfilling was the most economical form of disposal and, at the same time, offered a method that produced reclaimed land.[26] Land reclaimed in this manner, while suitable for parks, recreational areas, and even parking lots, was not acceptable for residential and commercial sites. A dramatic example occurred in the East Bronx. Row houses were erected on filled land in 1959. Within six months, cracks developed in the walls, and by 1965 the floors were badly tilted. A year later the housing commissioner condemned the dwellings and ordered them demolished.[27]

Another risk of landfilling was potential groundwater contamination. After World War II, several state departments of health issued warnings about possible groundwater pollution from sanitary landfills. By 1970, many states enacted regulations that required field investigations of groundwater locations prior to siting landfills. The most immediate problems focused on older sites, which may not have been well designed or had not taken leaching of the soil into account.[28]

As an alternative, incineration found some strong support in large cities. One expert noted, "Incineration of community refuse has now completed a cycle of favor, disfavor, and return to favor."[29] In cases where incinerators were in competition with landfills, debates focused on cost and convenience. Incineration continued to develop as a large-city disposal option in the 1950s either because landfill sites were not available or the cost of hauling wastes to distant sites was prohibitive.[30]

Waste destruction remained the primary objective of incineration.[31] Although the demand for economy in city government led to more investigations of heat-recovery systems, relatively few were built or operated successfully in the period. Obstacles to incineration development did not rest heavily on environmental costs. Many engineers still believed that if properly designed and operated, incinerators need not be offensive. The greatest objection was the traffic of collection vehicles moving to and from disposal sites. But incineration was far from a universal solution to the waste problem.[32]

The 1960s witnessed a great number of incinerator abandonments. One estimate suggested that a third of cities with incinerators (about 175) discontinued them in favor of other methods—primarily sanitary landfills. Problems arose in many cases because the plants were operated beyond design capacities. Most significant, the relationship between incineration and air pollution drew increasing attention. Even well-designed incinerators produced smoke when overloaded. A 1969 study of particulate emissions into the New York City atmosphere indicated that 19.3 percent were derived from municipal incineration and another 18.4 percent from on-site incineration, with the remainder from space heating, power generation, and industrial and mobile sources.[33]

Until the 1960s, knowledge of refuse furnace technology was in the hands of a few pioneer incinerator builders and some engineers. Increasing recognition of environmental problems linked to burning undermined what the experts had been saying about the safety of incinerators. In several states and cities, more stringent air-pollution control regulations emerged. Nevertheless, some engineers still insisted that unsatisfactory results from incinerator operation were due to economic compromises rather than lack of technical expertise. The APWA estimated in 1970 that 75 percent of incinerator facilities had inadequate air-pollution control provisions.[34]

Other disposal options had a less significant impact than landfilling and incineration between 1945 and 1970. Feeding garbage to swine continued until the 1950s. Between 1953 and 1955, owing to the rapid spread of vesicular exanthema (a swine disease that led to the slaughter of more than 400,000 pigs), regulations were instituted forbidding the feeding of raw garbage to hogs. Although cooking the waste solved the problem, hog feeding steadily declined because the method became economically prohibitive.[35]

Dumping of municipal waste at sea was terminated by the U.S. Supreme Court in 1934, but industrial and some commercial wastes were exempt from the ruling. At the end of the 1960s, it was estimated that each year from 50 to 62 million tons of waste (60 percent of which were dredging spills) were dumped into the ocean. In fact, industrial-waste dumping in the Northeast more than doubled between 1959 and 1968. New York City continued to dump sewage sludge, which was described by critics as a "dead sea of muck and black goo." In the mid-1970s, there were nearly 120 ocean sites for waste disposal, supervised by the U.S. Coast Guard.[36]

Various forms of recycling and reuse were just beginning as disposal options in the postwar years. Composting, or biochemical degradation of organic substances producing humus, was carried on in Europe and the Asian subcontinent. Since 1960, approximately 2,600 composting plants were in use outside the United States, 2,500 of which were small plants in India. By the 1950s, no American city employed composting of organic waste, but several years later a very few plants were built by private companies. High transport costs and the inability to sell large quantities of the by-products kept the business from expanding. By the end of 1967, of thirteen compost operations, only three were active—in St. Petersburg, Florida; Mobile, Alabama; and Houston, Texas.[37]

Other forms of recycling, although promoted by environmentally sensitive individuals in the late 1960s, made scant headway. The recycling rate for paper steadily declined between 1945 and 1970. In 1969, 58.5 million tons of paper were consumed in the United States; only 17.8 percent was recovered through recycling, compared with 23.1 percent in 1960 and 27.4 percent in 1950. The largest sources of wastepaper were corrugated boxes and newspaper. The decline in the recovery ratio and the rise in consumption made the problem of wastepaper increasingly significant.[38]

Until the mid-1960s, collection and disposal were considered functions of local government. However, growing recognition that solid waste was part of a more general national environmental problem, and that collection and disposal of wastes were straining local resources, led to the assumption that control limited to the local level would be inadequate to deal with an issue that extended beyond the city limits of every community.[39]

Change in perspective about solid waste as "third pollution" was pivotal. An HEW report stated that high rates of production and consumption of goods, plus rising urban population, had created "a refuse disposal problem that far outstrips the waste handling resources and facilities of virtually every community in the nation." It added that the results of the problem were "obvious and appalling—billowing clouds of smoke drifting from hundreds of thousands of antiquated and over-burdened incinerators, open fires at city dumps, wholesale on-site burning of demolition refuse . . . , acres of abandoned automobiles that blight the outskirts of our greatest cities, and veritable mountains of smoldering wastes abandoned at mining sites."[40]

Acknowledging the poor state of solid-waste research and urged on by demand for new legislation, the APWA and the USPHS sponsored the first National Conference on Solid Waste Research in Chicago in December 1963. Soon thereafter, the APWA established the Institute for Solid Wastes (1966).[41]

In a special message on conservation and restoration of national beauty in 1965, President Lyndon Johnson called for "better solutions to the disposal of solid waste" and recommended federal legislation to assist state governments in developing comprehensive disposal programs and to provide research and development funds. In that year Congress passed the Solid Waste Disposal Act as Title 2 of the 1965 amendments to the Clean Air Act.[42] It was the first piece of federal legislation to recognize refuse as a national problem. It also was the first to involve the federal government in solid-waste management through efforts to "initiate and accelerate" a national research and development program and to provide technical and financial assistance to state and local governments and interstate agencies in the "planning, development, and conduct" of disposal programs.[43]

The act focused on demonstration projects for new methods of solid-waste collection, storage, processing, and disposal. In Virginia Beach, Virginia, a vertical sanitary landfill was constructed. The Maryland State Department of Health attempted to determine the feasibility of using abandoned strip mines for landfills. In San Diego, experiments were conducted on converting waste into sterile and transportable fuel. The APWA studied rail haul for transporting wastes out of cities. Oregon State University investigated the chemical conversion of wastes to reduce volume and to discover salable materials. And the University of Maryland conducted research to derive a protein concentrate from the waste of food-processing companies.[44]

The new law was incomplete in its assessment of the waste problem and did not mandate a regulatory authority to deal with broader refuse issues. In addition, the primary focus was on disposal, not collection or street cleaning. Although the act was a significant step in federal involvement, there were scant reliable data on the extent of the solid-waste problem. President Johnson called for a special study to assess the scale of the nation's waste, and the National Survey of Community Solid Waste Practices (1968) became the first truly national study of its kind in the twentieth century. Although less than half of 6,000 communities participated, the survey helped to fill the "data gap."[45]

Initially, the enforcement of the 1965 act fell to the USPHS (municipal wastes) and to the Bureau of Mines (mining and fossil-fuel waste from power plants and industrial steam plants). With the creation of the EPA in 1970, responsibility for most refuse activities was transferred to it.[46] To refine the 1965 act, Congress passed the Resource Recovery Act in 1970, which shifted the emphasis from disposal to recycling, resource recovery, and the conversion of waste to energy. It created the National Commission on Materials Policy to develop a national policy on materials requirements, supply, use, recovery, and disposal. Another feature was the stipulation that a national system be implemented for storing and disposing of hazardous wastes.[47]

Although the federal legislation was incomplete, both the 1965 and 1970 acts caused the states to become more deeply involved in solid-waste collection and disposal. At the time of the passage of the Solid Waste Disposal Act of 1965, there were no state-level solid-waste agencies. Responding to federal pressure or incentives, states began to enact solid-waste management statutes and to designate one of their agencies as the state's solid-waste management office. Four-and-a-half years after the first grant of technical assistance was awarded, forty-four states had active programs. The most significant immediate result was the development of solid-waste management plans—a prerequisite for receiving federal funds. The state programs, however, demonstrated little uniformity.[48]

The key change in solid-waste management between 1945 and 1970 was the recognition of refuse as a national environmental problem. Local authorities welcomed state and federal aid in their battles against mounting hills of garbage, but they also felt resentment as new legislation mandated actions that required shifting local priorities and committing local resources.[49] Rising NIMBYism also increased citizen pressure for solutions to chronic problems. As with water supply and sewerage, public debate over an escalating "garbage crisis" would occur in the 1970s. Solid-waste management systems were no longer the "orphan child of sanitary engineering," but shared equal status with other technologies of sanitation.

From Earth Day to Infrastructure Crisis

FORCES SHAPING THE NEW SANITARY CITY

Implementation of new sanitary services and maintenance of existing systems faced serious challenges after 1970 as metropolitan growth became more complex, urban fiscal problems deepened, and environmental concerns intensified. By the early 1980s, there was much talk about a looming "infrastructure crisis."

Historians David R. Goldfield and Blaine A. Brownell have stated, "A new era of urbanization emerged after 1970, though few Americans noticed it at the time. Culminating a trend begun in 1920, the 1970 census announced that we had become a suburban nation."[1] Historian Jon C. Teaford added a new dimension to this observation by suggesting that the urban/suburban relationship was something quite different from past experience:

Metropolitan America was no longer organized around single dominant centers, as it had been in 1945; neither was it truly polycentered, with a few readily identifiable hubs. To identify metropolitan regions by the names of supposed centers was an anachronism, for metropolitan America was increasingly centerless. Yet it was not a featureless sprawl of indistinguishable elements or a uniform expanse of low-density settlement. . . . Instead it was rich in diversity, a historical accretion of settlement patterns and lifestyles that reflected the felt needs of millions of Americans of the past and present. . . . Metropolitan America was edgeless and centerless.[2]

Core cities continued to lose population and have their economic base erode. Metropolises developed multiple, albeit not always clearly defined, centers; a variety of self-contained communities arose on the city's malleable periphery; and nonmetropolitan growth challenged traditional suburban expansion. Southern and southwestern cities grew at the expense of older urban centers in the Northeast and Midwest.[3]

The metropolitan complex was home to the majority of Americans after 1970. In 1975, 73 percent of the U.S. population (213 million) lived in metropolitan counties; in the 1980s, the number jumped to 86 percent (248.7 million). The 1990 census indicated that California was the most urban state at 93 percent, while the southern region had the largest metropolitan population. In 2000, 80.3 percent of Americans (226 of 281.4 million) lived in metropolitan areas. Since 1990, population within metropolitan areas increased by 14 percent, while the nonmetropolitan population grew by 10 percent.[4]

Despite the anticipated return of some middle-class Americans to the central city through gentrification, population at the core continued to decline late into the twentieth century.[5] In 1986, only about 30 percent of the American population lived in central cities; a disproportionate number were poor African Americans and poor Latinos.[6] In addition, urban population density declined from 2,766 people per square mile in 1970 to 2,141 in 1990, indicating movement from central cities toward satellite communities and "out-towns." The decade of the 1970s was easily the period of greatest loss of population in central cities. However, core cities with boundaries similar to those in 1950 experienced a 2.2 percent increase since 1990, suggesting a modest renaissance for core cities into the next century. As one observer noted "There is virtually no chance that they will ever recover to their peak levels, but at least the losses have been stopped."[7]

The central city ceased to be the axis upon which the city turned, but it remained the focus of a service economy dominated by banking, finance, medicine, and education. The central city "had simply become one more suburb, yet another fragment of metropolitan America serving the special needs of certain classes of urban dwellers."[8] Within the metropolis, increasing numbers of urbanites lived outside central cities; by 1970, more than half of the metropolitan residents lived in suburban communities (including 20 to 40 percent of the poor in some areas).[9]

The term "suburb" became less useful in describing what was taking place on the urban fringe. The bedroom communities that provided workers for the central cities were increasingly self-sufficient, attracting new jobs and new business establishments, as well as cultural and recreational activities once reserved for downtown. Commentators searched for new designations to replace "suburbs," such as out-towns, mall-towns, edge cities, boomburbs, and technoburbs.[10]

In the years after 1970, American cities—especially in the East and Midwest—faced the so-called enduring plight characterized by financial stress, lack of a coherent federal urban policy, and rising social and physical problems.[11] Federal and state governments mandated that a city perform a wide array of functions for its residents, and it also had to finance services of benefit to nonresidents, especially commuters and those taking advantage of cultural and recreational events. Despite sporadic urban rehabilitation and back-to-the-city efforts in some locations, out-migration of middle-income residents and the influx of lower-income groups meant that many cities had less capacity to raise revenue and to provide satisfactory services.[12]

The costs of providing services tended to be higher in core cities than in the surrounding communities. First, the increasing concentration of low-income people at the core resulted in higher demand for welfare, health care, and hospitals. In 1978, 15.4 percent of central city populations had income below the poverty line; in 1992, it was 20.5 percent. In 2000, about 11 million poor people lived in the suburbs, compared with 13 million in central cities. Yet the rate of poverty was approximately twice as high in central cities as in suburbs (16.1 percent as compared to 7.8 percent).[13] Second, core cities were substantially older than suburbs, and thus the cost of maintaining or replacing existing infrastructure was high. And third, the cost of providing services to nonresidents reduced funds for other purposes.[14]

Beyond structural problems, fiscal woes were real and widespread. Sources of general revenue were in decline as white flight continued, and as reluctance to increase taxes led officials to find other means to balance their budgets. To meet service obligations, several departments borrowed from the city's cash flow, intending to pay off the debt with future revenues. Before long, cities were accumulating substantial internal debt.[15]

New York City's financial collapse in 1975 was the classic case, which brought nationwide attention to the municipal fiscal tailspin. The Big Apple's problems were at least ten years in the making. By 1975, the city had accumulated $2.6 billion in debt as a result of borrowing large sums during the economic expansion of the 1960s. New York City's expenditures were not conventional since it financed several services that were provided by the states, counties, and special districts in other communities. When the Nixon administration sharply cut federal aid to the city, local officials postponed payments on previous debts through short-term borrowing. Interest payments were becoming the fastest growing budget item in many cities. When New York City's largest banks withdrew from the bond market in late 1974, the financial juggling could not continue. The city's budget was in disarray, and more than 500,000 jobs were lost. A congressional bailout of the city restored some financial stability in the late 1970s, but other cities faced similar financial crises, such as Baltimore, Cincinnati, and Philadelphia. Some cities, especially those in the Sun Belt, such as Houston, Dallas, and Phoenix, avoided the same plight.[16]

Cities had long since turned to Washington and their state capitals for financial aid. In 1960, federal grants represented 3.9 percent of the general revenue of cities; in 1977, 16.3 percent. In several cities, one-quarter to one-half of operating revenues came from the federal government. Between 1960 and 1977, the percentage of local taxes used for general revenue dropped from 61.1 to 42.8 percent.[17] As local dependence upon federal largesse increased, the manner in which those funds were dispersed changed dramatically. In the early 1970s, Richard Nixon was particularly interested in dismantling many Great Society programs that gave priority to neighborhood participation over the authority of elected officials. Declining populations of northeastern and midwestern cities, and growth in the Sun Belt (with accompanying political prospects for Republicans), made the new administration less enthusiastic about financing decaying older central cities. Nixon's New Federalism called for greater responsibility of local government in controlling the spending of funds from Washington.[18]

The State and Local Assistance Act of 1972, a program of revenue sharing, dispersed approximately $30 billion over five years to 3,900 local government units. Two additional revenue-sharing acts followed. One of them, the Housing and Community Development Act (1974) passed under Gerald Ford, was considered an assistance program for central cities, but in practice revenue-sharing enactments became a program of suburban aid. Under the 1974 act, funds were sent automatically to more than eighty "urban counties" and to all cities in metropolitan areas with 50,000 or more population. Supreme Court–mandated reapportionment of state legislatures in 1962 had increased suburban political strength to vie for funds with central cities. Outward metropolitan growth promoted the development of independent administrative authorities—such as special districts—many of which qualified for revenue sharing. Central cities had to spend a large portion of their grants to avoid major budget cuts and to sustain existing programs. Suburban communities could use federal funds to keep tax rates low, to raise municipal salaries, and to underwrite new projects.[19]

The Democratic administration of Jimmy Carter continued the local-control approach of the Nixon-Ford years.[20] With the election of Ronald Reagan, the New Federalism veered sharply away from any major commitment to urban affairs in the 1980s. The Republican administration reshuffled priorities, strongly emphasizing federal support for national security and rebuilding of the nation's defense system at the expense of many domestic programs. While Reagan supported a "safety net" for the poor with Social Security, he advocated sweeping reductions in federal aid to cities, including an end to revenue sharing; support for incentives for private investment; and the privatization of many programs and services. Welfare payments were cut, half of the federally subsidized housing units were eliminated, rents charged to public housing tenants increased, and job-training programs were reduced. The only truly urban program in these years was the promotion of enterprise zones, championed by New York congressman Jack Kemp.[21]

Combined with the shrinking city tax base, this drastic change in national urban policy—described as a "nonurban urban policy"—initially created "havoc among metropolitan governments"; resulted in the closing of schools, parks, and libraries; and forestalled repairs to streets, sewers, and bridges. The dire predictions about the Reagan budget cuts were somewhat exaggerated since the Democratic-controlled Congress restored funds for several programs. State and local governments responded by establishing emergency funds and raising taxes to capture lost revenue. Nonetheless, funding for specifically urban programs declined 23 percent in 1982 from what had been projected under the old budget, and revenue sharing made way for no coherent federal urban policy at all. In addition, state aid did not take up the slack from lost federal support.[22]

The program of federal "disinvestment" was particularly difficult for cities already racked by scores of economic, social, and physical problems. Total federal spending in nondistressed cities increased by almost 66 percent (in current dollars) between 1980 and 1986; in declining cities, federal spending grew by only 5 percent. Reagan's New Federalism severely strained the nation's older cities, but the blows at least were not fatal. The disparity between central city decline and peripheral growth remained.[23] In 2003–4, the intergovernmental revenue of all cities in the United States exceeded $430 million (other sources, including property and sales taxes, were about $665 million), of which about $51 million (about 12 percent) came from the federal government.[24]

The early 1980s also witnessed public recognition of long-standing deterioration of the existing urban infrastructure and uneven investment in new infrastructure. The short-term picture revealed mixed results, leaving the impression that the breakdown was quite recent. Government spending on public works, discounting for inflation, rose from $60 billion in 1960 to $97 billion in 1984. This represented a decreasing share of the GNP from 3.7 to 2.7 percent. In 1977, the share funded by federal government programs stood at 53 percent of state and local public works capital investment, due primarily to healthy grants for transit, airport projects, and wastewater-treatment facilities. By 1982, it had declined to about 40 percent.[25]

In 1981, Pat Choate, a policy analyst at TRW Inc., and Susan Walter, manager for state government issues at General Electric, published *America in Ruins.* The book ignited wide debate over the state of the nation's public works. According to Choate and Walter, public facilities were wearing out faster than they were being replaced. In a time of tight budgets and inflation, maintenance was being deferred, replacement of obsolete public works was being postponed, and new construction was canceled. The most acute implication of this infrastructure crisis, they argued, was its impact as "a severe bottleneck to national economic renewal."[26] They painted a dismal picture of existing conditions: the Interstate Highway System was deteriorating at a rate requiring reconstruction of 2,000 miles of road per year; one of five bridges needed rehabilitation or reconstruction; and between $75 billion and $110 billion would be needed in the next twenty years to maintain urban water systems. They called for the federal government to help create a coherent public works policy and to analyze and budget for the nation's infrastructure needs.[27]

While not everyone agreed upon the extent of the "infrastructure crisis" or the actual costs required to alleviate it, few denied that a massive problem existed. In 1982, a Community and Economic Development Task Force supported by HUD noted that the nation had "discovered" its urban infrastructure problem especially apparent in the oldest urban centers.[28] A study completed by the Municipal Finance Officers Association (1983) flatly stated that "America's infrastructure is in trouble," emphasizing "disinvestment" in basic public facilities to the tune of from $500 million to $3 trillion. In the same year, a study completed at the request of the Senate Committee on the Budget observed that "a continual emphasis on new construction and replacement can induce states and localities to neglect needed repairs." A privately conducted study noted that since 1965, the percentage of the GNP spent on public works at all levels of government declined from 4.1 to 2.3 percent as priorities changed from these "less visible problems" to a concern over social issues.[29]

Studies in the mid-1980s, while still promoting the idea of a crumbling infrastructure, suggested refinements to the earlier projections. A 1984 workshop on alternatives for urban infrastructure stated what was becoming obvious: "There is . . . no consensus as to the extent of the problem" and no apparent single cause. But "business as usual" was impossible, and the situation was "crying out for innovation" and cooperation among administrative authorities.[30]

The Public Works Improvement Act of 1984 created the Council on National Public Works Improvement, directed to report to the president and Congress on the state of the nation's infrastructure. In its first report, the council stated that the infrastructure issue had matured since "the first sensational news reports." It added that many systems were in "sufficient disrepair," that the problem was widespread, and that the cost of meeting future needs "will be very high."[31]

In its much-anticipated 1988 report, *Fragile Foundations,* the council took a positive approach, finding "much that is good" about what was being done in the planning, building, operating, and maintaining of the nation's infrastructure. "But on balance, the Council concludes that our infrastructure is inadequate to sustain a stable and growing economy. As a nation, we need to renew our commitment to the future and make significant investments of our own to add to those of past generations." The council issued its "Report Card on the Nation's Public Works." The highest grade (B) went to water resources, based on the hope that the Water Resources Act of 1986 would make cost sharing mandatory for many water projects. Water supply and aviation scored grades of B-, the former because it stood out as "an effective, locally-operated program" and the latter because it handled "rapid increases in demand safely and effectively." Highways got a C+ for better maintenance and improved pavements; wastewater a C because of wide coverage, but little improvement in water quality; mass transit a C- for an overall decline in the system; solid waste a C- for more rigorous testing, monitoring, and development of alternative disposal options, but movement toward higher costs. Hazardous waste received the lowest grade (D) because progress was slower than expected for cleanup.[32]

The essence of the report, and several that came before it, was that the task ahead was formidable but not impossible. *Fragile Foundations* noted that capital investment in public works had peaked in the 1970s. Current levels of spending were regarded as "barely enough" to offset annual depreciation let alone meet new demand."[33] Into the 1990s, despite the promises of President Clinton and others to promote efforts to "Rebuild America," many questions remained. From a political perspective, public works still lacked a strong, unified constituency, and had to confront the vagaries of electoral politics.[34] The ASCE's 2005 "Report Card for America's Infrastructure" showed little improvement from that stated in *Fragile Foundations* in 1988. Aviation was rated as D+; roads, D; transit, D+; wastewater, D-; drinking water, an abysmal D-; with only solid waste (C+) and hazardous waste (D) graded the same. Granted, the evaluators were different—and possibly a little self-serving—and the criteria might have been different, but the ASCE's estimate that it would take $1.6 trillion over five years for investment in infrastructure was, nevertheless, a daunting number.[35]

While the specter of an infrastructure crisis hung over the maintenance and development of sanitary services in the late twentieth century and beyond, the momentum of the environmental movement further changed the context in which those services were viewed. What made the environmental movement so remarkable was the speed with which it gained national attention beginning in the late 1960s. Nothing epitomized that appeal better than Earth Day. The idea began as a "teach-in" on the model of an anti–Vietnam War tactic. Across the country, on 2,000 college campuses, in 10,000 high schools, and in parks and various open areas, as many as 20 million people celebrated purportedly "the largest, cleanest, most peaceful demonstration in America's history." In form, Earth Day was so much like a 1960s-style peace demonstration that the Daughters of the American Revolution (DAR) insisted that it must be subversive. In fact, it was pitched at moderate activists such as the Sierra Club and the Audubon Society. As a symbol of the new enthusiasm for environmental matters, and as a public recognition of a trend already well under way, Earth Day served its purpose.[36]

The Nixon administration gave its blessings to Earth Day, and on January 1, 1970, the president signed the National Environmental Policy Act of 1969 (NEPA). While opposing the bill until it cleared the congressional conferees, the Nixon administration ultimately embraced NEPA as its own. Going on record against "clean air, clean water, and open spaces" served no purpose. By identifying his administration with environmentalism, Nixon was able to address the issue on his own terms, most especially by focusing on antipollution problems through technical solutions.[37]

NEPA was far from "the Magna Carta of environmental protection" that some people proclaimed, but it called for a new national responsibility for the environment. NEPA was not simply a restatement of resource management; it also promoted efforts to preserve and enhance the environment. It particularly emphasized the application of science and technology in the decision-making process. The provision mandating action required federal agencies to prepare environmental impact statements (EISs) assessing the environmental effects of proposed projects and legislation.[38]

NEPA provided substantial opportunity for citizen participation, especially through access to information in agency files. It established the Council on Environmental Quality (CEQ) to review government activities pertaining to the environment, to develop impact-statement guidelines, and to advise the president on environmental matters. The CEQ was essentially a presidential instrument, and governmental environmental programs remained widely dispersed.[39]

After a plan to codify environmental programs into one department failed, it was announced that pollution control programs and the evaluation of impact statements would be the responsibility of a new body—the EPA. The EPA began operations in December 1970 under the direction of William Ruckelshaus. Initially, it included divisions of water pollution, air pollution, pesticides, solid waste, and radiation. But the EPA did not have overall statutory authority for environmental protection; it simply administered a series of specific statutes directed at particular environmental problems. The agency nevertheless was soon inundated by its regulatory responsibility and pressured by efforts of industry to skirt around those regulations.[40]

By the mid-1970s, environmentalism was a solidly fixed national issue. Mainstream environmental groups responded by taking greater initiative in helping to draft new legislation, pressing for the implementation of existing legislation, focusing on the environmental-impact review process, and monitoring government agencies. In addition, the courts became a battleground as more litigation tested key regulatory provisions.[41]

In the 1980s, the Reagan administration redirected the national government as a major force in formulating and implementing environmental regulations. Reagan supporters hoped that the new administration would reduce the adverse effects of environmental legislation on business, promote economic development, and reduce the government's role in environmental affairs. The administration's detractors sensed the onset of an era of antienvironmentalism. The selection of James Watt as secretary of the interior and Anne Gorsuch as administrator of the EPA seemed to confirm the views of both sides. The Reagan administration, according to some observers, applied the test of "economic development over environmental constraint" to much of its policy, preferring "administrative discretion rather than legislation as the preferred instrument of change."[42]

James Watt announced plans to develop resources on public lands in the West and step up offshore oil drilling. At the EPA, Gorsuch proposed a budget about one-quarter as large as that of the Carter administration, and seriously weakened the enforcement division. David Stockman, at the Office of Management and Budget, demanded cost-benefit analysis of all environmental regulations. And President Reagan fired the sitting members of the CEQ.[43]

The environmental momentum of the 1970s was stalled rather than derailed by the mid-1980s. As Watt's antienvironmentalism became more outrageous, membership in major environmental organizations increased and continued to do so even after his resignation in October 1983. In the same year, Gorsuch resigned under a cloud of wrongdoing and potential criminal conduct at the EPA. In an attempt to repair the political damage caused by the crippling of the EPA, Reagan persuaded William Ruckelshaus to return as administrator. In 1985, he was replaced by career government manager Lee Thomas. During George H. W. Bush's administration, William K. Reilly, head of the Conservation Foundation and World Wildlife Fund, became EPA administrator, but the administration's overall record on the environment was not strong.[44]

By the end of the decade, progress in environmental regulation and compliance had to be measured incrementally. The desire to balance economic development against environmental protection remained popular in political circles. In the environmental community, the concept of "sustainable development"—the notion that economic development can be maintained but not at the expense of an equitable standard of living and fulfilling of basic needs for all people while also taking steps to avoid damage to the environment—became the watchword.[45] However, a commitment to environmental affairs in general remained a national priority, manifest in entrenched environmental lobbyists in Washington, the emergence of a variety of grassroots organizations, and public battles over specific issues.

By the 1980s, mainstream environmentalists faced challenges not only from critics in politics and the private sector but also from competing environmental viewpoints. Sometimes acrimonious, these challenges nonetheless began to broaden the base of environmentalism. The environmental justice movement emerged in the mid-1980s to confront what it perceived as the growing threat of environmental racism. Some connected class and race, but many others viewed racism as the prime culprit. Rev. Benjamin E. Chavis Jr., former head of the National Association for the Advancement of Colored People (NAACP), is credited with coining the term "environmental racism" while executive director of the United Church of Christ's Commission for Racial Justice (CRJ). Chavis became interested in the connection between race and pollution in 1982 when the predominantly African American residents of Warren County, North Carolina, asked the CRJ for help in resisting the siting of a polychlorinated biphenyl (PCB) dump in their community. The protest proved unsuccessful, resulting in the arrest of more than 500 people, including Chavis. While the roots of the environmental justice movement go back into the 1970s, the Warren County case was a cause célèbre.[46]

The environmental justice movement found its strength at the grassroots level, especially among low-income people of color who faced environmental threats from toxics and hazardous wastes. According to sociologist Andrew Szasz, "The issue of toxic, hazardous industrial wastes has been arguably the most dynamic environmental issue of the past two decades."[47] The reaction of local groups to toxics and hazardous wastes began as NIM-

BYism, but evolved into something different. Lois Marie Gibbs, of the Citizens Clearinghouse for Hazardous Wastes and a well-known leader at Love Canal, stated, "Our movement started as Not In My Backyard (NIMBY) but quickly turned into Not In Anyone's Backyard (NIABY) which includes Mexico and other less developed countries."[48] What emerged, according to Szasz, was a radical environmental populism–ecopopulism within the larger tradition of American radicalism. Almost 4,700 local groups voiced public outcries against toxics by 1988, and a vibrant and networked social movement emerged, led at the local level in large part by women.[49]

Grassroots resistance to environmental threats, advocates say, is simply the reaction to more fundamental injustices brought on by long-term economic and social trends. Cynthia Hamilton, of California State University–Los Angeles, stated that the consequences of industrialization "have forced an increasing number of African Americans to become environmentalists. This is particularly the case for those who live in central cities where they are overburdened with the residue, debris, and decay of industrial production."[50] The struggles against "environmental injustice," as sociologist Robert D. Bullard has noted, are "not unlike the civil rights battles waged to dismantle the legacy of Jim Crow."[51]

In October 1991, a multiracial group of more than 600 met in Washington, D.C., for the first National People of Color Environmental Leadership Summit. In its "Principles of Environmental Justice" conference, participants asserted the hope "to begin to build a national and international movement of all peoples of color to fight the destruction and taking of our lands and communities."[52] In some respects, the civil rights movement that faltered in the late 1970s and 1980s was seeking resurgence through environmental justice. By embracing historic roots in civil rights activism, the environmental justice movement also disavowed connection to mainstream environmentalism, which it perceived as white, often male, middle and upper class, primarily concerned with wilderness preservation and conservation, and insensitive to interests of the minorities.[53]

The CRJ produced *Toxic Wastes and Race in the United States* in 1987, which was the first comprehensive national study of the demographic patterns associated with the location of hazardous-waste sites. The findings stressed that the racial composition of a community was the single variable best able to predict siting of commercial hazardous-waste facilities and that minorities were overrepresented in such settings. The report concluded that it was "virtually impossible" that these facilities were distributed by chance.[54] Its strong inference of deliberate targeting of communities because of race gave powerful ammunition to those interested in further developing the movement for environmental justice. It also set off a controversy over the importance of race as the central variable in siting hazardous-waste facilities, leading to charges of shoddy research and faulty logic by those advocating the findings of the CRJ report.[55]

Despite the controversy, environmental justice movement leaders saw PIBBY (Place in Blacks' Backyard) replacing NIMBY. Bullard argues, "Since affluent, middle-income, and poor African Americans live within close proximity of one another, the question of environmental justice can hardly be reduced to a poverty issue."[56] The government was also a culprit in deliberate targeting, he concluded, because it institutionalized unequal enforcement of laws and regulations, favored polluting industries over "victims," and delayed cleanups.[57]

Efforts by the federal government to address concerns over environmental racism were viewed with skepticism by those within the movement. A 1992 EPA report supported some of the claims of the exposure of racial minorities to high levels of pollution, but it linked race and class together in most cases. A study conducted by the *National Law Journal* in that same year questioned the EPA's environmental equity record, pointing out that in the administering of the Superfund program, disparities existed in dealing with hazardous-waste sites in minority communities as compared with white neighborhoods. EPA director William Reilly was strongly criticized by movement members for not attending the People of Color Environmental Summit. And despite convincing President Clinton to sign an executive order to "focus Federal attention on the environmental and human health conditions in minority communities and low-income communities with the goal of achieving environmental justice," there was disappointment because an environmental justice act never passed Congress.[58]

Among people of color, criticism of mainstream environmentalism and government action was not meant to leave the impression that minorities had little or no concern for environmental issues. Some in the environmental justice movement even advocated cooperating with mainstream environmental groups in areas of common interest.

Most important, the environmental justice movement has reintroduced, and in many ways broadened, the issue of "equity" as it relates to environmentalism in the United States. It persuaded—or possibly forced—environmental groups, government, and the private sector to consider race and class as central features of environmental concern. It helped to elevate toxics and hazardous-waste issues to central importance among a vast array of environmental problems. It shifted attention to urban blight, public health, and urban living conditions to a greater degree than earlier efforts had done. And it questioned demands for economic growth at the expense of human welfare. The movement, however, was not without its limitations. Its stances were sometimes inconsistent, especially on the issue of race versus class. It also sometimes underestimated its friends and mischaracterized its foes. This was particularly true because the movement, especially in its early years, was seeking political redress for the problems it raised.[59]

Broadening of the environmental movement also is obvious in the increasingly central role of women in the antitoxics movement and other grassroots reforms in the 1980s. With the exception of some antinuke campaigns and a few others, women had limited leadership roles in traditional environmental groups. In 1987, forty-three women met at the Women in Toxics Organizing Conference in Arlington, Virginia, to discuss strategy. By that time, ecofeminism—a combination of theoretical constructs and social activism—was beginning to provide new approaches to action alongside the growing influence of a variety of grassroots efforts.[60]

Despite skepticism among environmental justice leaders, the election of Bill Clinton as president in 1992, according to Richard N. L. Andrews, "raised widespread expectations of a return to vigorous presidential leadership in environmental policy."[61] But while initially defining a strong agenda and asserting the administration's commitment to leadership in international environmental policy, attention was diverted to an array of nonenvironmental controversies that pitted Clinton against his conservative Republican opposition. During the presidency of George W. Bush, the leadership of the United States in international environmental policy making waned, attempts increased to relax or reverse several major environmental policies at home, and efforts to utilize environmental resources such as new oil exploration stepped up.[62] Andrews queried: "While environmental issues remained visible and controversial . . . one might ask whether the environmental era itself had finally ended, or whether it had at least entered a new phase fundamentally different from the period of broad-based consensus priorities and politics that produced it."[63]

In the age of the New Ecology, science had come to play an important role in helping to formulate environmental issues, seeking antidotes, and contesting them in the political arena. In the 1970s, ecology grew from a branch of biology to become a separate discipline. Because its scope was so broad, it splintered into subdivisions such as population genetics, conservation ecology, systems ecology, and ecological economics. Environmental science, while embracing ecological principles, focused on resources, climate, and earth sciences, as well as on pollution and technology assessment.[64]

In the 1980s, the study of ecological risks grew rapidly. Although similar in approach to human health risk assessment, it went beyond individuals to examine communities, whole populations, resources, biodiversity, and ecosystem recovery.[65] Risk became the primary means of evaluating the interface between humans and the physical environment. Controversies over the levels of proof of risk and harm were inevitable. Yet in bringing environmental issues to the public and in helping to fashion regulations, scientific research provided tools for assessing risk or offering new technologies to control pollution. At the same time, scientific inquiry uncovered unanticipated environmental hazards, introduced a level of complexity not previously realized, and initiated debate over the accuracy of data into the political process.[66]

The new regulatory apparatus in the 1970s, advances in ecological sciences, and public discussion over deteriorating infrastructure influenced public works in major ways. The engineering community, in particular, underwent considerable readjustment. By the late 1960s, "sanitary engineers" had become "environmental engineers" who were employed in heavy industries and utility companies, many public works construction projects, federal agencies including the EPA, state and local governments, and consulting firms.[67] Environmental engineers "deal with structures, equipment, and systems that are designed to protect and enhance the quality of the environment and to protect public health and welfare."[68] They balance a concern for the physical environment with a human-centered view focusing on protection of individual health and well-being, and thus have moved away from an earlier predisposition to manage nature.

Historian Jeffrey K. Stine suggests a pragmatic objective for establishing the new field of environmental engineering: "By pushing environmental engineering, engineers sought to expand their constituencies."[69] The text *Environmental Engineering* reinforced that perspective, suggesting that while environmentally sound engineering "adds the dimension of social benefit to the engineer's job," pollution control engineering became "an exceedingly profitable venture."[70]

The commitment to technological solutions, however, remained widely accepted among engineers, and extended beyond their community. EPA administrator Russell E. Train wrote in 1975 that "in controlling pollution, whether by establishing discharge standards for new sources or compliance schedules for existing facilities, improvements in technology must and will be a driving force in achieving our environmental goals."[71]

A parallel, but decidedly different, approach to civil engineering was proposed in the 1960s—ecological engineering. Crediting H. T. Odum and Chinese ecological literature in the twentieth century, William J. Mitsch, professor of natural resources and environmental biology at Ohio State University, defined ecological engineering—or "ecotechnology"—as "the design of human society with its natural environment for the benefit of both." By contrast with environmental engineering, he added, ecological engineering is involved in "identifying those ecosystems that are most adaptable to human needs and in recognizing the multiple values of these systems." Unlike other forms of engineering and technology, "ecological engineering has as its raison d'etre the design of human society with its natural environment, instead of trying to conquer it."[72] While ecological science and ecological engineering were shaped largely by the biological sciences, environmental science and environmental engineering were shaped largely by the physical sciences. Principles of ecological engineering and ecological science were applied most notably to the treatment of wastewater.[73]

After 1970, the maintenance of existing sanitary systems and the development of new ones were caught between two forces: diminishing fiscal resources and environmental concerns among technical and nontechnical people. What constituted the sanitary city of the late twentieth century was a far cry from its origins in the nineteenth. Yet it remained to be seen if years of developing and maintaining the various technologies of sanitation resulted in systems resilient enough to cope with the demands of the modern metropolis.

19

Beyond Broken Pipes and Tired Treatment Plants

WATER SUPPLY, WASTEWATER, AND POLLUTION SINCE 1970

I n the wake of the "infrastructure crisis," water-supply and wastewater systems were spared dire predictions about deterioration in several major studies. A 1987 report of the National Council on Public Works Improvement (NCPWI) stated that a national water-supply "infrastructure gap" of the magnitude that "would require a substantial federal subsidy" did not exist. Urban water-supply systems as a whole, it concluded, "do not constitute a national problem," although a national problem did exist for small water-supply systems.[1]

This assessment was based on comparisons with other components of the nation's infrastructure. Water and wastewater needs appeared modest when compared with highway repair and replacement.[2] The relatively small, but hardly insignificant, number masked problems that had been building for years. Many drinking-water systems were outdated, faced massive leaks, were poorly maintained, and relied on pipes 100 or more years old. Wastewater treatment systems also were inadequate.[3]

After 1970, the idea of coupling water-supply and wastewater systems retained its advocates, but practical attempts were few. Some believed that systems of various sizes could not mesh or that long-standing habits resisted change.[4] Nevertheless, rising concern over new threats to water quality—especially groundwater deterioration and nonpoint pollution—intertwined water supply and wastewater conceptually. Broadening of federal regulatory authority over water pollution and tightening water quality standards were the first steps in recognizing the severity of water pollution in the era of the New Ecology. Since operation and maintenance of water-supply and wastewater systems largely remained local responsibilities, and water quality regulation occupied federal authorities, fragmentation threatened the safeguarding of pure water and effective disposal of effluent.

On the whole, availability of water supplies seemed adequate, but several concerns existed at the local and regional levels. Some urban areas experienced major droughts and chronic water shortages. In the mid-1980s, New York City prepared for water restrictions, Denver made plans for reusing wastewater, and shortages occurred in several California communities.[5]

The 1993 report of the Council on Environmental Quality (CEQ) concluded that combined withdrawals of water were continuing to meet needs. Between 1985 and 1990, the U.S. population increased 4 percent, while withdrawal and consumption of water only increased 2 percent. In fact, the rate of increase for water consumption was slowing in the 1970s compared to the previous decade. The CEQ reported that water-conservation measures, improved efficiency of water use, and utilization of water-reuse technology contributed to the decline.[6] For 2000, water use estimates for the United States, according to the U.S. Geological Survey, were about 408 billion gallons per day for all uses. The total varied less than 3 percent since 1985 largely because withdrawals stabilized for the two largest uses: irrigation and thermoelectric power. California, Texas, and Florida accounted for one-quarter of all water withdrawals in 2000.[7]

Figures for the mid-1980s indicate that there were 206,300 public water supplies in the United States. Of these, 71.7 percent were small, noncommunity supplies, and 28.3 percent were largely small community supplies serving residential areas. About 85 percent of the U.S. population received drinking water from public suppliers in 2000, compared to 62 percent in 1950. Approximately 82 percent of the urban water systems serving 50,000 or more population were publicly owned.[8]

Local circumstances continued to influence the operation and performance of the water-supply systems late into the twentieth century. Water issues tended to be defined in technical terms by experts operating in allegedly apolitical agencies that faced little public review. Since the Progressive Era, local bureaucracies were often dominated by appointed administrators and career bureaucrats who wielded more power in making and implementing policy than any group in local politics. Public works bodies were repositories of expertise and were active in mobilizing client groups (vendors, suppliers, and equipment manufacturers) in support of their work. The growth of urban service professions, such as civil engineering, strongly influenced the approaches taken in providing services.

While water supply and wastewater treatment and disposal remained largely public ventures in the United States by 2000, privatization made inroads nevertheless. Privatization of municipal water and wastewater systems could mean selling or leasing facilities to private companies, contracting to operate services, or other arrangements to shift some responsibility to private entities. Until 1997, private involvement in public water and wastewater systems was generally limited to three to five years. This limit was imposed by the Internal Revenue Service (IRS), which prohibited operating contracts of longer than five years for municipal ventures that were financed with tax-exempt bonds. The IRS changed the rule in 1997 because of lobbying efforts by the water and wastewater service industry.[9] By the first decade of the new century, privatization of water systems was much more widespread in Europe than in the United States. In 2003, only 5 percent of the water systems in the United States were privately owned, and only about 15 percent of the population was served by corporate water. Between 1997 and 2003, however, the number of publicly owned systems operated under long-term contracts by private companies increased from 400 to 1,100. The Center for Public Integrity—a nonprofit advocacy group based in Washington, D.C.—estimated that before 2020, 65 to 75 percent of public waterworks in Europe and North America would be controlled by private companies, with Africa and Asia not far behind.[10]

Some old battles persisted. The city of Los Angeles continued to clash with the residents of the Owens Valley. After the tensions of the 1920s and before 1970, remaining valley residents catered to vacationers taking advantage of one of the few undeveloped areas close to Los Angeles. Ironically, diversion of water to the city and to the Imperial Valley had spared the Owens Valley from extensive irrigated farming. After 1970, the completion of a second aqueduct led to a conflict with the Owens Valley residents over Mono Lake, which was followed by Los Angeles's efforts to begin a capacious groundwater pumping program that promised to dry up the Owens Valley for good. Unable to get a compromise, valley residents sought an injunction to stop new pumping, but failed. Tensions reached a peak in 1976 when a section of the aqueduct was blown up. Not until 1989 was a formula arrived at for determining future groundwater pumping, but even then not all issues were resolved. Restricted pumping in the valley led Los Angeles to increase its dependence on the Metropolitan Water District, and hence water from northern California and the Colorado River. The City of Angels had yet to find a solution to one of its most pressing problems, and the people of the Owens Valley continued to wonder what the future held.[11]

Regionalization of the water industry attracted considerable attention, especially the Metropolitan Water District in California and the Metropolitan Sanitary District of Greater Chicago. As in the United Kingdom and the Netherlands, a large number of uneconomical operating units delivering poor service existed in the United States. "Local parochialism" in water-supply systems undermined change, especially in the wake of the "infrastructure crisis" in the 1980s.[12]

Few municipal wastewater systems achieved regional status either, although service areas expanded. In 1986, 19,300 sewer systems and 16,000 public wastewater facilities served 70 percent of the U.S. population and 160,000 industrial sites. In 1996, the number of wastewater treatment facilities had modestly increased to 16,024, with 71.8 percent of the population served. In the late 1980s, treatment removed about 85 percent of the pollutants from the 37 billion gallons of wastewater flowing through the facilities each day. However, a substantial portion of raw waste was being discharged into surface waters by industries, possibly at as many as 39,000 sites.[13]

In the early 1980s, about half of the communities in the United States had wastewater facilities operating at full capacity, were unable to support further residential or industrial development, and could fail to meet water quality standards. A state sometimes used a sewer moratorium as a temporary growth-control measure to restrict new sewer lines, connections on existing lines, building permits, and rezoning. Limiting the extension of water and sewage lines curtailed growth, and thus necessary improvements in the systems were likely to be addressed quickly. A survey in late 1976 indicated that over 4,500 square miles were under moratoriums, affecting a population of approximately 9 million.[14]

While overloading of treatment facilities was a serious problem after 1970, the actual quality of treatment improved. In 1988, 88 percent of all plant capacity included secondary treatment or greater (in 1996 it was about the same), and more than three-quarters of all Americans were being served by centralized sewage-treatment facilities.[15]

Despite broadening the reach of sewage treatment, experts continued to seek technical improvements. These included mechanical removal of sludge from sediment tanks, more effective control of coagulation to permit shortened settling time and higher rates of filtration, and computerization of plant functions. Land treatment made a comeback in an era when recycling was becoming more popular. By the late 1980s, approximately 25 percent of municipal wastewater-treatment sludge was spread on land at more than

2,600 sites. But most industrial wastewater and sludge were less suitable for land treatment.[16]

Financing of water-supply and wastewater systems also demonstrates the persistent local nature of sanitary service delivery after 1970. While the federal government was significant in developing water quality standards and regulations, it provided a relatively small amount of funding for constructing water-supply systems and wastewater treatment facilities, other than assisting communities in developing projects. Statistics from the early 1980s indicate that state and local governments were primarily responsible for 83 percent of the expenditures for municipal water supplies, 92 percent for storm and combined sewer systems, and 80 percent for wastewater treatment. Water-resource investment was concentrated in Army Corps of Engineers' construction projects, flood control, recreation, hydroelectric power, and the development of water supplies. From 1991 to 2000, states contributed approximately $10 billion to match capitalization grants from the EPA for drinking water and wastewater projects, and made about $13.5 billion available in grants and loans and through bonds.[17]

The funding of wastewater-treatment facilities fell somewhere in the middle in terms of public attention, with water supplies clearly regarded as a local issue and pollution abatement rising as a major national concern. Federal spending on wastewater facilities peaked in the late 1970s. In 1984, federal grants for construction accounted for about 25 percent of public expenditures for wastewater treatment. States contributed about 5 percent toward construction, and local outlays covered the operating costs and about half of the construction costs. State aid to local governments to finance wastewater facilities was relatively new, prodded by the efforts of the Reagan administration to roll back federal spending on nonmilitary projects. By 1981, forty-one states had developed programs to provide grants or loans to help meet the 25 percent share of capital costs required by the EPA to obtain federal grants. Ironically, state and local funding also declined in response to the EPA's financial commitment in the early 1970s, but only increased slightly as EPA dollars disappeared. By 1989, funding for wastewater treatment, like water supply, was increasingly in the hands of local and state authorities.[18]

Reduction in federal money in recent years for the construction of wastewater collection and treatment systems led municipalities to be more cost-effective and to consider alternatives to conventional systems. One approach has been decentralized wastewater management (DVM), defined as "the collection, treatment and reuse of wastewater at or near its source of generation." The method of treating waste on-site or near the site, rather than transporting it to a centralized facility, has had to be developed with attention to the public's aversion to conspicuous waste technologies, or honoring the "out of sight, out of mind" perspective. In 2000, decentralized systems served about 25 percent of the U.S. population and about 37 percent of new development.[19]

In an era of heightened concern over a range of environmental risks threatening the nation's water supplies and complicating the process of wastewater treatment, outbreaks of waterborne diseases continued to lose ground. Diseases such as cholera and typhoid fever had been virtually eliminated, and the few disease outbreaks primarily occurred in small community or noncommunity systems. Periodically, a large outbreak could arise, as in the case of cryptosporidiosis in Georgia in 1987 affecting 13,000 people, or a major attack of gastroenteritis affecting 403,000 people in Milwaukee in 1993. Contamination in distribution systems and inadequately disinfected water were significant causes of outbreaks.[20]

Disinfection continued to be regarded as one of the most important steps in water treatment. In 1970, chlorine accounted for 95 percent of the disinfectants used to treat potable water in cities. Aside from increasing costs, chloramines were believed to have a limited impact on viral agents in the water and thus attention turned to alternative disinfectants such as other halogens, bromine, and iodine. Chlorine disinfection also produced some undesirable by-products, including trihalomethane (THM). To control THM, officials took a new look at ozone because of its high germicidal effectiveness; its ability to combat odor, taste, and color problems; and its benign decomposition. There had been resistance to ozone because it was produced electrically and could not be stored, was difficult to adjust to variations in water quality, and was not universal in its action as a disinfectant. No consensus emerged to abandon chlorination, and the EPA still believed that it was the most effective commonly used additive for controlling waterborne diseases.[21]

Despite its impressive record, chlorination came under severe criticism. A 1974 report alleged a causal relationship between the use of Mississippi River water (containing chlorinated sewage effluent) and the incidence of cancer. While the study's validity was later questioned, it led to an increased awareness of chemical contaminants in water supplies and the impact of chlorination. In April 1975, EPA reports of cancer-causing chemicals in the water supplies of seventy-nine cities splashed on the front pages of many newspapers. Water scares resulted in cities such as New Orleans and Duluth, Minnesota. Ultimately, chlorine treatments were found to produce chloroform and other trihalogenated methanes. In March 1976, the National Cancer Institute issued a report indicating that chloroform was carcinogenic. The EPA set the first trihalomethane limits in 1979.[22]

Fluoridation did not suffer the same scrutiny as disinfection, but debate persisted. In 1990, fluoride was being added to the water supplies of 57 percent of the American people, and was endorsed as a tooth-decay preventative by most major health organizations. Other groups such as the National Health Action Committee and the Safe Water Foundation actively challenged community water fluoridation and fluoride mouth-rinse programs in the schools. Proponents charged the opposition with employing scare tactics, spreading half-truths, and even attempting to link fluoridation with aging and AIDS.

Some research studies raised doubts about the ability of fluoride to curb tooth decay, suggesting that there were no appreciable differences in groups exposed to fluoride and those who were not. Some concern also mounted about incidences of dental fluorosis—in the most extreme cases leading to brown stains and pitting of the teeth. The EPA concluded that conflicting studies produced inadequate evidence to restrict fluoride's use.[23]

As in the case of waterborne disease, there appeared to be measurable improvement in surface-water quality in several locations in the 1970s and 1980s. An NCPWI report noted net improvement in stream and estuary water quality, but net degradation in lake water quality between 1972 and 1982.[24] Yet serious pollution problems remained. Since 1973, asbestos fibers were found in water supplies throughout the United States, and sewage sludge continued to be a particularly difficult problem to solve.[25] With over 300,000 factories in the United States using water, a variety of toxic materials found their way into surface supplies. The EPA claimed that industrial discharges impaired the use of 11 percent of the nation's stream miles and 10 percent of the lake area. One assessment estimated that industrial discharges contributed approximately one-third of the total point source loading of oxygen-demanding wastes. By the end of the 1980s, dissolved-oxygen concentrations had increased, dissolved solids had decreased, and some concentrations of toxics (such as arsenic, cadmium, lead, chlordane, dieldrin, DDT, and toxaphene) had declined, suggesting that point source controls were having a positive effect.[26] As of 1998 the EPA estimated that 40 percent of the nation's freshwater did not meet water quality goals and about half of the 2,000 major watersheds had "water quality problems."[27]

While surface water continued to be a major source of supply for municipal and industrial uses, for irrigation, and for generating electricity—74 percent of total freshwater use in 2000—groundwater use after 1950 was clearly on the rise, especially for community water systems (CWSs) in smaller and medium-size cities. Groundwater is the source of about half of the drinking water for all Americans and virtually all of the rural population. In 2000, groundwater withdrawals accounted for 37 percent of the public supply, over 98 percent of the self-supplied domestic water, and 42 percent of irrigation water, not to mention important uses in mining, livestock raising, and industry. In total, groundwater represented 26 percent of freshwater use.[28] The maximum average well yields were located in the Columbia Lava Plateau (Washington, Oregon, California, Nevada, and Idaho) and in the Southeastern Coastal Plain. In the East and South, groundwater was used primarily for domestic and industrial purposes; in the West, it was used essentially for irrigation.

Agricultural use—about 50 billion gallons per day in 2000—was the primary source of depletion in the Texas-Gulf region and parts of the West. In the High Plains of Texas, the Ogallala aquifer is almost depleted, and Texas lost 1,435 million acres of irrigated cropland between 1982 and 1997. From 1950 to 1975, groundwater withdrawals overall increased about two times as fast as surface water withdrawals, due especially to irrigation. Even in urban areas, overdrafting of wells—pumping water faster than it is replaced—was a problem.[29]

Land subsidence was a consequence of overdrafting, which occurs when too much groundwater is removed from a particular location, resulting in the surrounding clay collapsing and then compacting. On the surface, land subsidence occurs; below the surface, freshwater cannot replenish the structure. Subsidence also may break pipes and clog sewers, encourage saltwater encroachment into an aquifer, and increase flooding. Significant subsidence due to declining groundwater levels occurred in Louisiana, Texas, Arizona, Nevada, and California. In the area surrounding Houston's Ship Channel, for example, subsidence of up to ten feet due to groundwater withdrawals was measured in 1978.[30]

While groundwater depletion was a chronic concern, the question of quality had rarely been an issue. Beginning in the 1970s, the hallowed belief that groundwater was intrinsically purer than surface water came into question. The sources of groundwater contamination included waste-disposal practices, irrigation return flows, spills, abandoned oil and gas wells, and leaks from buried tanks and pipelines. One study referred to groundwater contamination as "slow, insidious degradation."[31] In 1984, water in 8,000 public and private wells was reported degraded or unusable. That year the Office of Technology Assessment listed approximately 175 organic chemicals, more than fifty inorganic chemicals, radio nuclides, and several biological organisms in groundwater. Between 1971 and 1985, 245 groundwater-related disease outbreaks were reported in the United States. In 1988, the EPA noted the presence of seventy-four pesticides in the groundwater in thirty-eight states, and DDT was found in many sources—even after it was banned. Once contaminated, groundwater is much more difficult to restore to high quality than surface water.[32]

Increase in bottled-water purchases was an indicator of growing concern over the quality of drinking water. The bottled-water industry in the United

States grew at a rate of 10 to 15 percent annually in the 1980s, and between 8.2 and 11.8 percent between 2000 and 2006. In 1990, more than 700 different brands of water were on the market in the United States. Approximately one of fifteen Americans consumed bottled water (one of three in Southern California). In 2005, more than 7,500 million gallons were sold. Purchases of in-home water-treatment devices also were on the rise.[33]

The 1970s also witnessed an increased awareness of nonpoint source pollution. By 1990, it was estimated that more than 50 percent of America's total water quality problems were attributed to agricultural, industrial, and residential nonpoint sources. Runoff into watercourses contained asbestos, heavy metals, oil and grease, salts, manures, pesticides and herbicides, construction-site pollutants, bacterial and viral contaminants, hydrocarbons, and topsoil.[34] A NCPWI study stated, "In some regions of the country, nonpoint source pollution is so bad that eliminating all point sources entirely would still not result in significantly cleaner water."[35] The EPA reported that "the latest *National Water Quality Inventory* [in the late 1990s or 2000] indicates that agriculture is the leading contributor to water quality impairments, degrading 60 percent of the impaired river miles and half of the impaired lake acreage surveyed by states, territories, and tribes. Runoff from urban areas is the largest source of water quality impairments to surveyed estuaries (areas near the coast where seawater mixes with freshwater)."[36] In the late 1980s, pesticides were detected in groundwater in at least twenty-six states as a result of routine agricultural activity. For cities, urban stormwater runoff is an immediate and critical nonpoint pollution problem as well.[37]

Some local studies detected a pattern in the polluting characteristics of urban runoff in the 1970s and 1980s.[38] A Milwaukee project discovered CSO in the major source of fecal coliforms in the Milwaukee River. Urban runoff was determined to be the sole source of degradation in Lake Eola, Florida. Lead concentrations two to three times greater than in nonurban samples were found in urban samples of algae, crawfish, and cattails in some areas. CSOs were a major pollution concern for approximately 770 cities with combined sewer systems in the United States. In 1989, the EPA published its *National CSO Central Strategy,* which focused attention on CSOs but did not address the site-specific nature of the CSOs, the cost of controls, or a complete range of possible solutions. In April 1994, however, the EPA published its CSO Control Policy, calling it "the national framework for control of CSOs" through the National Pollutant Discharge Elimination System (NPDES) permitting program. The NPDES, negotiated "among municipal organizations, environmental groups, and State agencies," was meant to help municipalities and permitting authorities to effectively meet the Clean Water Act's pollution control goals.[39]

The federal role in water-supply and wastewater issues was greatest after 1970, which included broadening and deepening water quality and effluent standards. In 1970, Congress passed the Water Quality Improvement Act, which established liability for owners of vessels that spilled oil, regulated thermal pollution, and instituted a system of permits for activities that might cause violations of water quality standards. In 1971, the EPA was given responsibility for water quality, and it set standards for twenty-two contaminants. Because of the limits on the success of previous legislation, the 1972 Federal Water Pollution Control Act (renamed the Clean Water Act) ultimately was passed with overwhelming support and became a critical turning point in national water quality legislation.[40]

The intent of the law was to place the federal government in a leadership role in dealing with water pollution, with the assistance of the states. The act was meant to reduce the quantity of pollutants discharged into surface waters by setting two goals: to attain water quality in navigable waters suitable for fisheries and for swimming (the so-called fishable-swimmable goal) by 1983 and require that all publicly owned wastewater-treatment facilities provide secondary treatment by 1988; and to achieve a zero pollutants-discharge goal by 1995. These goals were meant to change the strategy of federal water quality management by replacing in-stream (ambient) water quality standards with limits on the discharge of pollutants from industrial point source discharges and municipal wastewater-treatment plants (technology-based standards).

States were required to develop water quality management plans, including the identification of point and nonpoint sources of pollution. Every discharger of point source pollution had to obtain a federal permit to control the specific pollutants entering watercourses. An older, unworkable permit system was replaced by the National Pollution Discharge Elimination System (NPDES) to be administered by the states and territories when their own permit programs met the rigorous new federal standards. The permits contained effluent limitations as well as monitoring and reporting requirements. By 1994, thirty-nine of the fifty-four states and territories had taken responsibility for the NPDES, although compliance was uncertain.[41]

The focus on technology-based standards—or "technology-forcing"—meant that all municipal wastewater-treatment plants had to meet uniform effluent standards based on secondary treatment, and all industrial sources of pollution had to install the "best practicable technology" (BPT) by 1977, and the "best available technology" (BAT) by 1983.[42] This goal has yet to be fully met.

The 1977 amendments to the Clean Water Act continued the emphasis on technology-based standards. They also attempted to deal with a zero-discharge goal by modifying the deadlines and specifying which pollutants should receive the most attention. The federal financial and managerial roles were reduced, and all programs were to revert to the states with EPA oversight. One explanation for this shift was that the 1977 amendments were shaped more by the states, municipalities, and businesses than the 1972 legislation, which was strongly influenced by public interest groups. As a consequence, the problems of smaller jurisdictions and rural areas received more attention in the amendments.[43]

Despite its role in the improvement of water quality, the Clean Water legislation did not meet its original timetable, did not operate with the authorized funds, and did not achieve its antipollution targets. Some critics charged that both the Clean Water Act and the Clean Air Act were heavy with provisions that "can be understood as accommodations, appeasements, or buy-offs of this group or that one."[44] Others blamed the Reagan administration for weakening pollution control laws in the early 1980s by claiming that the existing regulatory system was inefficient, and that the level of control had a severe impact on economic productivity.[45]

These arguments have merit. Industry fears of overly stringent standards influenced the shaping of the legislation and the Reagan administration's response to it. As *Business Week* noted in 1977, the 1972 act "has been a beauty for engineers, contractors, and equipment manufacturers who supply sewage treatment systems for cities and industry. But for those companies that will fail to meet the act's July 1, 1977, deadline for 'best practicable technology'—and who see the even stiffer 1983 rules looming ahead—the same law is a menacing beast."[46] The mirror image is that the difficulty with the 1972 law was EPA's inability to secure compliance. In addition, the faith in new technology to reduce pollutants entering watercourses was reinforced in the 1977 amendments.[47]

The Clean Water Act and its amendments were imperfect mechanisms for addressing a complex set of problems, especially those that relied so heavily on technology forcing. Legal expert William H. Rodgers Jr. asserted that (like the Clean Air Act) the Clean Water Act "offers dramatic collisions between absolute rights and utilitarianism, often dressed in the garb of economic efficiency." "These two missions impossible," he added, "especially the no-discharge goal, are among the most thoroughly denounced actions taken by any twentieth century Congress."[48] An urban water expert noted that the zero-discharge goal was "the object of many jokes among engineers."[49]

Advocates of technology forcing argued that there was potential for redesigning systems to reduce the amount of water used and the discharge levels, and even to recover products for reuse.[50] Technology forcing also had its limits. The requirement that all publicly owned wastewater facilities adopt secondary treatment was an arbitrary way to move toward zero discharge or for setting quality standards. Effluent standards based on BAT or BPT and requiring permits did not take into account the quality of the receiving water or the effects of the discharge on the environment. Instead, they relied on the permits themselves and faith in effectiveness of treatment technologies to eliminate point sources of pollution. In addition, the process of defining BAT and BPT was quite complex, and when industry opposed them, it was sometimes simpler for the EPA to weaken the standards further than try to justify them or to fight the battle in the courts.[51] The most serious indictment of the Clean Water Act was that it had been ineffective in reducing pollution. At best, the zero-discharge goal was commendable, but vague and unachievable. The debate over the Clean Water Act battled on as proponents and detractors utilized the act as a point of controversy for focusing on the nation's struggle with water pollution.[52]

The new legislation was guided by powerful historic forces in dealing with effluent, that is, a dependence on highly centralized, capital-intensive water-supply filtration and wastewater-treatment facilities designed to capture point sources of pollutants entering watercourses or returning to consumers via those watercourses. These mechanisms alone were not capable of addressing water pollution in all its manifestations, and did little to address the problem of preventing pollution at its place of origin.

Federal protection of groundwater could be identified in bits and pieces throughout a variety of laws enacted in the 1960s and later, but these laws usually dealt with groundwater remediation rather than groundwater quality.[53] In the 1980s, however, Congress and some federal agencies made efforts to bolster groundwater protection authority. In 1984, the EPA established the Office of Ground Water Protection and released its groundwater-protection strategy. This was not enough to appease those concerned with the problem. The following year, the Conservation Foundation and the National Governors' Association organized a National Groundwater Policy forum that produced recommendations for a national groundwater-protection plan.[54] In 1987, Congressman James L. Oberstar (D-Minn.) introduced the Groundwater Protection Act, meant to amend federal pesticide law. For some time the public interest group Clean Water Action had lobbied to restrict the use of cancer-causing pesticides that contaminated groundwater. However, regional divisions thwarted consensus on the proposed legislation. Support came from northeastern states and southern states, but states in the Midwest, especially where pesticide use was extensive, were less enthusiastic about the Oberstar bill. Little action was taken, and a comprehensive federal groundwater law awaits the future.[55]

Nonpoint source pollution proved to be most perplexing because its causes were so diverse, existing treatment systems were ill suited to deal with it, and prevailing regulations had long ignored it. In the 1970s, the only federal program to address nonpoint pollution directly was Section 208 of the Clean Water Act, which called for planning agencies to examine waste-management alternatives and to develop plans to reduce point and nonpoint sources of pollution.[56] Finally, in the 1987 amendments to the Clean Water Act, the EPA and state officials were directed to supplement technology-based standards with a water quality–based approach to attempt to control nonpoint source pollution and a variety of toxic pollutants. Emphasis on technology forcing was modified in an attempt to deal with very knotty problems that had previously defied easy resolution.[57]

Section 402(p) of the 1987 amendments provided a system of regulatory controls specifically related to nonpoint source discharges and stormwater. This action was a start in the right direction, but it did not resolve all issues concerning nonpoint pollution. Attention quickly shifted to the EPA, and controversy arose over the implementation of the stormwater provisions in Section 402(p). Rules for general stormwater permits were promulgated, but other issues of stormwater discharge were yet to be fully addressed, such as unresolved issues with respect to CSOs and sanitary sewer overflows (SSOs). Aside from decentralized waste management, serious consideration has been given to programs of wastewater reuse and to better understand the impact of wet-weather flow (WWF), or wet-weather induced discharges, that add to the pollution load of sewers. Much remains to be done.[58]

The legislative history of water-pollution abatement after 1970 demonstrates the difficulty in resolving a wide array of pollution problems. Establishing quality standards also proved perplexing. The Clean Water Act (1972) did not significantly address drinking-water standards, particularly because it did little to confront the protection of groundwater or nonpoint source pollution. Previous laws dealing with drinking-water standards were limited to water supplied by interstate carriers, focused on contaminants causing communicable disease, and made little provision for underground sources. The Safe Drinking Water Act (SDWA) of 1974 led to the National Primary Drinking Water Regulations, which established maximum contaminant levels (MCLs) for biological, chemical, and physical contaminants, and for residuals remaining in drinking water. The states had primary enforcement authority.[59] Unlike in past years, when biological pollutants gained primacy, the 1974 act recognized a wide array of threats to potable water. According to environmental engineer Daniel Okun, "Considering its elephantine gestation period, the Safe Drinking Water Act of 1974 . . . promised the birth of a new era in public water supply in the United States."[60]

The most immediate impetus for new national standards began with the 1970 USPHS Water Supply Study, which stated that many water-supply systems failed to meet the 1962 USPHS Drinking Water Standards. Revelations that carcinogens were identified in the water supplies of New Orleans and other cities using the Mississippi River as a source for drinking water came to attention through findings of the Environmental Defense Fund and the EPA in November 1973, and a CBS special program, "Drinking Water May Be Dangerous to Your Health," broadcast in 1974. After much wrangling, a compromise bill was signed by President Ford in December 1974.[61]

Implementation of the SDWA proved difficult. The EPA was slow to carry out the requirements, in part because it was reluctant to use its enforcement powers against local government units, the most common providers of drinking water. Although reauthorization was difficult, the SDWA was revised extensively in 1986 to step up the pace of enforcement, and to urge the EPA to carry out its regulatory responsibilities. The amendments significantly expanded EPA requirements to regulate the levels of contaminants in drinking water, setting out eighty-three individual substances for which enforceable limits had to be established. The revision also resulted in two major groundwater-protection programs. Under the 1974 act and the 1986 amendments, communities that did not treat their surface-water supplies were now required to do so.[62] An EPA attorney noted that the amendments received little fanfare: "Perhaps the lack of attention shouldn't be surprising. The Safe Drinking Water Act always has been the Rodney Dangerfield of federal environmental law as far as getting respect goes."[63]

Outside of Washington, D.C., the SDWA came under attack from critics. Overriding issues were "How safe is safe?" and "How clean is clean?" Local authorities saw the act as an additional burden imposed by the federal government. State governments balked at having to enforce the law, and some considered returning responsibility to the federal government. Others argued that the deadlines for compliance were unrealistic. Still others contended that the law did not deal adequately with the problems for people in small towns and rural communities. In the early 1990s, efforts were made by Senator Pete Domenici (R-N.M.) to place a moratorium on enforcement of parts of the standards, while the Clinton administration sought ways to provide federal assistance to communities for carrying out the SDWA.[64]

As part of the *Newspapers in Education* program for students from grades seven to twelve, a lesson, "The Clean Water Act Turns 30," was made available on October 22, 2002, on a Web site created for the *Cincinnati Enquirer and Post*. It aptly noted:

Friday, October 18, 2002, signified the 30th anniversary of the Clean Water Act—a critical law for ensuring the health of our nation's citizens. The 1972 act was created because severe health problems . . . linked to polluted drinking water had become a huge concern. . . .

Along with the Safe Drinking Water Act of 1974, the Clean Water Act has allowed the Environmental Protection Agency (EPA) to give out about $80 million in wastewater treatment assistance across the U.S. These efforts have helped get 79 million more citizens connected to modern sewage treatment facilities than there were in 1968. The proportion of reported disease outbreaks that can be attributed to problems at public water treatment systems has steadily declined, from 73% in 1989–1990 to 30% in 1995–1996.

Still, in 1998, close to 1,000 community drinking water supply systems, affecting about 18 million people, violated EPA's Surface Water Treatment Rule, which was aimed at guarding against a microorganism known as Giardia and viruses in drinking water supplies. Plus, the day before the Act's anniversary, the U.S. Public Interest Research Group (USPIRG) reported that between 1999 and 2001, four of five wastewater treatment plants and chemical and industrial facilities in the United States polluted waterways beyond what their federal permits allow.[65]

Change was coming, but slowly. Whatever its value as a general critique of the state of public works in the United States, the "infrastructure crisis" was not sufficiently broad to characterize accurately water-supply and wastewater systems after 1970. Granted, these technologies of sanitation demonstrated problems of aging and deterioration, but they were never designed to anticipate problems such as groundwater contamination and nonpoint pollution. Particularly in the case of the latter, the framework within which those systems were designed and built focused on problems of epidemic disease and (eventually) an array of point source pollutants, but did not allow for—and arguably could not anticipate—the kinds of environmental problems evident by the late twentieth century. If there was an "infrastructure crisis" for water-supply and wastewater systems, it was acknowledged by some, but not fully confronted.

Out of State, Out of Mind

THE GARBAGE CRISIS IN AMERICA

On March 22, 1987, the garbage barge *Mobro* left Islip, New York, looking for a landfill that would take its disagreeable cargo. Over two months, five states (North Carolina, Louisiana, Alabama, Mississippi, and Florida) and three countries (Mexico, Belize, and the Bahamas) banned the barge from unloading. Reluctantly, the captain turned the *Mobro* toward home, where it received a dispirited welcome. Somewhat ironically it was allowed to dump its 3,100 tons of waste where it started.[1] The garbage barge story was a journalist's delight, highlighting New York's chronic problems of service delivery and exposing the insensitivity of the Sun Belt for the plight of the Snow Belt. It also graphically demonstrated the so-called garbage crisis in America—a predicament much more immediate and personal than the "infrastructure crisis."

Talk of a general garbage crisis began in the early 1970s, although examples can be found much earlier.[2] In 1973, the National League of Cities and the U.S. Conference of Mayors issued a report in which they noted the "skyrocketing" volume of solid waste and the "sharp decline" in available urban land for disposal sites. The notion of a crisis in waste volume and a crisis in landfill space persisted into the following decades.[3]

In a widely discussed December 1989 article in *Atlantic Monthly*, archaeologist-turned-garbologist William Rathje rightly questioned the standard perception of the crisis:

> The press in recent years has paid much attention to the filling up (and therefore the closing down) of landfills, to the potential dangers of incinerators, and to the apparent inadequacy of our recycling efforts. The use of the word "crisis" in these contexts has become routine. For all the publicity, however, the precise state of affairs is not known. It may be that the lack of reliable information and the persistence of misinformation constitute the real garbage crisis.[4]

The idea of a "garbage crisis," as characterized, is a call for action but also a convenient label for a complex set of issues. It is convenient because the notion of "crisis" confers upon the problem relatively tangible properties, which might be resolved through equally concrete solutions, such as new technology, effective management, or popular will. In some sense "crisis" denies the complexity of the problem and ignores its persistence over time, failing to question whether it is chronic, recurrent, or temporary. Reflecting a more a positive response, the *Economist* stated in May 1993 that "in many wealthy countries, waste is the environmental problem that people care about most. Government policies are beginning to reflect that fact." While waste was once a public health problem, goals have shifted to focus on the reduction of waste and to recycle more. The issue had seemed to rise in importance because "waste tends to be the pollution problem to which other problems are eventually reduced."[5]

There is little doubt that the vast quantities of wastes generated in the United States contributed significantly to the problems of collection and disposal. Annual per capita waste generation has been high in the United States, especially in comparison with other countries. Between 1970 to 1990, MSW increased 61.6 percent, from 121.9 million tons per year to 198 million tons per year. Population increases, as much as changes in consumption patterns, were responsible for the steady rise in aggregate discards. In 2005, residences, businesses, and institutions in the United States produced more than 245.7 million tons of MSW (a decrease of 1.6 million tons from 2004, however), or approximately 4.54 pounds per person per day (down 1.5 percent from 2004).[6]

While the quantity of waste has been critical in the garbage problem, it has not been the only issue. Composition of materials in the waste stream also contributed significantly. The generation of waste called into question the rationale for consumption of scarce resources and at the same time led to demands for quick and efficient disposal of society's discards. The waste stream in the years after 1970 included a complex mix of hard-to-replace as well as recyclable materials, and a vast array of toxic substances.

In recent years, paper has made up the largest percentage of municipal waste (34.2 percent before recycling in 2005). Yard and food wastes, the major organic discards (13.1 percent before recycling in 2005), have steadily declined. Plastics represented an increasing portion of the total (11.9 percent before recycling in 2005). Inorganic materials were scant in terms of percentage, but included household chemicals with potentially serious environmental implications.[7]

The nature of the waste stream suggested important "front end" problems. Of all the discarded items, paper, plastics, textiles, and aluminum have grown steadily in thousands of tons if not in percentage of the total MSW.[8] The use of these materials, with the exception of textiles, reflected a substantial increase in packaging waste and a wide variety of additional uses for paper. Few would disagree that rampant consumerism has been a major contributing factor to the solid-waste problem. Yet excessive packaging has not been the only problem. Between 1958 and 1976, packaging consumption rose by 63 percent, but rapid growth in aggregate packaging waste did not continue into the 1980s and 1990s. Between 1970 and 1986, packaging as a component of MSW increased only 9 percent, while durable items increased 35 percent; clothing/footwear, 88 percent; nondurables, 300 percent; newspapers/magazines/books, 40 percent; and office and commercial paper, 69 percent. In addition, by 2005 recovery of paper and paperboard for recycling was the highest rate overall of any material in the municipal waste stream. About 50 percent of postconsumer paper and paperboard were recovered in that year compared to 42.8 percent in 2000.[9]

A critical issue with packaging was changes in composition. Paperboard cartons, plastic jugs, and plastic jars replaced glass milk bottles and other glass containers; aluminum beverage cans replaced steel cans; and plastic grocery bags began to replace paper bags. In 1985, 47 billion pounds of plastics were produced in the United States. Of the 39 billion pounds consumed domestically, 33 percent was used for packaging, and over half of the discarded plastics was from packaging. In 2005, plastics generated for durable and nondurable goods and for packaging and other items were almost 29 million tons, of which 5.7 percent was recovered. The dependence on plastics required extensive use of petrochemical and other feedstocks, resulting in discards of substantial bulk in landfills; posed a health hazard if burned in incinerators; and demanded complex recycling methods. Additional problems arose if packages were composites of a combination of materials.[10]

As suggested in chapter 17, not only did the amount of discards increase but the uses changed as well. Some packaging was no longer dependent on paper, but office paper and commercial printing papers began mounting up. Newsprint as a percentage of total MSW was increasingly less significant. In 2005, more than 12 million tons of newspapers (including newsprint and groundwood inserts) were generated, but almost 89 percent was recovered, leaving a little more than 1 million tons discarded. The rise in the use of computer printouts and facsimile paper, paper for direct-mail advertising, and the output from high-speed copiers was particularly significant. So much for the "paperless office" in the age of computers.

The potential for recycling paper was great, although paper and paperboard of various types were often mixed in landfills and thus had to be recycled at the source. Even with source separation, effective techniques for converting all but newsprint and clean, white paper into new products were not very promising. Significantly, paper adds to the volume of waste in a greater proportion than weight. Since most projections about aggregate discards are calculated in terms of weight, paper's contribution to the disposal problem is often underestimated.[11]

Questions of excessive waste generation and the squandering of natural resources became important issues in the age of the New Ecology. The composition of materials in the waste stream depended on several external issues, including patterns of consumption and production, the availability and use of new materials and new technologies, and the obsolescence of older materials and older technologies. Without careful attention to the types of materials entering the waste stream, and without recognition that composition was generally in a state of flux, traditional collection and disposal practices sometimes proved ineffective and obsolete.

Collection of solid waste had long been regarded as a cumbersome, labor-intensive, and expensive venture, but not a particularly glaring part of the "garbage crisis." Surveys estimated that from 70 to 90 percent of the cost of collection and disposal in the United States went for collection. Conservative estimates placed the annual cost of solid-waste management in 1974 at $5 billion, with collection and delivery of wastes to disposal sites accounting for nearly $4 billion.[12]

Into the twenty-first century, the cost of collection remains high. Refuse collection suffered as metropolitan areas expanded. Longer hauls to disposal sites or transfer stations meant significantly increased costs in fuel and in wear and tear on vehicles. Reliance on a mechanized fleet to collect refuse also contributed to greater amounts of air-pollution problems, more traffic congestion, and a growing number of accidents.[13]

In recent years, collection also resurrected some old issues. The debate over privatization of service delivery increasingly raged in some cities. For consumers, frequency of service remained a key concern. Rising interest in recycling reopened the question of source separation versus mixed-refuse collection. In general, collection problems exposed the significance of external factors influencing the quality of service. Modification in the structure and size of cities, changes in population distribution, and increasing environmental awareness all had powerful effects on collection.

If there has been a garbage crisis, it has been most closely linked with disposal, especially shrinking of landfill sites in the eastern United States. In 1988, Karen Tumulty of the *Los Angeles Times* wrote that "Fresh Kills, New York's only remaining big landfill, also symbolizes a crisis that faces communities throughout the Northeast. When it runs out of space around the turn of the century, 7.1 million people will have to find something else to do with three quarters of their trash."[14]

Fresh Kills Landfill, owned by New York City and operated by the Department of Sanitation, was the great symbol of the end of the sanitary landfill era. Located on a salt marsh on the western shore of Staten Island, the largest landfill in the world spread over more than 2,100 acres, with four mounds ranging in height from 90 to 500 feet. It is so big, as the oft-repeated observation goes, that as the highest point on the eastern seaboard south of Maine, it could be seen from space. Before it closed, barges from nine marine transfer stations operated around the clock, six days a week, to deliver approximately 11,000 tons of refuse daily, or approximately 2.7 million tons of solid waste and incinerator residue annually.[15]

Beyond its "gee-whiz" quality, Fresh Kills was fraught with controversy almost from its inception in 1948. Residents of Staten Island never forgave the decision makers who singled them out to carry the burden of New York City's waste disposal. In addition, like all landfills of its time, Fresh Kills had few environmental controls, including no bottom liner to protect the surrounding area from leachate (toxic liquids). And, despite its size, it did not even accommodate half of the city's daily solid waste production.[16]

Celebrations on Staten Island broke out in March 2001 when a garbage barge from Queens departed for Fresh Kills on the last of 400,000 such trips to deposit its load at the landfill. New York governor George E. Pataki, New York City mayor Rudolph W. Giuliani, and Staten Island Borough president Guy V. Molinari were on hand to mark the occasion. "This is a glorious day for Staten Island," Molinari stated. The landfill was "the most notorious environmental burden in Staten Island history."[17] The euphoria was short-lived, however. In the wake of the shocking attacks of September 11, 2001, and the destruction of the World Trade Center, Fresh Kills was opened to receive most of the estimated 1.2 million tons of debris from Lower Manhattan. As several people pointed out, Fresh Kills also became the final resting place for remains of many of the victims of that tragic event.[18] What had been a reviled disposal site became hallowed ground.

One solution to the landfill shortage was exporting wastes. In the not-so-distant past, doing so smacked of desperation by those who have given up on local solutions. In the late 1980s, three northeastern states—New Jersey, Pennsylvania, and New York—were exporting 8 million tons of garbage a year, with much of it deposited in the Midwest, where landfill sites were more plentiful and tipping fees were lower.[19] In the 1990s, interstate movement of solid waste was commonplace, particularly because it was relatively inexpensive and a simple alternative to other disposal options. In 1995, all states either imported or exported solid waste. In that year, 252 interactions (movement of solid waste between two states or countries) occurred. This represented a 91 percent increase over the 1990 figure.[20]

In 2000, 32 million tons of MSW crossed state lines, an increase of 13 percent since 1998. New York, by far, was the biggest net exporter with 6.3 million tons, followed by New Jersey, Maryland, Missouri, and Illinois. Imported waste shipments were concentrated in the Midwest and along the East Coast, with Pennsylvania the biggest net importer by a wide margin (9.2 million tons compared with Virginia, in second place, with 3.7 million tons).[21]

Several factors accounted for the movement of waste across state lines. Most cities relied on the closest and most readily available disposal capacity, which in some cases meant crossing state lines. St. Louis and Kansas City, Missouri, for example, delivered waste across state lines because disposal sites were closest there, and transport was not restricted by geographic barriers. Institutional changes, such as regionalization of service delivery and the consolidation of the solid-waste management industry, removed state lines as impenetrable boundaries. In addition, as landfills became larger and fewer, available options for dumping waste narrowed.[22]

The trend toward interstate transport concerned state governments in many respects. State and local governments passed flow-control ordinances that designated where MSW was taken for processing, treatment, or disposal. However, because of flow controls, certain designated facilities—such as incinerators—could maintain monopolies on local sources of solid waste or recoverable materials. Relationships between private solid-waste companies and municipal authorities were strained by the flow-control issue. In the early 1990s, the EPA found that thirty-five states, the District of Columbia, and the Virgin Islands directly authorized flow control, and four other states authorized it through various administrative mechanisms. The agency concluded that flow controls were efficient tools for solid-waste management, but not essential for developing a new management capacity or achieving recycling goals. The courts thought differently. In the *Carbone* decision (1994), the U.S. Supreme Court ruled that flow-control ordinances like the one passed in Spokane County, Washington, to ensure a supply of waste for incineration violated the Constitution's interstate commerce clause. For its part, Congress continued to examine the merits of flow control, while supporters hoped for a reversal of the 1994 decision.

Battle lines were being drawn along state borders. In February 1999, inspectors from several states began an "East Coast garbage truck inspection blitz." On one day alone, 417 trucks were stopped in Maryland, the District of Columbia, and New Jersey. Of those, 37 were ordered off the road because they might be carrying potential hazards. The legislative debate over flow control and the battles in the courts and in individual states made it unclear as to whether local governments would be able to rely on flow control to direct refuse to designated facilities in the twenty-first century.[23]

By the 1970s, solid-waste professionals and others already began to doubt the adequacy of sanitary landfills to serve the future needs of cities. Competition from recycling programs raised questions about the economic feasibility of burying resources in the ground. The case for more elaborate recycling, the preservation of virgin materials, the use of MSW as an alternative source of fuel, and the high costs of construction and maintenance were stripping sanitary landfilling of its role as the preeminent disposal option. Environmental concerns about leachate pollution of groundwater and unmonitored methane production also played a part in discrediting the sanitary landfill.

The most immediate point of discussion was the problem in acquiring adequate space. Siting new landfills became problematic in some sections of the country. Many communities did not set aside land specifically designated for waste disposal, partly because marginal land was no longer available. Consensus favoring sanitary landfills also meant that other disposal options received less attention and, in many cases, languished.[24]

Landfill siting was treacherous business because of citizen resistance and intensifying environmental standards. NIMBYism received wider press coverage because of attempts to site landfills along the urban fringe, where the population was not characteristically poor or heavily populated with people of color. In Contra Costa County, California, the proposed sites for new landfill space were to be located on the predominantly blue-collar east side of the county. The organizer of an east county coalition, WHEW (We Have Enough Waste), stated that "whenever there are undesirable things to be located, if there is a dump or a jail, where do you look? East county!"[25]

Once passive or politically neutered neighborhoods were fighting back, unwilling to provide dumping space for the whole city. In the 1980s, grassroots organizations that had been dealing with toxics also began confronting landfills and incinerators. The Citizens Clearinghouse for Hazardous Wastes (CCHW) based in Falls Church, Virginia, established the Solid Waste Organizing Project in 1986 to stop unsafe disposal practices, to promote alternatives such as recycling and composting, and to eliminate products that added to the solid-waste problems. The CCHW claimed victories from Georgia to California.[26] Along with protesting the siting of toxic facilities, members of the environmental justice movement also decried the risk of hazardous substances in landfills and in releases from incinerators.[27] The question of proper refuse disposal was moving beyond what was technically feasible, economically possible, and environmentally sound to what was socially and politically acceptable.

Growing concerns that sanitary landfills were not as sanitary as their name implied gave citizens' groups weapons to resist new sitings. Birds, insects, and rodents frequenting landfills carried pathogens back to people. Five of six existing MSW landfills in the late 1980s were not lined, thus posing a threat to groundwater and surface water, especially through leachate. There was also concern about contamination from various hazardous materials and incinerator ash that found their way into the landfills.[28]

Until the promulgation of strict federal laws in the early 1990s, more than 75 percent of MSW ended up in landfills. However, the amounts directed to landfills still remained high at the turn of the twentieth century. Conversely, the number of sanitary landfills declined rapidly since the 1980s. Figures vary widely on what constitutes a landfill for purposes of statistical evaluation, but all estimates clearly indicate a substantial shrinking of the total number of landfills in the United States. One study set the number of municipal landfills at 15,577 in 1980; another reported a drop to less than 8,000 by 1989. Between 1990 and 2000, the number of landfills declined from approximately 7,300 to 2,200 (some assessments are even as low as 1,967). In 2005, *BioCycle* reported that there were 1,654 solid-waste landfills in the contiguous United States.[29]

What was a crisis for some communities was a distant danger for others, especially since determining landfill capacity was difficult. Regional differences were more obvious. In 2005, the largest number were in the South (581) and the fewest in the Northeast (133). The steep rise in tipping fees in the Northeast between 1985 and 1990 demonstrated how serious the landfill problem had become at that time.[30]

The cost of disposal, environmental risk factors, management problems, compliance with regulations, and squandering of resources combined to exacerbate the problem of landfills. If sanitary landfills deserved crisis status, it was because American cities had depended on a one-dimensional disposal system for at least forty years and had not adequately prepared the way for viable alternatives.

Of the available alternatives, incineration had the strongest following. By 1966, the total number of plants in operation had declined to 265, and by

1975 there were only about 160. In 1988, only 3 percent of MSW was incinerated.[31] These figures do not speak well for the second most popular disposal
technology, but incineration continued to appear and disappear. Its ability
to occupy relatively small spaces and to reduce waste volume was offset by
high capital costs and air-pollution problems.

Interest in waste-to-energy (WTE) incinerators intensified in the 1970s
due to the search for alternative fuels in the wake of the energy crisis. Over
the years, interest in mass burn and refuse-derived fuels (RDFs) appeared periodically, but incineration never rivaled the sanitary landfill. Nonetheless,
supporters of WTE incinerators in the early 1990s believed that municipal
waste combustion remained "a necessary and integral part of overall solid
waste management strategies."[32] Two types of waste-combustion facilities
continued in use. The first—typically a mass burn unit—only burned waste
to reduce the volume for landfilling and usually were built before 1975; the
second produced steam for the production of electricity or for sale directly
to users of thermal energy. A WTE plant employed mass burn techniques or
could be an RDF facility. An RDF plant isolated burnable material, with the
resulting combustible substance often turned into pellets for use at a later
time. Despite the sophistication of modern incinerators, only 134 waste-
reduction-only (38) and WTE (96) facilities were in operation in 1988. Of that
number, only 18 were new WTE plants, and the highest number of these
facilities were east of the Mississippi River.[33]

The economic feasibility of incineration was a concern among engineers
and city officials. Improvements in ash-disposal practices and equipment
to control air pollution added substantially to construction and operating
costs. Several time-specific problems were also important, including the
assumption that incinerators could not overcome their environmental liabilities; they met only specific disposal needs, and the production of usable
by-products did not outweigh other liabilities.

By the early 1970s, pollution emissions in particular were placing incinerators in ill favor with the public, but their unique ability to reduce the volume of waste kept them from disappearing entirely. Retrofitting plants with
scrubbers and precipitators reduced air pollution, and the renewed promise
of raising steam to produce energy—plus the development of RDFs—played
well late in the decade. Yet the cost of adding pollution controls often was
prohibitive, and incinerators rarely could compete directly with sanitary
landfills in the 1970s. Even the more hopeful prospect of producing energy
was hampered because of the difficulty in selling the steam.[34]

Resource recovery plants got a modest boost in 1979 because of the Public Utility Regulatory Policies Act (PURPA), which provided guaranteed
markets for electricity sales, despite the high cost of generating electricity

from garbage. The potential for energy generation made resource recovery systems intriguing because they held out the possibility of return on investment. PURPA, however, did not give resource recovery plants a strong competitive role in the disposal arena.[35]

Some communities adopted mass burning units in the 1980s. Mass burning, however, remained expensive, and alarm over the production of dioxin and furans, acid gas, and heavy metal emissions weakened interest.[36] WTE was the byword of the day, but opponents claimed that resource recovery plants were unreliable, unable to produce energy in sufficient quantity to offset costs, and produced dioxin emissions. As with landfills, grassroots groups and environmental justice advocates fought the siting of incinerators. Concerned Citizens of South-Central Los Angeles (CC-SCLA), for example, was successful in the 1980s in scuttling a plan to build a huge incinerator in a predominantly black, inner-city neighborhood. Battles over proposed incinerator facilities increased substantially during the mid- to late 1980s.[37]

By the mid-1980s, the environmental debate over disposal began to push aside enthusiasm for WTE facilities. Promoters of WTE had to make the dual case that the system they embraced could generate sufficient energy to offset costs and could do so without serious threats to the air or groundwater. As the debate intensified, it was apparent that public standards were changing.[38] The decade of support that the EPA and the Department of Energy provided for planning, research, demonstration, and commercialization of incinerators also was undermined by federal budget cuts and a shift in interest away from MSW to hazardous and toxic wastes, as popularized by the revelations of severe toxic pollution at Love Canal in New York (1978). Furthermore, the method of financing incineration systems was changing, which affected local management decisions.[39]

Conditions began to improve somewhat for incineration in the early 1990s. Further constriction of sanitary landfills was a major reason. A bit of a counterweight were national standards for municipal waste combusters, promulgated by the EPA in 1991, which placed restrictions on existing incinerators. The 1990 Clean Air Act also limited incinerator emissions. The hope was that cleaner technologies would prevail to meet federal standards and would encourage more efficient energy production through new WTE units. Also, promotion of "integrated solid waste management" by governmental entities included a place for incineration.[40]

In recent years, promoters of WTE developed new strategies to increase its use.[41] The continuing debate over incineration was a classic confrontation between those with faith in technology to overcome social and economic problems and those who were suspicious of the "technical fix." In theory, WTE offered an efficient means to reduce the volume of wastes and also to provide a valuable by-product. Proponents did not deny the potential environmental hazards associated with burning waste, but were inclined to believe that newer WTE facilities could stand the performance test.[42] Critics questioned the efficiency of the plants, bemoaned the squandering of resources that provided the feedstock, and feared the environmental impact of their use. Less subtle in his criticism than others, environmentalist Barry Commoner referred to the new generation of incinerators as "dioxin-producing factories."[43] The Institute for Local Self-Reliance, a research group based in Washington, D.C., strongly criticized incineration and also rejected sanitary landfilling as a viable disposal option, instead supporting recycling and waste minimization. It linked cost factors and environmental implications, arguing that "because of unresolved, environmental regulatory requirements, the ultimate cost of mass burn facilities [and presumably sanitary landfills] is not fully known."[44] A 1990 Greenpeace report charged that communities with existing incinerators had populations of people of color 89 percent higher than the national average, and 60 percent higher in communities with proposed incinerators.[45]

Despite trends indicating that incineration in the early 1990s was capturing from 15 to 20 percent of the total MSW, it remained a volatile disposal technology whose stock continued to rise and fall depending on a variety of externalities. At the beginning of the twenty-first century, most incineration involves recovery of energy products, although there has been a small trend in burning source-separated municipal solid waste. In 2000, there were 102 WTE facilities, but only 88 in 2005 (39 in the Northeast).[46]

Recycling, once regarded as a grassroots method of source reduction and an innocuous protest against overconsumption in the 1960s, became an alternative disposal strategy to incineration and landfilling by the 1980s. As an Oregon engineer stated, "Once ridiculed as an ineffective hobby of environmentalists," recycling "is now regarded as an essential component of solid-waste management and a cost-effective way to reduce dependence on landfills."[47]

Recycling in the United States was "at the takeoff stage" in the 1980s. Many questions needed to be answered: What incentives should be used to foster compliance among householders, businesses, and manufacturers? What about mandatory laws? Should there be a governmental procurement policy? How much attention should be paid to recycling literacy? And most important, can markets be found for the increasing volume of recyclables?[48]

As solid-waste managers, private industry, and citizens' groups struggled with these questions, pressure to find answers came in the form of recycling programs popping up across the country. Before 1980, fewer than 140 communities in the United States had door-to-door recycling collection service. Estimates for the 1990s put the total at over 1,000. In 1989, there were more than 10,000 recycling drop-off and buy-back centers in operation and more than 7,000 scrap processors.[49]

A major national goal was to increase the recycling rate, which stood at 10 percent in the late 1980s. In 1988, the EPA called for a national recycling goal of 25 percent by 1992 (although it was not achieved, EPA estimates for 1990 were 16 percent and 29 percent for 2000) and 35 percent by 2005 (the EPA estimated a 32.1 recovery rate in that year for recycling, including composting, which translates into 1.46 pounds per person per day). Highest rates of recovery were achieved for yard trimmings, metal products, and paper and paperboard products. Several states in the 1980s began aggressive recycling programs, including Connecticut, New Jersey, New York, Pennsylvania, and Rhode Island in the East; Florida in the South; and Oregon in the West. Some communities were claiming achievements beyond the national standard.[50]

Skeptics of recycling were concerned about cost, the lack of stable markets, and what they believed to be inaccurate portrayal of the success of many recycling programs. And criticism persisted well beyond the 1980s. In 1996, John Tierney wrote an article for the *New York Times Magazine* in which he stated that as a result of the *Mobro* incident in 1987, "The citizens of the richest society in the history of the planet suddenly became obsessed with personally handling their own waste." He said that as a result of the concern about filling up of landfills, Americans saw recycling as their only option, but failed to recognize that the crisis of 1987 was a false alarm and that mandatory recycling programs "aren't good for posterity" and diverted money from "genuine social and environmental problems." He concluded that "Americans have embraced recycling as a Transcendental experience, an act of moral redemption."[51]

The piece was selective in its use of data, and Tierney had thrown out the proverbial baby with the bathwater for the sake of a story: He characterized the recycling craze as new, and therefore impulsive, rather than as part of a long-standing dialogue going back many years; he criticized recycling supporters for not being grounded in the realities of the costs-benefits of recycling, but did not provide careful documentation of recycling's failures; he depicted advocates as representing a clearly identifiable social or political fringe rather than as a diverse group; he regarded recycling as independent of other waste-disposal issues, instead of viewing it as a component in a large and more complex system; and he painted the history of recycling as if it already had played itself out.

Critics of recycling were often guilty of overstating their case, and proponents were sometimes too optimistic about the success of their programs. In some instances cautionary notes were appended to enthusiastic reports. Just as the sanitary landfill came to be viewed as a disposal panacea at midcentury—only to be later discredited—the euphoria over recycling needed to be tempered with a strong record of tangible results, but through careful analysis rather than shoot-from-the-hip journalism.[52]

Recycling was not going to go away. In 1968, Madison, Wisconsin, may have been the first city to begin curbside recycling (of newspapers). In that same year, the aluminum industry in the United States began to recycle discarded aluminum products. In 1971, Oregon was the first state to enact a "bottle bill," placing a five-cent refund on every beer and soft-drink container unless it could be reused by more than one bottler (in which case it would have a two-cent refund value) and outlawed pull-tab cans.[53]

Recycling centers in several communities experienced modest success, but the inconvenience of dropping off materials at a centralized location severely limited participation. Curbside collection programs, while not inexpensive to run, became the most effective method of gathering recyclables, achieving the highest diversion rates from the waste stream of any option available. From the modest beginnings in Madison and a few other locations, citywide and regularly routed curbside programs increased to 218—located primarily in California and the Northeast—by 1978. In 1989, it was estimated that there were 1,600 full-scale and pilot curbside recycling programs in the United States, with participation rates estimated from 49 to 92 percent.

As the 1990s began, recycling was a growth industry, and by mid-decade markets for recovered materials was on the rise. By 1991, the number of curbside programs soared to 2,711, with forty-seven of the fifty largest cities having such programs in 1992. In 2005, there were about 8,550 curbside recycling programs, although the number rises and falls depending on who is counting. Also, by the early 1990s there were materials recovery facilities (MRFs) operating in twenty-four states. A 1993 Congressional Research Service report stated that "in the urban States of the Northeast and Pacific Coast . . . curbside programs are now so common that areas without them are becoming the exception rather than the rule." By later in the decade, this was also true for other parts of the country with large urban populations.[54]

Recycling had developed strong political and social appeal in the 1980s. While not inexpensive, recycling put policy makers on the side of conservation of resources, as well as giving concerned citizens a way to participate in confronting the solid-waste dilemma specifically and environmental problems in general. The perception of recycling as an answer to the nation's disposal issues gave it strong momentum. Its linkage to waste reduction and waste minimization as a means to conserve resources and reduce pollution attracted new followers.[55]

"There is nothing as crucial to an alderman as garbage." So spoke Alderman Roman Pucinski of Chicago in 1984.[56] Many people agreed, and many argued that interest-group politics alone explained the garbage crisis in America. It is difficult to deny the confusion and complexities that the surfeit of interest groups brought to the garbage problem—and the harrowing task of political leaders to sort out the relevant issues. But political circumstance alone did not explain priority setting for individual cities, nor did it address the means to confront the so-called garbage crisis.

With the ban on backyard burning of trash in many cities during the 1970s, trash collecting became mandatory along with garbage collecting. The increased volume of collectible wastes led several cities to contract with private firms for servicing some of the routes. Dissatisfaction with municipal service encouraged several private collectors to claim that they could compete favorably with the city.[57] By 1974, large firms held contracts in over 300 communities. Between 1964 and 1973, 65 percent of cities surveyed operated some sort of collection service, but the proportion of cities with exclusively municipal collection declined from 45 to 39 percent.[58]

Debate about privatization of collection services led to confrontation over several solid-waste issues. The major point was who had control of collection and who made decisions about the disposition of the wastes themselves. In many ways, the shift to private services happened more easily than anticipated. A 1981 study of ten cities (ranging in size from 35,000 to 593,000 people) stated that the change to private collection was widely supported. Substantial reductions in the cost of service were achieved in the first year of the contract. Key factors contributing to support for private contracts included: shifts in federal or state regulations that compelled cities to reexamine their collection systems; a reduction in city resources; and the efforts of private firms to promote privatization. These factors and city size proved most important. Large cities with entrenched, centralized, and professional public works systems tended to resist contracting with private firms for collection much more vigorously than smaller cities.[59]

The trend toward more privatized collection continued into the 1980s and 1990s. In 1988, the NSWMA stated that over 80 percent of the nation's garbage was collected by private companies under government contract or working directly for local residents. Private control of disposal facilities was much less dramatic. In 1987, only 22 percent of municipal governments had contracted with private landfills or combustion facilities, although 32 percent had plans to hire a contractor in 1989. In 1988, only 15 percent of the country's landfills, but about half of the resource recovery plants, were privately owned.[60]

With increased privatization came consolidation within the solid-waste industry in the 1970s. "Historically," as financial consultant Anne Hartman noted in *Waste Age*, "the solid waste control industry as a whole has been fragmented, undercapitalized, and unsophisticated." The industry, she added, was now "assuming a new stature, and new management methods are being brought to bear on the long-standing and relatively simple problem of disposing of wastes."[61] Three companies that dominated the U.S. industry in 1980 had been formed as recently as the late 1960s and early 1970s—Waste Management, Inc., Browning-Ferris Industries, and SCA Services of Boston. While they only handled 15 percent of the nation's solid waste, they had extensive influence in the industry because of their capital, their size, and their management teams.[62]

Into the 1980s, consolidation in the industry was producing record growth in privatized solid-waste services. Growth also brought charges of violating federal antitrust laws and mismanagement of disposal sites, including alleged violations of federal environmental statutes. In response to the variety of charges, Browning Ferris, Waste Management, and the other conglomerates sought to bolster their image, and some attempted corporate makeover campaigns. Criticism of the waste giants did not disappear, and in some cases periodic drops in stock value reflected unease with the consolidation and conduct of the solid-waste industry.[63]

Consolidation accounted for much of the industry's growth in the 1990s, with some companies becoming international businesses. In 1996, the top six companies posted total revenue of more than $19 billion. In 1995–96, large numbers of consolidations took place, with 28 of *Waste Age's* top 100 companies being acquired or merging with other top 100 companies. The acquisitions were largely part of a plan for vertical integration, that is, companies controlling the waste stream and then dumping the waste into their own landfills.[64] One expert noted that "big has bought small, big has bought big, and small may be merging with small."[65] Smaller haulers were willing to sell to bigger companies largely because government regulations had driven up the cost of doing business. Traditional family-owned companies often felt unprepared to deal with an industry becoming increasingly more complex. In some cases, larger companies were willing to maintain the original name of the smaller hauling company and involve the original owners in management. And, of course, profit opportunities or some pressure from the bigger firms encouraged several small companies to sell out.[66]

In March 1998, USA Waste Services, Inc.—at the time, the third largest waste disposal company in the country—acquired industry leader, but ailing, Waste Management, Inc. to become the largest waste disposal corporation. The "new" WMI now controlled more than 20 percent of the national waste stream. Of the top 100 companies ranked by revenue in 1997, 15 were acquired in 1998, with consolidations continuing along the whole spectrum of the industry. *Waste News* reported on January 12, 1998, "Industry observers see no fast finish for the current consolidation trend in the trash-handling and scrap metal markets." Indeed, the very next year Allied Waste

Industries Inc. of Scottsdale, Arizona (then number three in the industry), announced that it would acquire Houston's Browning-Ferris Industries, then ranked number two behind WMI.[67]

The transformation of the solid-waste industry since the 1960s, or the consolidations in the 1980s and 1990s, did not stabilize the industry or make it immune from continued criticism. As one journalist noted in 1990, "Waste handlers have provided skeptics with every conceivable reason to distrust them, including technological fiascos, bid rigging and other antitrust violations, mismanagement and links to organized crime." Exposés on the waste giants and recurrent stories of price-fixing and mob involvement dogged the industry.[68] Critic Harold Crooks asserted that "the fragmented and disorganized nature of the garbage business left most firms extremely vulnerable to the hazards of the open market, and so like other big-city industries in similar situations, it often resorted to coercive means of self-protection."[69] Even after the rise of the new agglomerates, the industry was not free of its rough-and-tumble history, its intense competition, its internal battles, and its brushes with government agencies and the courts. But as Crooks also noted somewhat ironically, "the mobster has become a dinosaur in the waste disposal industry. Financial muscle had replaced the physical kind."[70]

The privatization trend suggests a dramatic swing in the operations phase of solid-waste management. With respect to ultimate responsibility for collection and disposal services, privatization was simply a complicating feature. Competition rather than the type of ownership appeared to be the most important factor in efficiency of service. The trend toward consolidation in the waste-disposal industry could strongly affect the nature of competition, and thus the relationship with municipal governments in setting collection and disposal policies.[71]

Competition also existed among various governmental entities through jurisdictional rivalries. A variety of new arrangements emerged out of the realization that the garbage problem was infrequently contained within the city limits. Some areas tried interlocal agreements, countywide systems, multicounty corporations, or intrastate planning bodies. Often born of necessity, these arrangements not only spoke to the complex problems of collection and disposal but also required substantial attention to determining lines of authority in the decision-making process. The question of "Who is in charge?" is particularly significant because of the commitment of both public and private entities to "integrated waste management" systems.[72]

However, proponents of integrated waste management may not have given sufficient attention to the fact that attempts to implement it came with the risk of sharpening the traditional rivalry between public and private forces seeking to control collection and disposal systems. The EPA's promotion of the concept suggested a significant new role for the federal government not only in setting a national agenda for solid waste but also in setting an expanded regulatory role.

Especially since the 1960s, the federal government broadened the scope of the waste problem, stressing its significance as an environmental issue with national repercussions. The enlargement in federal authority in solid-waste management in the late 1980s was a return to a role it had put aside. With the creation of the EPA in 1970, responsibility for most refuse activities was transferred to it, and soon the Office of Solid Waste (OSW) acquired the authority to conduct special studies, award grants, and publish guidelines. The OSW provided a degree of stability for federal solid-waste programs, but not without controversy. Since a primary function of the EPA was to aid in the control and elimination of pollution, several officials preferred to concentrate on the problem of hazardous wastes and deemphasized solid-waste management issues as envisioned in the 1965 and 1970 acts.

In the mid-1970s, the EPA proposed a drastic cutback in the federal solid-waste program and recommended that activities be limited to regulating hazardous wastes. When Congress and state and local groups balked, the EPA backed away from this extreme position and announced its willingness to continue to develop and promote resource recovery systems and technology. But when the immediate threat of the energy crisis had passed and belt-tightening provisions cut into the EPA's budget in the early 1980s, the agency all but turned away from MSW.[73] With public attention alerted to a pending "garbage crisis," and with the strengthening of the provisions of the Resource Conservation and Recovery Act (RCRA) affecting MSW in the mid-1980s, the EPA was back in the middle of the MSW debate, along with other federal agencies.[74]

There were some who applauded the resolve of the federal government to join the battle over the "garbage crisis." Others argued that increase in regulatory authority was not matched with action. William Kovacs, chief counsel of the House of Representatives Subcommittee on Transportation and Commerce, observed that the "EPA's implementation of RCRA can only be described as tardy, fragmented, at times nonexistent, and consistently inconsistent."[75]

For optimists, the expanded role of the federal government into issues involving solid waste had the potential to mobilize Americans to action. For pessimists, it simply added one additional discordant voice to a tone-deaf choir. At the very least, the likely outcome produced a more middle-of-the-road effect.[76]

As stated earlier, the Solid Waste Disposal Act (1965) and the Resource Recovery Act (1970) caused the states to become more deeply involved in local issues related to solid waste. In 1976, RCRA passed Congress and was reauthorized as the Hazardous and Solid Waste Amendments in 1984. The new legislation added significantly to the federal government's role in solid-waste management and was the first comprehensive framework for hazardous-waste management in the United States.[77] It completely changed the language of RCRA, redefining solid waste to include hazardous waste. RCRA continued provisions on solid waste and on resource recovery; ordered the EPA to require "cradle to grave" tracking of hazardous waste and controls on hazardous-waste facilities; closed most open dumps; and set minimum standards for waste-disposal facilities. The EPA did not give major attention to carrying out the provisions dealing with hazardous wastes until publicity over Love Canal in 1978.[78]

Of particular significance for MSW was Subtitle D of RCRA, which provided for environmentally sound disposal methods and the protection of groundwater with new landfill standards. According to one analyst: "Simply put, the premise of RCRA is that the solid waste problem is a result of our industrial society. As such, the true costs of disposal should be borne by those benefiting from the products that generate the waste. Therefore, once the subsidy of the cheap landfill is removed, other more advanced technologies will develop."[79]

Under RCRA, the EPA acquired extensive regulatory authority over MSW, especially in the design and operation of landfills and incinerators. Congress directed the EPA through RCRA to develop landfill criteria that included a prohibition on open dumping. The EPA issued the criteria in 1979. The 1984 amendments attempted to tighten standards with respect to landfills, and revised Subtitle D regulations in 1991 set minimum national standards for landfills. The cost of meeting the Subtitle D criteria for existing and new landfills was extremely high and resulted in further constriction in the landfill supply nationwide, thus favoring the development of larger regional landfills to lower per-unit costs or moving communities toward alternative disposal methods. In addition, landfills were more frequently becoming privately held because the costs were too high for communities to bear alone. In 1984, only 17 percent of the landfills in the United States were privately owned; in 1998, at least 40 percent, if not more, were privately owned.[80]

The modern solid-waste problem in the United States was too complex to regard it as a crisis. Implied in this usage of "crisis" is the assumption that society had reached a point beyond which nothing less than a dangerous outcome could be expected. Much of what had come to be considered the "garbage crisis" was a series of chronic conditions, interrelated in such a way as to defy a simple solution. The difficult task was to determine what needed immediate action and what needed steadfast action. Things that fell into the category of chronic conditions included the volume of waste, inherent problems with collection, the impulse to depend on a single disposal option, emphasis on "back end" solutions to "front end" problems, debate over public versus private operation and management, and jurisdictional disputes over regulation. Those issues especially in need of immediate attention included increasing amounts of toxic materials in the municipal waste stream, the squandering of limited resources, lack of viable landfill space, and increased levels of air and water pollution. Some of these problems varied regionally and varied in intensity. Others were more universal, as in the case of consumption of scarce or otherwise valuable resources.

In some respects, dealing with the chronic conditions may prove more difficult than confronting those deemed to be at a crisis point. Clearly, the broad acceptance of a "garbage crisis" derived much of its power from the fact that, since the 1970s at least, the problem of waste was given greater credit for environmental degradation than earlier. What had been a nuisance in the nineteenth century became a major environmental blight in the twentieth. In a perverse sort of way, solid waste had reached parity with water supply and sewerage as a significant public works challenge for the future.

Epilogue

By the twentieth century, most major American cities and many smaller ones had achieved the goal of establishing permanent, centralized citywide sanitary (environmental) services. This was a major accomplishment. Water was readily available for home and business use, for combating fires, and for tasks too numerous to mention. Effluent and refuse flowed away or was hauled away in a systematic and usually efficient manner. Cities were being spared the most serious consequences of many communicable diseases. As we have seen, these vital services, which were intertwined and sometimes hardly visible in the maze of urban infrastructure, were intimately connected to the prevailing health and environmental views and values of their day. While sanitarians, engineers, and city officials could take justifiable pride in providing urban residents with the makings of a sanitary city, these technologies of sanitation also contributed to, or failed to address, an array of new and different environmental problems. And despite their dramatic overall impact, their availability and use were not always equitable across class and racial lines.

Beginning in the early nineteenth century, building sanitary systems originated with the goal of providing pure and plentiful supplies of water as public goods. With the major exception of Philadelphia's waterworks in 1801, few modern citywide systems appeared before 1830. By the late nineteenth century, waterworks were generally regarded as a public enterprise, justified as such because of the need to protect the public health and to supply water on a citywide basis. As with wastewater systems and solid-waste collection and disposal in later years, modern citywide water-supply systems were conceived in an Age of Miasmas. The structure of those systems and their functions were linked inextricably to the goals of environmental sanitation, that is, to utilize the prevailing sensory tests of purity to deliver a product that would not only be free of disease but also would be utilized to alleviate disease. Bountiful supplies of water, for example, could help to flush away stench-ridden wastes.

Even though massive increases in piped-in water proved to be a major reason for wastewater systems, water supply and sewerage were addressed separately in the nineteenth century. The justification for wastewater systems was also graphically linked to the precepts of environmental sanitation. The leap of logic from liquid wastes to solid wastes was made possible by the adherence to environmental sanitation, to a point of view that almost etched in stone the primary directive that wastes had to be removed from the presence of humans as quickly as possible to preserve the public health. In essence, sanitary services in this period were little more than elaborate transportation networks. The underground sewer was a logical embodiment of the goals of environmental sanitation, as were the sanitary landfill and the incinerator. These technologies were meant to distance humans from their wastes and discards—materials that presumably had embedded in them the threat of disease.

Whole systems were designed to meet the ends of environmental sanitation. Water-supply systems began with a protected source—or one that could be purified through filtration or treatment. The new distribution system of pipes and pumps removed from the individual responsibility for filling containers at a public well or local watercourse, and made the waterworks—private and public—responsible for bringing water directly to each consumer. Implicit in this system was a guarantee that the supply met the prevailing standards of purity. With respect to wastewater systems, citizen accountability also was at a minimum. Combined or separate sewer pipes whisked effluent from homes and businesses, and placed responsibility for disposal in the hands of the city. The objectives of environmental sanitation had been met because human contact with the waste—at least at the source—was dramatically reduced. In the case of refuse, on-site pickup served the same purpose as water or sewer pipes.

Within the context of nineteenth-century environmental sanitation, sanitary services were at their best in the areas of collection and delivery of water, and collection of effluent and solid wastes. Insofar as pure water was delivered efficiently and initial human contact with wastes was minimized, these services fulfilled their objective. A change in environmental paradigm—from miasmas to bacteria—did not disrupt these methods of collection or influence major changes in "front of the pipe" components in these technologies of sanitation. The new age of bacteriology did help to identify and ultimately confront "end of the pipe" problems. The major weakness of environmental sanitation as a concept was its limited attention to disposal and treatment of effluent and refuse after they had been directed away from homes and businesses.

With the onset of the bacteriological revolution, water pollution, in particular, received greater and more pointed attention. Scientists, physicians, and engineers now had a much better idea of what they were looking for in the fight against communicable disease, and a clearer idea on how to combat biological pollutants. Older methods, such as dilution, had value under proper circumstances, but the primary focus centered on treatment. The goals of environmental sanitation were no longer regarded as sufficiently broad to deal with the several confounding problems of waste disposal—or to assure the purity of a water supply. Without attempting to change the basic structure of the technologies of sanitation, experts introduced new methods of water testing and focused increasing attention on refining treatment technologies. Never in question was the basic precept of designing permanent, citywide sanitary systems; the basic designs originating in the Age of Miasmas were not substantively changed in the twentieth century. In essence, bad science led to good technology.

While bacteriology contributed effectively to eradicating biological pollution, it offered less in addressing other environmental problems, particularly industrial wastes and new toxic chemicals entering water supplies and land-disposal sites. The New Ecology brought into clearer focus the extent of water and land pollution related to the technologies of sanitation. Methods of measuring pollutants from point sources in water, in landfills, and from other sanitary service processes became more sophisticated and holistic. Again, the basic structure of the technologies of sanitation did not change, but a better understanding of their capabilities did.

The New Ecology offered greater awareness of environmental inputs and outputs, which could help maximize the value of sanitary services. Through continual refinements and effective maintenance of the technologies of sanitation, the best of those systems delivered valuable service to metropolitan communities. The traditional emphasis on permanent, centralized systems, however, ultimately exposed limitations in their functions. All technologies of sanitation were capital intensive, and they required continual maintenance and repair. The publicly acknowledged infrastructure crisis in the 1970s and 1980s demonstrated a lack of commitment to maintaining existing systems adequately. Other priorities in city budgets strongly influenced the fate of these sanitary systems, as did a variety of fiscal constraints, brought on by economic depression or a change in political leadership and ideology. Sanitary services, like other infrastructure issues, also were caught between the choice of investing in existing systems or expanding to meet new demand. The connection between these services and the nature and rate of urban growth also was closely linked.

The results of path dependence—initial choices constraining future options—complicated the ability of cities to maintain or expand their sanitary systems or to develop new ones. The decision to invest in a distant water supply, pumping stations, filtration and treatment facilities, incinerators, and landfills gave a certain degree of comfort to city leaders in the past who believed they were creating monuments to good sanitation and effective service delivery. The emphasis on project design as opposed to long-range planning often meant that future generations could not choose to abandon these systems and begin again, but must maintain or expand them—even if inadequate—or face extraordinary costs. It was not so much that flawed technologies were chosen initially, but that systems were designed to be permanent, to resist change in order to justify their worth to the contemporary community. In essence, the systems lacked flexibility, that is, they were not capable of substantial alteration due to changes in technology, fiscal conditions, or urban growth patterns. They also lacked the ability to combine functions with other systems, such as unifying water supplies and wastewater systems. More recent demands for "modular flexibility" in the design of infrastructure is a response to the realization that past practices in the construction and use of technical systems have severely limited choices for contemporary decision makers.

With respect to problems of pollution, it is clear that the technologies of sanitation—especially water-supply and wastewater systems—have limited value in dealing with certain kinds of pollutants not present or not considered significant in the initial design phase of the systems. Of particular note are runoff and nonpoint source pollution. Designers of the systems and decision makers in the nineteenth and early twentieth centuries cannot be faulted for failing to anticipate these problems; after all, these decisions took place during the "century of technological enthusiasm," as coined by historian Thomas P. Hughes in *American Genesis*, and during a period when the centralizing tendencies of government were strong. However, the commitment to permanent, centralized technical systems to resolve sanitary problems in fact left little room for adapting those systems to meet new and serious challenges. In other words, the systems built in permanence but not resilience. Historical circumstances, therefore, strongly influence the current choices available for sanitary services, whether decision makers like it or not.

The technologies of sanitation, originating in the nineteenth century and continuing to the present, have a mixed record of achievement. As the circulatory system of the city, sanitary services have contributed mightily to the environmental well-being of urbanites throughout the country. They were and remain an important component of the city-building process—necessary elements in the growth of urban areas. They can be growth mechanisms for extending development outward from the urban core, or, by withholding them, city leaders can use them as tools for limiting growth. Yet sanitary services also are sources of pollution—shifted from one location to another, and from one jurisdiction to another. In all instances, they are intimately connected with the prevailing health and environmental views of their day, remaining central to understanding many of the implications of urban growth. Taken as a group, water supply, wastewater, and solid-waste collection and disposal are basic and fundamental to the urban experience. To function effectively the American city has to be a sanitary city.

ABBREVIATIONS

AAAS   American Association for the Advancement of Science

ACIR   U.S. Advisory Commission on Intergovernmental Relations

AJPH   *American Journal of Public Health*

APHA   American Public Health Association

APWA   American Public Works Association

ASCE   American Society of Civil Engineers

ASMI   American Society for Municipal Improvements

AWWA   American Water Works Association

CEQ   Council on Environmental Quality

CSO   combined sewer overflow

CWS   community water systems

EPA   U.S. Environmental Protection Agency

GAO   U.S. General Accounting Office

gcd   gallons per capita per day

HEW   U.S. Department of Health, Education, and Welfare

HUD   U.S. Department of Housing and Urban Development

JAWWA   *Journal of the American Water Works Association*

mgd   million gallons per day

MSW   municipal solid waste

NCPWI   National Council on Public Works Improvement

NPDES   National Pollution Discharge Elimination System

NSWMA   National Solid Waste Management Association

RCRA   Resource Conservation and Recovery Act

SDWA   Safe Drinking Water Act

SMSA   Standard Metropolitan Statistical Area

USPHS   U.S. Public Health Service

NOTES

Introduction

1. Bryan D. Jones et al., *Service Delivery in the City* (New York: Longman, 1980), 2.

2. "A National Movement for Cleaner Cities," *American Journal of Public Health* 20 (Mar. 1930): 296–97; George W. Cox, "Sanitary Services of Municipalities," *Texas Municipalities* 26 (August 1939): 218.

3. Martin V. Melosi, *Garbage in the Cities*, rev. ed. (Pittsburgh: University of Pittsburgh Press, 2005).

4. Joel A. Tarr and Gabriel Dupuy, *Technology and the Rise of the Networked City in Europe and America* (Philadelphia: Temple University Press, 1988), xiii.

5. Martin V. Melosi, "Cities, Technical Systems and the Environment," *Environmental History Review* 4 (Spring/Summer 1990): 45–50.

6. Jon Peterson, "Environment and Technology in the Great City Era of American History," *Journal of Urban History* 8 (May 1982): 344.

7. Martin V. Melosi "Path Dependence and Urban History: Is a Marriage Possible?" in *Resources of the City*, ed. Dieter Schott, Bill Luckin, and Genevieve Massard-Guilbaud (Hampshire, U.K.: Ashgate, 2005), 262–75.

Chapter 1. Sanitation Practices in Pre-Chadwickian America

1. Sam Bass Warner Jr., *The Urban Wilderness* (New York: Harper and Row, 1972), 158.

2. Ernest S. Griffith and Charles R. Adrian, *A History of American City Government, 1775–1870* (1976; reprint, Washington, D.C.: University Press of America, 1983), 218; Cady Staley and George S. Pierson, *The Separate System of Sewerage*, 2nd ed. (New York, 1891), 53; Zane L. Miller and Patricia M. Melvin, *The Urbanization of Modern America*, 2nd ed. (San Diego, Calif.: Harcourt Brace Jovanovich, 1987), 20–21; Blake McKelvey, *American Urbanization* (Glenview, Ill.: Scott, Foresman, 1973), 14.

3. Carl Bridenbaugh, *Cities in the Wilderness*, 2nd ed. (New York: Knopf, 1955), 18, 85–86.

4. John Duffy, *The Sanitarians* (Urbana: University of Illinois Press, 1990), 33; Melosi, *Garbage in the Cities*, 13–15.

5. Duffy, *Sanitarians*, 13, 30; Joel A. Tarr, "Urban Pollution: Many Long Years Ago," *American Heritage*, Oct. 1971, 64–69, 106.

6. Duffy, *Sanitarians*, 20–23.

7. John J. Hanlon, *Principles of Public Health Administration*, 4th ed. (St. Louis: C. V. Mosby, 1964), 48; Duffy, *Sanitarians*, 10–11, 35.

8. John Duffy, "Yellow Fever in the Continental United States during the Nineteenth Century," *Bulletin of the New York Academy of Medicine* 44 (June 1968): 687–88; George Rosen, *A History of Public Health* (New York: MD Publications, 1958), 234; Harrison P. Eddy, "Sewerage and Drainage of Towns," *Proceedings of the ASCE* 53 (Sept. 1927): 1604; David R. Goldfield and Blaine A. Brownell, *Urban America*, 2nd ed. (Boston: Houghton Mifflin, 1990), 152.

9. Sam Bass Warner Jr., *The Private City*, rev. ed. (Philadelphia: University of Pennsylvania Press, 1987), 99.

10. Goldfield and Brownell, *Urban America*, 66, 68–69; Eric H. Monkkonen, *America Becomes Urban* (Berkeley: University of California Press, 1988), 112–15.

11. Raymond A. Mohl, *Poverty in New York, 1783–1825* (New York: Oxford University Press, 1971), 10–13, 104–6.

12. Hanlon, *Principles of Public Health Administration*, 47; John B. Blake, "The Origins of Public Health in the United States," *AJPH* 38 (Nov. 1948): 1539; Duffy, *Sanitarians*, 15, 18.

13. Other cities contested Boston's claim, including Petersburg, Virginia (1780), Philadelphia (1794), Baltimore (1793), and New York (1796). See Hanlon, *Principles of Public Health Administration*, 48–49.

14. Edwin D. Kilbourne and Wilson G. Smillie, eds., *Human Ecology and Public Health*, 4th ed. (New York: Macmillan, 1969), 114; Hanlon, *Principles of Public Health Administration*, 47–48; Stanley K. Schultz, *Constructing Urban Culture* (Philadelphia: Temple University Press, 1989), 119–20; Duffy, *Sanitarians*, 62.

15. Rosen, *History of Public Health*, 234; Schultz, *Constructing Urban Culture*, 119–21, 140.

16. Duffy, *Sanitarians*, 11–12, 48–49, 57; Hanlon, *Principles of Public Health Administration*, 47.

17. See Martin V. Melosi, "Hazardous Waste and Environmental Liability," *Houston Law Review* 25 (July 1988): 761–63.

18. Schultz, *Constructing Urban Culture*, 43; Melosi, "Hazardous Waste and Environmental Liability," 763–64.

19. "Protosystem" is meant to connote an "original" system or "first in rank or time" as opposed to the idea of a "primitive" system.

20. Letty Anderson, "Hard Choices: Supplying Water to New England," *Journal of Interdisciplinary History* 15 (Autumn 1984): 211.

21. Jon C. Teaford, *The Municipal Revolution in America* (Chicago: University of Chicago Press, 1975), 104.

22. M. N. Baker, *The Quest for Pure Water: The History of Water Purification from the Earliest Records to the Twentieth Century* (1948; reprint, New York: AWWA, 1981), 903; Rosen, *History of Public Health*, 125; Jean-Pierre Goubert, *The Conquest of Water* (Princeton, N.J.: Princeton University Press, 1986), 34–40; F. E. Turneaure and H. L. Russell, *Public Water-Supplies* (New York: John Wiley and Sons, 1911), 6.

23. Earle Lytton Waterman, *Elements of Water Supply Engineering* (New York: Wiley, 1934), 4–5; Turneaure and Russell, *Public Water-Supplies* (1911), 6; Richard Shelton Kirby and Philip Gustave Laurson, *The Early Years of Modern Civil Engi-*

neering (New Haven, Conn.: Yale University Press, 1932), 81, 194; Edward S. Hopkins, ed., *Elements of Sanitation* (New York: D. Van Nostrand, 1939), 53; Rosen, *History of Public Health*, 124–25; Harold E. Babbitt and James J. Doland, *Water Supply Engineering*, 4th ed. (New York: McGraw-Hill, 1949), 3.

24. Kirby and Laurson, *Early Years*, 192, 212–14; Hopkins, *Elements of Sanitation*, 53; Daniel E. Lipschutz, "The Water Question in London, 1827–1831," *Bulletin of the History of Medicine* 42 (Sept./Oct. 1968): 510; Goubert, *Conquest of Water*, 22; Rosen, *History of Public Health*, 124–25.

25. Milo Roy Maltbie, "A Tale of Two Cities," *Municipal Affairs* 3 (June 1899): 193; Asok Kumar Mukhopadhyay, *Politics of Water Supply* (Calcutta: World Press Private, 1981), 1; J. J. Cosgrove, *History of Sanitation* (Pittsburgh: Standard Sanitary Manufacturing, 1909), 78, 82; W. S. Chevalier, *London's Water Supply, 1903–1953* (London: Staples Press, 1953), 1; W.H.G. Armytage, *A Social History of Engineering* (London: Faber and Faber, 1976), 71–72; William Freeman, *Water Supply and Drainage* (London: Sir Isaac Pitman and Sons, 1945), 13; Kirby and Laurson, *Early Years*, 187; Rosemary Weinstein, "New Urban Demands in Early Modern London," in *Living and Dying in London*, ed. W. F. Bynum and Roy Porter (London: Wellcome Institute for the History of Medicine, 1991), 34–36.

26. *The Sanitary Industry* (New York: Johns-Manville, 1944), w4; Rosen, *History of Public Health*, 124–25; Maltbie, "Tale of Two Cities," 193–94; Kirby and Laurson, *Early Years*, 188–91; C. W. Hutt and H. Hyslop Thompson, eds., *Principles and Practices of Preventive Medicine*, vol. 1 (London: Methuen, 1935), 6; Turneaure and Russell, *Public Water-Supplies* (1911), 7.

27. Eric E. Lampard, "The Urbanizing World," in *The Victorian City*, ed. H. J. Dyos and Michael Wolff (London: Routledge and Kegan Paul, 1973), 1, 4, 10–13, 21–22; H. J. Habakkuk and M. Postan, eds., *The Industrial Revolutions and After*, vol. 6 of *The Cambridge Economic History of Europe* (Cambridge: Cambridge University Press, 1966), 274; William Oswald Skeat, ed., *Manual of British Water Engineering Practices*, vol. 1 (Cambridge: Heffer, 1969), 2; Lewis Mumford, *City in History* (New York: Harcourt Brace and World, 1961), 461–65.

28. J. A. Hassan, "The Growth and Impact of the British Water Industry in the Nineteenth Century," *Economic History Review* 38 (Nov. 1985): 532.

29. Ibid., 531–34.

30. Ibid., 532.

31. H. W. Dickinson, *Water Supply of Greater London* (London: Newcomen Society at the Courier Press, 1954), 103; Chevalier, *London's Water Supply*, 4.

32. Samuel Rideal and Eric K. Rideal, *Water Supplies* (London: Crosby Lockwood and Son, 1914), 97; ASCE, *Pure and Wholesome* (New York: ASCE, 1982), 2; Skeat, *Manual of British Water Engineering Practices*, 30.

33. Skeat, *Manual of British Water Engineering Practice*, 5–6.

34. Brian Read, *Healthy Cities* (Glasgow: Blackie, 1970), 42–43; George W. Fuller, "Progress in Water Purification," *JAWWA* 25 (Oct. 1933): 1566; Chevalier, *London's Water Supply*, 7; Rideal and Rideal, *Water Supplies*, 97; M. N. Baker, "Sketch of the History of Water Treatment," *JAWWA* 26 (July 1934): 904.

35. Anne Hardy, "Parish Pump to Private Pipes," in Bynum and Porter, *Living and Dying in London*, 77–80.

36. Stephen F. Ginsberg, "The History of Fire Protection in New York City,

1800–1842" (Ph.D. diss., New York University, 1968), 318; Letty Donaldson Anderson, "The Diffusion of Technology in the Nineteenth-Century American City: Municipal Water Supply Investments" (Ph.D. diss., Northwestern University, 1980), 87–88.

37. Ginsberg, "History of Fire Protection," 338.

38. U.S. Bureau of Census, *Census of Population: 1960*, vol. 1, *Characteristics of the Population* (Washington, D.C.: Department of Commerce, 1961), pt. A, 1-14–15, Table 8; Waterman, *Elements of Water Supply Engineering*, 6; Harrison P. Eddy, "Water Purification—A Century of Progress," *Civil Engineering* 2 (Feb. 1932): 82; J.J.R. Croes, *Statistical Tables from the History and Statistics of American Water Works* (New York, 1885), 4–69.

39. Joel A. Tarr, James McCurley, and Terry F. Yosie, "The Development and Impact of Urban Wastewater Technology," in *Pollution and Reform in American Cities, 1870–1930*, ed. Martin V. Melosi (Austin: University of Texas Press, 1980), 59–60.

40. Other cities claimed to be the first to establish a municipal waterworks, but they had not constructed a "system" on the scale of Philadelphia's. See George S. Davison, "A Century and a Half of American Engineering," *Transactions of the ASCE* (Oct. 5, 1926): 560.

41. John C. Trautwine Jr., "A Glance at the Water Supply of Philadelphia," *Journal of the New England Water Works Association* 22 (Dec. 1908): 421.

42. Michal McMahon, "Fairmount," *American Heritage*, Apr./May 1979, 100–101; Donald C. Jackson, "'The Fairmount Waterworks, 1812–1911,' at the Philadelphia Museum of Art," *Technology and Culture* 30 (July 1989): 635; City of Philadelphia, Department of Public Works, Bureau of Water, "Description of the Filtration Works and Pumping Stations, Also Brief Historical Review of the Water Supply, 1789–1900" (1909), 57–59; Michal McMahon, "Makeshift Technology," *Environmental Review* 12 (Winter 1988): 24.

43. Edward C. Carter II, "Benjamin Henry Latrobe and Public Works," *Essays in Public Works History* (Washington, D.C.: Public Works Historical Society, 1976); McMahon, "Fairmount," 100–101.

44. Jane Mork Gibson, "The Fairmount Waterworks," *Bulletin of the Philadelphia Museum of Art* 84 (Summer 1988): 9.

45. Ibid.

46. Ibid., 10–12; Jackson, "Fairmount Waterworks," 635; McMahon, "Makeshift Technology," 25–26; Goubert, *Conquest of Water*, 56; Turneaure and Russell, *Public Water-Supplies* (1911), 9–11; George W. Fuller, "Water-Works," *Proceedings of the ASCE* 53 (Sept. 1927): 1594; Trautwine, "Glance at the Water Supply," 425; Frederick P. Stearns, "The Development of Water Supplies and Water-Supply Engineering," *Transactions of the ASCE* 56 (June 1906): 455–56; F. E. Turneaure and H. L. Russell, *Public Water-Supplies*, 4th ed. (New York: John Wiley and Sons, 1948), 8.

47. Ellis Armstrong, Michael Robinson, and Suellen Hoy, eds., *History of Public Works in the United States* (Chicago: APWA, 1976), 232; Griffith and Adrian, *History of American City Government*, 73.

48. M. J. McLaughlin, "142 Years of Water Distribution," *American City* 58 (Dec. 1943): 50; Armstrong et al., *History of Public Works*, 233; Stearns, "Devel-

opment of Water Supplies," 455; Turneaure and Russell, *Public Water-Supplies* (1948), 7.

49. Joel A. Tarr, "The Evolution of the Urban Infrastructure in the Nineteenth and Twentieth Centuries," in *Perspectives on Urban Infrastructure*, ed. Royce Hanson (Washington, D.C.: National Academy Press, 1984), 19; "Golden Decade for Philadelphia Water," *Engineering News-Record* 159 (Sept. 19, 1957): 37.

50. Martin J. McLaughlin, "Philadelphia's Water Works from 1798 to 1944," *American City* 59 (Oct. 1944): 86–87.

51. U.S. Bureau of Census, *Census of Population: 1960*, vol. 1, *Characteristics of the Population*, pt. A, 1-14–15, Table 8; Waterman, *Elements of Water Supply Engineering*, 6.

52. Anderson, "Diffusion of Technology," 1.

53. Charles Jacobson, Steven Klepper, and Joel A. Tarr, "Water, Electricity, and Cable Television," *Technology and the Future of Our Cities* 3 (Fall 1985): 9; Anderson, "Diffusion of Technology," 103–8.

54. Waterman, *Elements of Water Supply Engineering*, 6.

55. Nelson Manfred Blake, *Water for the Cities* (Syracuse, N.Y.: Syracuse University Press, 1956), 44–62, 101–20; J. Michael LaNier, "Historical Development of Municipal Water Systems in the United States, 1776–1976," *JAWWA* 68 (Apr. 1976): 174–75; Gustavus Myers, "History of Public Franchises in New York City," *Municipal Affairs* 4 (Mar. 1900): 85–87; Ginsberg, "History of Fire Protection," 318ff.

56. Fern L. Nesson, *Great Waters* (Hanover, N.H.: University Press of New England, 1983), 1.

57. Blake, *Water for the Cities*, 172–98; LaNier, "Municipal Water Systems," 174; John B. Blake, "Lemuel Shattuck and the Boston Water Supply," *Bulletin of the History of Medicine* 29 (1955): 554–62; Griffith and Adrian, *History of American City Government*, 70–71.

58. John W. Hill, "The Cincinnati Water Works," *JAWWA* 2 (Mar. 1915): 42–53; Bert L. Baldwin, "Development of Cincinnati's Water Supply," *Military Engineer* 22 (July/Aug. 1930): 320–21; Richard Wade, *The Urban Frontier* (Cambridge, Mass.: Harvard University Press, 1959), 294–95.

59. Wade, *Urban Frontier*, 297; LaNier, "Municipal Water Systems," 176; Gordon G. Black, "The Construction and Reconstruction of Compton Hill Reservoir," *Journal of the Engineers' Club of St. Louis* 2 (Jan. 2, 1917): 4–8.

60. Gary A. Donaldson, "Bringing Water to the Crescent City," *Louisiana History* 28 (Fall 1987): 381–96.

61. Wade, *Urban Frontier*, 294–95; James C. O'Connell, "Chicago's Quest for Pure Water," *Essays in Public Works History* 1 (Washington, D.C.: Public Works Historical Society, 1976), 3; Tarr, "Evolution of the Urban Infrastructure," 14; Tarr et al., "Development and Impact," 60.

62. Warner, *Urban Wilderness*, 202.

63. Fred B. Welch, "History of Sanitation," paper presented at the First General Meeting of the Wisconsin Section of the National Association of Sanitarians, Inc., Milwaukee, Dec. 1944, 39, 41; Hopkins, *Elements of Sanitation*, 51–52, 104; *Sanitary Industry*, w1–w3; Frederick Charles Krepp, *The Sewage Question* (London,

1867), 7; Benjamin Freedman, *Sanitarian's Handbook* (New Orleans: Peerless, 1957), 2–4; Baldwin Latham, *Sanitary Engineering* (London, 1878), 21–22.

64. Barrie M. Ratcliffe, "Cities and Environmental Decline," *Planning Perspectives* 5 (1990): 190–91; Kirby and Laurson, *Early Years*, 227–28; Krepp, *Sewage Question*, 7; Cosgrove, *History of Sanitation*, 85–86, 91; Mansfield Merriman, *Elements of Sanitary Engineering*, 4th ed. (New York: John Wiley and Sons, 1918), 141.

65. Weinstein, "New Urban Demands in Early Modern London," 29–31.

66. Kirby and Laurson, *Early Years*, 230.

67. Krepp, *Sewage Question*, 12; Henry L. Jephson, *The Sanitary Evolution of London* (London: John Wiley and Sons, 1907), 14; Charles J. Merdinger, "Civil Engineering through the Ages," *Transactions of the ASCE CT* (1953): 95; Leonard P. Kinnicutt, C.-E. A. Winslow, and R. Winthrop Pratt, *Sewage Disposal* (New York: John Wiley and Sons, 1919), 8; Latham, *Sanitary Engineering*, 35; H. B. Hommon, "Brief History of Sewage and Waste Disposal," *Pacific Municipalities* 42 (May 1928): 161; "The London Water Supply," *Engineering Magazine* 2 (Jan. 1870): 82.

68. Melosi, *Garbage in the Cities*, 5–6.

69. Tarr et al., "Development and Impact," 59–60.

70. Eddy, "Sewerage and Drainage," 1603.

71. Tarr et al., "Development and Impact," 60–61; Duffy, *Sanitarians*, 13, 29.

72. Samuel A. Greeley, "Street Cleaning and the Collection and Disposal of Refuse," *Proceedings of the ASCE* 53 (Sept. 1927): 1621.

73. Howard P. Chudacoff and Judith E. Smith, *The Evolution of American Urban Society*, 4th ed. (Englewood Cliffs, N.J.: Prentice-Hall, 1994), 11.

74. Melosi, *Garbage in the Cities*, 111–13; Duffy, *Sanitarians*, 16, 28.

Chapter 2. Bringing the Serpent's Tail into the Serpent's Mouth

1. C.-E. A. Winslow, *The Conquest of Epidemic Disease* (1943; reprint, New York: Hafner, 1967), 243.

2. Ann F. La Berge, *Mission and Method* (New York: Cambridge University Press, 1992), xiii, 3, 283–84.

3. Asa Briggs, *Victorian Cities* (1963; reprint, Berkeley: University of California Press, 1993), 16–17.

4. Anthony Brundage, *England's "Prussian Minister"* (University Park: Pennsylvania State University Press, 1988), 4–5.

5. D. D. Raphael, "Jeremy Bentham," *Collegiate Encyclopedia*, vol. 2 (New York: Grolier, 1971), 518–19.

6. Brundage, *England's "Prussian Minister,"* 7–11.

7. Ibid., 150–57.

8. Ibid., 35–77.

9. C. Fraser Brockington, *The Health of the Community*, 3rd ed. (London: J. and A. Churchill, 1965), 29, 32–33; Margaret Pelling, *Cholera, Fever, and English Medicine, 1825–1865* (London: Oxford University Press, 1978), 7; Winslow, *Conquest of Epidemic Disease*, 242–43; W. M. Frazer, *A History of English Public Health 1834–1939* (London: Bailliere, Tindall and Cox, 1950), 13, 15.

10. Brundage, *England's "Prussian Minister,"* 79–81.

11. Ibid., 83–84.

12. Briggs, *Victorian Cities*, 21; John H. Ellis, *Yellow Fever and Public Health in the New South* (Lexington: University Press of Kentucky, 1992), 3, 5–6.

13. Edwin Chadwick, *Report on the Sanitary Condition of the Labouring Population of Great Britain*, ed. with an introduction by M. W. Plinn (Edinburgh: University Press, 1965), 1; Anthony S. Wohl, *Endangered Lives* (Cambridge, Mass.: J. M. Dent, 1983), 7.

14. Roy Porter, "Cleaning Up the Great Wen," in Bynum and Porter, *Living and Dying in London*, 69.

15. D. B. Eaton, "Sanitary Legislation in England and New York," paper read before the Public Health Association of New York, 1872, 6; M. W. Flinn, *Public Health Reform in Britain* (London: Macmillan, 1968), 14.

16. Lipschutz, "Water Question in London," 510, 523–25; M. W. Flinn, introduction to Chadwick, *Report*, 9–10; Read, *Healthy Cities*, 9.

17. William Hobson, ed., *The Theory and Practice of Public Health* (New York: Oxford University Press, 1979), 4.

18. Bill Luckin, *Pollution and Control* (Bristol: Adam Hilger, 1986), 4.

19. Pelling, *Cholera, Fever, and English Medicine*, 10.

20. Christopher Hamlin, "Providence and Putrefaction," *Victorian Studies* 28 (Spring 1985): 381–86.

21. Brockington, *Health of the Community*, 30; Frazer, *History of English Public Health*, 14–15; Hobson, *Theory and Practice*, 4; Pelling, *Cholera, Fever, and English Medicine*, 12; Flinn, introduction, 58–67.

22. Christopher Hamlin, "Edwin Chadwick and the Engineers, 1842–1854," *Technology and Culture* 33 (Oct. 1992): 680.

23. S. E. Finer, *The Life and Times of Sir Edwin Chadwick* (London: Methuen, 1952), 226–29; Read, *Healthy Cities*, 1012; Armytage, *Social History of Engineering*, 140–41; R. A. Lewis, *Edwin Chadwick and the Public Health Movement, 1832–1854* (London: Longmans, 1952), 33, 52–53, 58–59, 105; Brundage, *England's "Prussian Minister,"* 81–82.

24. Quoted in Finer, *Life and Times*, 222.

25. Christopher Hamlin, "Muddling in Bumbledom," *Victorian Studies* 32 (Autumn 1988): 59–60, 78–83; Flinn, *Public Health Reform*, 31–32; Flinn, introduction, 1; Frazer, *History of English Public Health*, 108, 110, 135; Albert Palmberg, *A Treatise on Public Health and Its Applications in Different European Countries* (London, 1895), 7–8; Brockington, *Health of the Community*, 35–38; Arthur J. Martin, *The Work of the Sanitary Engineer* (London: MacDonald and Evans, 1935), 5; Eaton, "Sanitary Legislation," 15–18; Brundage, *England's "Prussian Minister,"* 113–33.

26. Brundage, *England's "Prussian Minister,"* 85, 101–2.

27. Rosen, *History of Public Health*, 232; George Newman, *The Building of a Nation's Health* (London: Macmillan, 1939), 24; Armytage, *Social History of Engineering*, 142, 244; Brockington, *Health of the Community*, 39–40; Martin, *Work of the Sanitary Engineer*, 6–8.

28. Hamlin, "Edwin Chadwick," 682, 695. See also Christopher Hamlin, *Public Health and Social Justice in the Age of Chadwick* (New York: Cambridge University Press, 1998).

29. Read, *Healthy Cities*, 57–60; Flinn, *Public Health Reform*, 11; John E. J. Sykes, *Public Health Problems* (London, 1892), 291; Palmberg, *Treatise*, 116–17, 209; John V. Pickstone, "Dearth, Dirt and Fever Epidemics," in *Epidemics and Ideas*, ed. Terrence Ranger and Paul Slack (Cambridge: Cambridge University Press, 1992), 137.

30. Flinn, *Public Health Reform*, 17–18.

31. Merdinger, "Civil Engineering through the Ages," 22, 98.

32. Hamlin, "Edwin Chadwick," 683–84.

33. Terry S. Reynolds, ed., *The Engineer in America* (Chicago: University of Chicago Press, 1991), 7–9; Merdinger, "Civil Engineering through the Ages," 3–4, 18–19; Martin, *Work of the Sanitary Engineer*, 24–26.

34. Hamlin, "Edwin Chadwick," 681–706.

35. In 1842, Hamburg became the first city in Germany to introduce a well-designed sewerage system, while the first modern main sewer was built in Paris in 1851. See William Paul Gerhard, *Sanitation and Sanitary Engineering* (New York: author, 1909), 100.

36. George W. Fuller and James R. McClintock, *Solving Sewage Problems* (New York: McGraw-Hill, 1926), 3, 22–23; Hamlin, "Providence and Putrefaction," 393; Gerhard, *Sanitation and Sanitary Engineering*, 100.

37. Flinn, *Public Health Reform*, 40–41, 44.

38. Kirby and Laurson, *Early Years of Modern Civil Engineering*, 231; Hommon, "Brief History," 161; Leonard Metcalf and Harrison P. Eddy, *American Sewerage Practice*, vol. 1 (New York: McGraw-Hill, 1914), 5, 10.

39. Read, *Healthy Cities*, 15, 20.

40. Armytage, *Social History of Engineering*, 141; L.T.C. Rolt, *Victorian Engineering* (London: Allen Lane, 1970), 143; Flinn, *Public Health Reform*, 37–40; Welch, "History of Sanitation," 43, 45; Harold Farnsworth Gray, "Sewerage in Ancient and Medieval Times," *Sewage Works Journal* 12 (Sept. 1940): 945.

41. S. H. Adams, *Modern Sewage Disposal and Hygienics* (London: E. and F. N. Spon, 1930), 52.

42. Nicholas Goddard, "Nineteenth-Century Recycling," *History Today* 31 (June 1981): 36.

43. Christopher Hamlin, "William Didbin and the Idea of Biological Sewage Treatment," *Technology and Culture* 29 (Apr. 1988): 191–92; Hamlin, "Providence and Putrefaction," 393–94.

44. Hamlin, "William Didbin," 189–218; Read, *Healthy Cities*, 26–33.

45. Cosgrove, *History of Sanitation*, 113.

46. Luckin, *Pollution and Control*, 49, 141–43.

47. Metcalf and Eddy, *American Sewerage Practice*, vol. 1, 1–2; Kirby and Laurson, *Early Years*, 235; T.H.P. Veal, *The Disposal of Sewage* (London: Chapman and Hall, 1956), 2–4, 16–17; Frazer, *History of English Public Health*, 225.

48. Elizabeth Porter, *Water Management in England and Wales* (Cambridge: Cambridge University Press, 1978), 26; F.T.K. Pentelow, *River Purification* (London: Edward Arnold, 1953), 9; Fuller and McClintock, *Solving Sewage Problems*, 29–30; Skeat, *Manual of British Water Engineering Practices*, 9; "London Water Supply," 82–83; Clement Higgins, *A Treatise on the Law Relating to the Pollution and Obstruction of Watercourses* (London, 1877), 1–2; Julius W. Adams, *Sewers*

and Drains for Populous Districts (New York, 1880), 39; W. Santo Crimp, *Sewage Disposal Works* (London, 1894), 7–30.

49. Dickinson, *Water Supply of Greater London*, 106.

50. Stuart Galishoff, "Triumph and Failure," in Melosi, *Pollution and Reform*, 38.

51. Baker, "Sketch of the History of Water Treatment," 905.

52. Luckin, *Pollution and Control*, 35–37, 41, 45, 48. For a careful study of water analysis, see Christopher Hamlin, *A Science of Impurity* (Berkeley: University of California Press, 1990).

53. Hassan, "Growth and Impact of the British Water Industry," 543.

Chapter 3. The "Sanitary Idea" Crosses the Atlantic

1. Miller and Melvin, *Urbanization of Modern America*, 31–32, 48–49, 57; McKelvey, *American Urbanization*, 26; Goldfield and Brownell, *Urban America*, 104–5.

2. Miller and Melvin, *Urbanization of Modern America*, 32.

3. Griffith and Adrian, *History of American City Government*, 19.

4. Charles E. Rosenberg, *The Cholera Years* (1962; reprint, Chicago: University of Chicago Press, 1987), 228.

5. Ibid., 7, 37–38, 55–62, 135–37.

6. Howard N. Rabinowitz, *Race Relations in the Urban South, 1865–1890* (Urbana: University of Illinois Press, 1980), 114–21.

7. David R. Goldfield, *Urban Growth in the Age of Sectionalism* (Baton Rouge: Louisiana State University Press, 1977), 152–53, 160; William H. Pease and Jane H. Pease, *The Web of Progress* (New York: Oxford University Press, 1985), 90, 93, 99.

8. Khaled J. Bloom, *The Mississippi Valley's Great Yellow Fever Epidemic of 1878* (Baton Rouge: Louisiana State University Press, 1993), 10–11.

9. Winslow, *Conquest of Epidemic Disease*, 266.

10. Howard D. Kramer, "The Germ Theory and the Public Health Program in the United States," *Bulletin of the History of Medicine* 22 (May/June 1948): 234–35.

11. J. K. Crellin, "The Dawn of the Germ Theory: Particles, Infection and Biology," in *Medicine and Science in the 1860s*, ed. F.N.L. Poynter (London: Wellcome Institute of the History of Medicine, 1968), 57–67, 71–74; J. K. Crellin, "Airborne Particles and the Germ Theory: 1860–1880," *Annals of Science* 22 (Mar. 1966): 49, 52, 56–57; Mazyck P. Ravenel, ed., *A Half Century of Public Health* (New York: APHA, 1921), 66–67; Erwin H. Ackernecht, "Anticontagionism between 1821 and 1867," *Bulletin of the History of Medicine* 22 (Sept./Oct. 1948): 567–68, 575–78, 580–82, 587–89.

12. Howard D. Kramer, "The Beginnings of the Public Health Movement in the United States," *Bulletin of the History of Medicine* 21 (May/June 1947): 354.

13. Schultz, *Constructing Urban Culture*, 132–33.

14. C.-E. A. Winslow, introduction to Lemuel Shattuck, *Report of the Sanitary Commission of Massachusetts, 1850,* (Cambridge, Mass.: Harvard University Press, 1948), 237; Duffy, *Sanitarians*, 96–97, 137; Ellis, *Yellow Fever and Public Health*, 7.

15. Charles E. Rosenberg, *Explaining Epidemics and Other Studies in the History of Medicine* (New York: Cambridge University Press, 1992), 126–27.

16. Schultz, *Constructing Urban Culture*, 137.

17. Barbara Gutmann Rosenkrantz, *Public Health and the State* (Cambridge, Mass.: Harvard University Press, 1972), 16, 19–20.

18. Winslow, introduction, vi–vii; Hanlon, *Principles of Public Health Administration*, 49–50; Schultz, *Constructing Urban Culture*, 131–32; Rosenkrantz, *Public Health and the State*, 14–23.

19. Kramer, "Beginnings of the Public Health Movement," 362; Hugo Muench, "Lemuel Shattauck—Still a Prophet: The Vitality of Vital Statistics," *AJPH* 39 (Feb. 1949): 152; Rosenkrantz, *Public Health and the State*, 31; Shattuck, *Report of the Sanitary Commission of Massachusetts*, 301–2.

20. Shattuck, *Report of the Sanitary Commission of Massachusetts*, vii–ix, 109; Blake, "Origins of Public Health," 1539; Hanlon, *Principles of Public Health Administration*, 50–51; Kilbourne and Smillie, *Human Ecology and Public Health*, 115; Abel Wolman, "Lemuel Shattuck—Still a Prophet: Sanitation of Yesterday—But What of Tomorrow?" *AJPH* 39 (Feb. 1949): 145; C.-E. A. Winslow, "Lemuel Shattuck—Still a Prophet: The Message of Lemuel Shattuck for 1948," *AJPH* 39 (Fall 1949): 158; Rosen, *History of Public Health*, 241–43.

21. Hanlon, *Principles of Public Health Administration*, 29, 51; Rosenkrantz, *Public Health and the State*, 36.

22. Duffy, *Sanitarians*, 99.

23. Kramer, "Beginnings of the Public Health Movement," 362–63.

24. Ibid., 370–71; John Duffy, "The American Medical Profession and Public Health," *Bulletin of the History of Medicine* 53 (Spring 1979): 2; Duffy, *Sanitarians*, 102–8; Henry I. Bowditch, *Public Hygiene in America* (Boston, 1877), 35; Schultz, *Constructing Urban Culture*, 144–46.

25. Harold M. Hyman, *A More Perfect Union* (New York: Knopf, 1973), 320–22, 331–36.

26. Citizens' Association of New York, *Report of the Council of Hygiene and Public Health of the Citizens' Association of New York, upon the Sanitary Condition of the City* (New York, 1866), cxlii–cxliii.

27. Ellis, *Yellow Fever and Public Health*, 9–12, 35–36.

28. Rosen, *History of Public Health*, 244–48; Gert H. Brieger, "Sanitary Reform in New York City," in *Sickness and Health in America*, ed. Judith Walzer Leavitt and Ronald L. Numbers (Madison: University of Wisconsin Press, 1985), 408; Blake, "Origins of Public Health," 1540; Duffy, "American Medical Profession," 3; Duffy, *Sanitarians*, 120; Schultz, *Constructing Urban Culture*, 121.

29. Rosenkrantz, *Public Health and the State*, 1, 37–73.

30. Hanlon, *Principles of Public Health Administration*, 49.

31. Sam Bass Warner Jr., "Public Health Reform and the Depression of 1873–1878," *Bulletin of the History of Medicine* 29 (Nov./Dec. 1955): 512–13; Kilbourne and Smillie, *Human Ecology and Public Health*, 115; Duffy, *Sanitarians*, 130–32.

32. See Ellis, *Yellow Fever and Public Health*, for a thorough treatment of the subject.

33. Warner, "Public Health Reform," 504, 515–16.

34. Kramer, "Germ Theory," 235–36.

35. Terry S. Reynolds, "The Engineer in Nineteenth-Century America," in Reynolds, *Engineer in America*, 15–17.

36. Ibid., 19–20, 23–24; Melosi, *Garbage in the Cities*, 78–79; Anderson, "Diffusion of Technology," 30–33.

37. Reynolds, "Engineer in Nineteenth-Century America," 25.

38. Joel A. Tarr, "Bringing Technology to the Cities," unpublished paper, 5.

39. Stuart Galishoff, *Newark* (New Brunswick, N.J.: Rutgers University Press, 1988), 16–17, 66–67; David R. Goldfield, "The Business of Health Planning," *Journal of Southern History* 42 (Nov. 1978): 557–70; Harriet E. Amos, *Cotton City* (Tuscaloosa: University of Alabama Press, 1985), 136–37.

40. See Alan I. Marcus, *Plague of Strangers* (Columbus: Ohio State University Press, 1991).

41. Kenneth Fox, *Better City Government* (Philadelphia: Temple University Press, 1977), 5–16; Griffith and Adrian, *History of American City Government*, 91–92, 133–38; Maury Klein and Harvey A. Kantor, *Prisoners of Progress* (New York: Macmillan, 1976), 338–64.

42. Chudacoff and Smith, *Evolution of American Urban Society*, 165–66.

43. Philip J. Ethington, *The Public City* (New York: Cambridge University Press, 1994), 24, 26–27; M. Craig Brown and Charles N. Halaby, "Machine Politics in America, 1870–1945," *Journal of Interdisciplinary History* 7 (Winter 1987): 587–88, 609–11; M. Craig Brown, "Bosses, Reform, and the Socioeconomic Bases of Urban Expenditure, 1890–1940," in *The Politics of Urban Fiscal Policy*, ed. Terrence J. McDonald and Sally K. Ward (Beverly Hills, Calif.: Sage, 1984), 69–70.

Chapter 4. Pure and Plentiful

1. Waterman, *Elements of Water Supply Engineering*, 6.

2. Anderson, "Diffusion of Technology," 102–4, 117; Anderson, "Hard Choices," 218; Tarr, "Evolution of the Urban Infrastructure," 30–31.

3. Griffith and Adrian, *History of American City Government*, 198–217.

4. Goldfield and Brownell, *Urban America*, 151–52.

5. McMahon, "Makeshift Technology," 30–33.

6. Alexander B. Callow Jr., *The Tweed Ring* (New York: Oxford University Press, 1965), 195.

7. M. N. Baker, "Public and Private Ownership of Water-Works," *Outlook*, May 7, 1898, 79.

8. Anderson, "Diffusion of Technology," 108.

9. Blake, *Water for the Cities*, 44–62, 101–20; LaNier, "Municipal Water Systems," 174–75; Myers, "History of Public Franchises," 85–87.

10. Blake, *Water for the Cities*, 172–98; LaNier, "Municipal Water Systems," 174; Blake, "Lemuel Shattuck," 554–62.

11. Nesson, *Great Waters*, 6–12.

12. Blake, *Water for the Cities*, 219.

13. O'Connell, "Chicago's Quest for Pure Water," 1–3; W. W. DeBerard, "Expansion of the Chicago, Ill., Water Supply," *Transactions of the ASCE CT* (1953): 588–93; LaNier, "Municipal Water Systems," 176.

14. Wade, *Urban Frontier*, 297; LaNier, "Municipal Water Systems," 176; Black, "Construction and Reconstruction of Compton Hill Reservoir," 4–8.

15. George C. Andrews, "The Buffalo Water Works," *JAWWA* 17 (Mar. 1927): 280; "History of the Buffalo Water Works," *Engineering Record* 38 (Sept. 24, 1898): 363–64.

16. Bruce Jordan, "Origins of the Milwaukee Water Works," *Milwaukee History* 9 (Spring 1986): 2–5; Elmer W. Becker, *A Century of Milwaukee Water* (Milwaukee: Milwaukee Water Works, 1974), 1–3.

17. John Ellis and Stuart Galishoff, "Atlanta's Water Supply, 1865–1918," *Maryland Historian* 8 (Spring 1977): 6–7; Ellis, *Yellow Fever and Public Health*, 29, 142; William Wright Sorrels, *Memphis' Greatest Debate* (Memphis, Tenn.: Memphis State University Press, 1970), 15–24.

18. Louis P. Cain, *Sanitation Strategy for a Lakefront Metropolis* (De Kalb: Northern Illinois University Press, 1978), 37–51; DeBerard, "Expansion of the Chicago, Ill., Water Supply," 593–97; Frank J. Piehl, "Chicago's Early Fight to 'Save Our Lake,'" *Chicago History* 5 (Winter 1976–77): 223–24; Samuel N. Karrick, "Protecting Chicago's Water Supply," *Civil Engineering* 9 (Sept. 1939): 547–48; John Ericson, *The Water Supply System of Chicago* (Chicago: Barnard and Miller, 1924), 11–13.

19. Larry D. Lankton, "1842," *Civil Engineering* 47 (Oct. 1977): 93–94.

20. Ibid., 95–96; Galishoff, "Triumph and Failure," 36.

21. Lankton, "1842," 90.

22. Eugene P. Moehring, *Public Works and the Patterns of Urban Real Estate Growth in Manhattan, 1835–1894* (New York: Arno Press, 1981), 31–32, 44–45, 47, 50.

23. Blake, *Water for the Cities*, 199–218; Nesson, *Great Waters*, 11–12; William R. Hutton, "The Washington Aqueduct, 1853–1898," *Engineering Record* 40 (July 29, 1899): 190–93.

24. Cited in Blake, "Origins of Public Health," 1541.

25. Galishoff, "Triumph and Failure," 37–38; Michael P. McCarthy, *Typhoid and the Politics of Public Health in Nineteenth-Century Philadelphia* (Philadelphia: American Philosophical Society, 1987), 1.

26. William P. Mason, *Water-Supply* (New York: John Wiley and Sons, 1897), 466.

27. "Community Water Supply," 236; Baker, "Sketch of the History of Water Treatment," 906–8; George E. Symons, "History of Water Supply 1850 to Present," *Water and Sewage Works* 100 (May 1953): 191; Baker, *Quest for Pure Water*, 127.

28. Baker, "Sketch of the History of Water Treatment," 908–10; Eddy, "Water Purification," 83; Baker, *Quest for Pure Water*, 133, 135.

29. George C. Whipple, "Fifty Years of Water Purification," in Ravenel, *Half Century of Public Health*, 163; Baker, *Quest for Pure Water*, 148.

30. J. Leland FitzGerald, "Comparison of Water Supply Systems from a Financial Point of View," *Transactions of the ASCE* 24 (Apr. 1891): 252–56.

31. Armstrong et al., *History of Public Works*, 232–33; Stearns, "Development of Water Supplies," 455; Turneaure and Russell, *Public Water-Supplies* (1911), 7–8; Goubert, *Conquest of Water*, 56–58; Anderson, "Diffusion of Technology," 10–14; Allen Hazen, "Public Water Supplies," *Engineering News-Record* 92 (Apr. 17,

1924): 696; John W. Alvord, "Recent Progress and Tendencies in Municipal Water Supply in the United States," *JAWWA* 4 (Sept. 1917): 291–92.

32. Warner, *Urban Wilderness*, 202.

33. Wade, *Urban Frontier*, 294–95; O'Connell, "Chicago's Quest for Pure Water," 3; Tarr, "Evolution of the Urban Infrastructure," 14; Tarr et al., "Development and Impact," 60.

Chapter 5. Subterranean Networks

1. Gerhard, *Sanitation and Sanitary Engineering*, 99.

2. Warner, "Public Health Reform," 507.

3. Edward K. Spann, *The New Metropolis* (New York: Columbia University Press, 1981), 133.

4. Metcalf and Eddy, *American Sewerage Practice*, vol. 1, 15.

5. Joel A. Tarr and Francis Clay McMichael, "Decisions about Wastewater Technology: 1850–1932," *Journal of the Water Resources Planning and Management Division* (May 1977): 48–51; Joel A. Tarr and Francis Clay McMichael, "Historic Turning Points in Municipal Water Supply and Wastewater Disposal, 1850–1932," *Civil Engineering—ASCE* 47 (Oct. 1977): 82–83.

6. See Tarr and McMichael, "Historic Turning Points," 83.

7. J. B. White, *The Design of Sewers and Sewage Treatment Works* (London: Edward Arnold, 1970), 3.

8. Joel A. Tarr, "The Separate vs. Combined Sewer Problem," *Journal of Urban History* 5 (May 1979): 312.

9. Tarr and McMichael, "Decisions about Wastewater Technology," 52,

10. Metcalf and Eddy, *American Sewerage Practice*, vol. 1, 24.

11. Jon A. Peterson, "The Impact of Sanitary Reform upon American Urban Planning, 1840–1890," *Journal of Social History* 13 (Fall 1979): 88; Gerhard, *Sanitation and Sanitary Engineering*, 102–3; Harrison P. Eddy, "Sewerage and Sewage Disposal," *Engineering News-Record* 92 (Apr. 17, 1924): 693; Metcalf and Eddy, *American Sewerage Practice*, vol. 1, 21; Charles Gilman Hyde, "A Review of Progress in Sewage Treatment during the Past Fifty Years in the United States," in *Modern Sewage Disposal*, ed. Langdon Pearse (New York: Federation of Sewage and Industrial Waste Associations, 1938), 1.

12. Brooklyn, Board of Water Commissioners, *The Brooklyn Water Works and Sewers, a Descriptive Memoir* (New York, 1867), 71–72.

13. Ibid.

14. Morris M. Cohn, *Sewers for Growing America* (Ambler, Pa.: Certain-teed, 1966), 46–47.

15. City of Boston, *Sewage of Boston* (1876), 1–2.

16. Ibid., 4, 7.

17. Eliot C. Clarke, *Main Drainage Works of the City of Boston* (Boston, 1885), 12–15. See William A. Newman and Wilfred E. Holton, *Boston's Back Bay* (Boston: Northeastern University Press, 2006).

18. Metcalf and Eddy, *American Sewerage Practice*, vol. 1, 15; Eddy, "Sewerage and Sewage Disposal," 693; Clarke, *Main Drainage Works*, 16; Warner, "Public Health Reform," 508.

19. Robin L. Einhorn, *Property Rules* (Chicago: University of Chicago Press, 1991), 137–38.

20. Chicago, Bureau of Engineering, Department of Public Works, *A Century of Progress in Water Works, 1833–1933* (Chicago: Department of Public Works, 1933), 7; Louis P. Cain, "Raising and Watering a City: Ellis Sylvester Chesbrough and Chicago's First Sanitation System," *Technology and Culture* 13 (July 1972): 354–56.

21. Cain, "Raising and Watering a City," 353–54, 356.

22. Before 1871, the flow of the Chicago River was reversed to prevent waste from entering the water supply, but abnormally heavy rains in 1879 forced the city to try to augment its pollution-control problem with the development of a canal. See Louis P. Cain, "The Creation of Chicago's Sanitary District and Construction of the Sanitary and Ship Canal," *Chicago History* 8 (Summer 1979): 98–110; Cain, "Raising and Watering a City," 358–59; Karrick, "Protecting Chicago's Water Supply," 547–48.

23. George C. D. Lenth, "The Chicago Sewer System," *Journal of the Western Society of Engineers* 28 (Apr. 1923): 103; C. D. Hill, "The Sewage Disposal Problem in Chicago," *AJPH* 8 (Nov. 1918): 834; Einhorn, *Property Rules*, 139.

24. Piehl, "Chicago's Early Fight," 225.

25. Cain, "Raising and Watering a City," 360–64.

26. Ibid., 365.

27. Einhorn, *Property Rules*, 139–40.

28. Joanne Abel Goldman, *Building New York's Sewers* (West Lafayette, Ind.: Purdue University Press, 1997).

29. See Ellis, *Yellow Fever and Public Health*, 142; Rabinowitz, *Race Relations*, 122–23.

30. Moehring, *Public Works*, 317–22.

31. Metcalf and Eddy, *American Sewerage Practice*, vol. 1, 29; Leonard Metcalf and Harrison P. Eddy, *American Sewerage Practice*, vol. 2 (New York: McGraw-Hill, 1915), 8; Eddy, "Sewerage and Sewage Disposal," 693; Hyde, "Review of Progress in Sewage Treatment," 1–2; C. D. Hill, "The Sewerage System of Chicago," *Journal of the Western Society of Engineers* 16 (Sept. 1911): 566.

Chapter 6. On the Cusp of the New Public Health

1. Tarr and Dupuy, *Technology and the Rise of the Networked City*, xiv–xvi.

2. McKelvey, *American Urbanization*, 73, 104.

3. Ernest S. Griffith, *A History of American City Government: The Conspicuous Failure, 1870–1900* (Washington, D.C.: University Press of America, 1974), 152; Miller and Melvin, *Urbanization of Modern America*, 72, 79; Chudacoff and Smith, *Evolution of American Urban Society*, 90.

4. Chudacoff and Smith, *Evolution of American Urban Society*, 105–7; Ernest S. Griffith, *A History of American City Government: The Progressive Years and Their Aftermath, 1900–1920* (1974; reprint, Washington, D.C.: University Press of America, 1983), 5; Goldfield and Brownell, *Urban America*, 180.

5. Kenneth Finegold, *Experts and Politicians* (Princeton, N.J.: Princeton University Press, 1995), 3, 11–12.

6. Jon C. Teaford, *The Unheralded Triumph* (Baltimore: Johns Hopkins University Press, 1984), 30.

7. Finegold, *Experts and Politicians*, 13–22.

8. Morton Keller, *Regulating a New Society* (Cambridge, Mass.: Harvard University Press, 1994), 4.

9. Ibid., 190–91; Goldfield and Brownell, *Urban America*, 230; William D. Miller, *Memphis during the Progressive Era, 1900–1917* (Memphis: Memphis State University Press, 1957), 113.

10. Martin V. Melosi, "Battling Pollution in the Progressive Era," *Landscape* 26 (1982): 36–37.

11. Henry W. Webber, "Civic Pride in New York City," *Forum* 53 (June 1915): 731.

12. M. Christine Boyer, *Dreaming the Rational City* (Cambridge, Mass.: MIT Press, 1983), 17.

13. Teaford, *Unheralded Triumph*, 7.

14. Finegold, *Experts and Politicians*, 15.

15. Griffith, *Conspicuous Failure*, 215; Griffith, *Progressive Years*, 124; Charles N. Glaab and A. Theodore Brown, *A History of Urban America*, 2nd ed. (New York: Macmillan, 1976), 174–76.

16. Teaford, *Unheralded Triumph*, 105, 122; Griffith, *Progressive Years*, 124–25, 128.

17. Alan D. Anderson, *The Origin and Resolution of the Urban Crisis* (Baltimore: Johns Hopkins University Press, 1977), 9, 13.

18. Griffith, *Conspicuous Failure*, 163.

19. Anderson, *Origin and Resolution*, 10; Griffith, *Progressive Years*, 171–76.

20. Griffith, *Progressive Years*, 189.

21. See Nancy Tomes, *The Gospel of Germs* (Cambridge, Mass.: Harvard University Press, 1998).

22. Crellin, "Dawn of the Germ Theory," 57–74.

23. Crellin, "Airborne Particles," 49, 52, 56–57, 59–60.

24. Hutt and Thompson, *Principles and Practices of Preventive Medicine*, 13–14; Carl E. McCombs, *City Health Administration* (New York: Macmillan, 1927), 6; Kramer, "Germ Theory," 234, 241; Ravenel, *Half Century of Public Health*, 69–71, 78; Duffy, *Sanitarians*, 193.

25. Kilbourne and Smillie, *Human Ecology and Public Health*, 116; Nancy Tomes, "The Private Side of Public Health," *Bulletin of the History of Medicine* 64 (Winter 1990): 509–39; Hibbert Winslow Hill, *The New Public Health* (1916; reprint, New York: Arno, 1977), 8–13.

26. Ernest McCullough, *Engineering Work in Towns and Small Cities* (Chicago: Technical Book Agency, 1906), 62.

27. Charles V. Chapin, "Sanitation in Providence," *Proceedings of the Providence Conference for Good Government and the 13th Annual Meeting of the National Municipal League* (Nov. 19–22, 1907), 325–26.

28. George C. Whipple, "Sanitation—Its Relation to Health and Life," *Transactions of the ASCE* 88 (1925): 94.

29. Ibid., 95.

30. George M. Price, *Handbook on Sanitation*, 2nd ed. (New York: John Wiley and Sons, 1905), 241; Barbara Gutmann Rosenkrantz, "Cart before Horse," *Journal of the History of Medicine and Allied Sciences* 29 (Jan. 1974): 62; Milton Terris, "Evolution of Public Health and Preventive Medicine in the United States," *AJPH* 65 (Feb. 1975): 165; Duffy, "American Medical Profession," 7, 16; Duffy, *Sanitarians*, 196–97; James G. Burrow, *Organized Medicine in the Progressive Era* (Baltimore: Johns Hopkins University Press, 1977), 88–89.

31. Blake, "Origins of Public Health," 1547–48; Duffy, *Sanitarians*, 205.

32. See Galishoff, *Newark*, 103; John Duffy, *A History of Public Health in New York City, 1866–1966* (New York: Russell Sage Foundation, 1974), 191; Carl V. Harris, *Political Power in Birmingham, 1871–1921* (Knoxville: University of Tennessee Press, 1977), 155–56, 238–39; Judith Walzer Leavitt, *The Healthiest City* (Princeton, N.J.: Princeton University Press, 1982), 35, 263; Joy J. Jackson, *New Orleans in the Gilded Age* (Baton Rouge: Louisiana State University Press, 1969), 183; Don H. Doyle, *Nashville in the New South, 1880–1930* (Knoxville: University of Tennessee Press, 1985), 86; Don H. Doyle, *New Men, New Cities, New South* (Chapel Hill: University of North Carolina Press, 1990), 280–81.

33. Kilbourne and Smillie, *Human Ecology and Public Health*, 115; Blake, "Origins of Public Health," 1545–47; Duffy, *Sanitarians*, 194, 201–2.

34. Duffy, *Sanitarians*, 239; Hanlon, *Principles of Public Health Administration*, 53–54; Manfred Waserman, "The Quest for a National Health Department in the Progressive Era," *Bulletin of the History of Medicine* 49 (Fall 1975): 355–57; Burrow, *Organized Medicine in the Progressive Era*, 100–102; Kilbourne and Smillie, *Human Ecology and Public Health*, 18; Ralph Chester Williams, *The United States Public Health Service, 1798–1950* (Washington, D.C.: Commissioned Officers Association of the USPHS, 1951), 441–42; Alfred Crosby, *America's Forgotten Pandemic* (1989; reprint, Cambridge: Cambridge University Press, 2003).

35. Reynolds, *Engineer in America*, 25–26, 169, 173–74, 178; David Noble, *America by Design* (New York: Knopf, 1977), 38–39; Schultz, *Constructing Urban Culture*, 187–89; Goldfield and Brownell, *Urban America*, 245.

36. Noble, *America by Design*, 44.

37. Edwin T. Layton Jr., *The Revolt of the Engineers* (1971; reprint, Baltimore: Johns Hopkins University Press, 1986), vii.

38. Stanley K. Schultz and Clay McShane, "To Engineer the Metropolis" *Journal of American History* 65 (Sept. 1978): 389–411.

39. "The Sanitary Engineer," *Scientific American*, Feb. 16, 1918, 142.

40. "The Sanitary Engineer—A New Social Profession," *Charities and the Commons (The Survey)* 16 (June 2, 1906): 286.

41. Gerhard, *Sanitation and Sanitary Engineering*, 58–59.

42. Melosi, *Garbage in the Cities*, 73–79; Pratt, "Industrial Need of Technically Trained Men," 150; Whipple, "Training of Sanitary Engineers," 803; W. B. Bizzell, "Sanitary Engineering as a Career," *AJPH* 15 (June 1925): 510; Abel Wolman, "Contributions of Engineering to Health Advancement," *Transactions of the ASCE CT* (1953): 583–84.

43. Mansfield Merriman, *Elements of Sanitary Engineering*, 2nd ed. (New York, 1899), 7.

Chapter 7. Water Supply as a Municipal Enterprise

1. Waterman, *Elements of Water Supply Engineering*, 6.

2. Tarr, "Evolution of the Urban Infrastructure," 14.

3. Hill, "Cincinnati Water Works," 56–60; Baldwin, "Development of Cincinnati's Water Supply," 321–23; C. R. Hebble, "A Few Interesting Things about the Cincinnati Water-Works," *American City* 2 (May 1915): 381–83; "Progress on the New Water-Works for Cincinnati, O.," *Engineering News* 40 (Dec. 8, 1898): 354; DeBerard, "Expansion of the Chicago, Ill., Water Supply," 593–97; Morris Knowles et al., "Lawrence Water Supply—Investigations and Construction," *Journal of the New England Water Works Association* 39 (Dec. 1925): 346–55; Terry S. Reynolds, "Cisterns and Fires: Shreveport, Louisiana, as a Case Study of the Emergence of Public Water Supply Systems in the South," *Louisiana History* 22 (Fall 1981): 348–66.

4. Tarr, "Evolution of the Urban Infrastructure," 26; Ellis and Galishoff, "Atlanta's Water Supply," 5–22.

5. Cornelius C. Vermeule, "New Jersey's Experience with State Regulation of Public Water Supplies," *American City* 16 (June 1917): 602.

6. Griffith, *Progressive Years*, 86–87; Maureen Ogle, "Redefining 'Public' Water Supplies, 1870–1890," *Annals of Iowa* 50 (Spring 1990): 507–30; Gregg R. Hennessey, "The Politics of Water in San Diego, 1895–1897," *Journal of San Diego History* 24 (Summer 1978): 367–83.

7. Committee on Municipal Administration, "Evolution of the City," *Municipal Affairs* 2 (Sept. 1898): 726–27; Griffith, *Conspicuous Failure*, 180; Anderson, "Diffusion of Technology," 106.

8. *The Manual of American Water-Works* (New York, 1897), f–g.

9. Anderson, "Diffusion of Technology," 106, 108, 112; Tarr, "Evolution of the Urban Infrastructure," 26, 30; Baker, "Public and Private Ownership," 78.

10. Anderson, "Diffusion of Technology" 122; Henry C. Hodgkins, "Franchises of Public Utilities as They Were and as They Are," *JAWWA* 2 (Dec. 1915): 743.

11. Anderson, "Diffusion of Technology," 115, 119, 121.

12. Myers, "History of Public Franchises," 88–91.

13. Todd A. Shallat, "Fresno's Water Rivalry," *Essays in Public Works History* 8 (1979): 9–13.

14. A. S. Baldwin, "Shall San Francisco Municipalize Its Water Supply?" *Municipal Affairs* (June 1900): 317–28; Clyde Arbuckle, *History of San Jose* (San Jose, Calif.: Memorabilia of San Jose, 1986), 301, 486–87, 501–9.

15. "The Recent History of Municipal Ownership in the United States," *Municipal Affairs* 6 (Winter 1902–3): 524, 529; Anderson, "Diffusion of Technology," 122–23; M. N. Baker, "Municipal Ownership and Operation of Water Works," *Annals of the American Academy* 57 (Jan. 1915): 281.

16. Anderson, "Diffusion of Technology," 123.

17. Baker, "Public and Private Ownership," 78.

18. Olivier Zunz, *The Changing Face of Inequality* (Chicago: University of Chicago Press, 1982), 118–19.

19. "The Relation of the Municipality to the Water Supply," *Annals of the American Academy* 30 (Nov. 1907): 557–92.

20. John Thomson, "A Memoir on Water Meters," *Transactions of the ASCE* 25 (July 1891): 40–65; Armstrong et al., *History of Public Works,* 234–35; Hazen, "Public Water Supplies," 697.

21. W. J. Chellew, "How Meters Promote Equity and Economy in the Distribution of Water," *American City* 6 (Apr. 1912): 665; Morris Knowles, "Equitable Water Rates the Result of Metering," *American City* 8 (Feb. 1913): 172.

22. "The Water-Supply of Cities," *North American Review* 136 (Apr. 1883): 373.

23. "Why Meter?" 522; C. J. Renner, "The Experience of a Small City with Water Meters and Water Rates," *American City* 11 (Dec. 1914): 474; Thomas H. Hooper, "Should Meters Be Owned and Controlled by the Municipality?" *American City* 20 (Feb. 1919): 183.

24. "Consumption of Water and Use of Meters," 62–63; "Water-Supply Statistics of Metered Cities," *American City* 23 (Dec. 1920): 614–20, and 24 (Jan. 1921): 42–49.

25. Armstrong et al., *History of Public Works,* 222–24; Edward Wegmann, *The Water-Supply of the City of New York, 1658–1895* (New York, 1896), 90; M. N. Baker, "Water Supply of Greater New York," *Municipal Affairs* 4 (Sept. 1900): 486–505; "The New Water Supply for New York City," *Scientific American,* Mar. 24, 1906, 250; William W. Brush, "City Aqueduct to Deliver Catskill Water Supply to the Five Boroughs of Greater New York," *Proceedings of the Brooklyn Engineers' Club* 95 (Jan. 1911): 76–114; Percey C. Barney, "Catskill Water System of New York City," *Proceedings of the Brooklyn Engineers' Club* 94 (Jan. 1911): 51–75.

26. William L. Kahrl, *Water and Power* (Berkeley: University of California Press, 1982); Abraham Hoffman, *Vision or Villainy* (College Station: Texas A&M University Press, 1981); Catherine Mulholland, *William Mulholland and the Rise of Los Angeles* (Berkeley: University of California Press, 2000).

27. Remi Nadeau, "The Water War," *American Heritage* 13 (Dec. 1961): 31–35, 103–7; William L. Kahrl, "The Politics of California Water," *California Historical Quarterly* 55 (Spring/Summer 1976): 98–119.

28. Nesson, *Great Waters,* 76–77; J. K. Finch, "A Hundred Years of American Civil Engineering, 1852–1952," *Transactions of the ASCE CT* (1952): 93; Willis J. Milner, "Some Difficulties in Obtaining a Water Supply," *City Government* 6 (May 1899): 105; "Mount Ayr, Iowa, Water Supply," *American Municipalities* 29 (Apr. 1915): 20–21.

29. "Water-Supply Statistics of Metered Cities," (Dec. 1920): 614–20, and (Jan. 1921): 41–49; Perry Hopkins, "Origin and Growth of Public Water-Supply: Part II," *American City* 36 (Jan. 1927): 51.

30. Alvord, "Recent Progress and Tendencies," 291; F. A. Barbour, "Leakage from Pipe Joints," *American City* 15 (Dec. 1916): 660–62; "The Use and Waste of Water," *Engineering Record* 42 (Sept. 1, 1900): 196–98.

31. Waterman, *Elements of Water Supply Engineering,* 254; Anderson, "Diffusion of Technology," 150; Tarr, "Evolution of the Urban Infrastructure," 32; Hazen, "Public Water Supplies," 695–96; Alvord, "Recent Progress and Tendencies," 288–97; Griffith, *Progressive Years,* 179; Babbitt and Doland, *Water Supply Engineering,* 6; H. M. Blomquist, "The Planning of Water-Works for Fire Protec-

tion," *American City* 21 (Sept. 1919): 242; George W. Booth, "Water Distribution Systems in Relation to Fire Protection," *American City* 14 (June 1916): 593–97.

32. LaNier, "Municipal Water Systems," 178; "Present Tendencies in Water-Works Practice," *Engineering News* 37 (Apr. 15, 1897): 233; J. F. Springer, "Water Pipes of Wood," *Scientific American*, Sept. 11, 1920, 250, 262, 264; Hazen, "Public Water Supplies," 696; Alvord, "Recent Progress and Tendencies," 291–92; Armstrong et al., *History of Public Works*, 233.

33. Amory Prescott Folwell, *Water-Supply Engineering* (New York: John Wiley and Sons, 1903), 45.

34. P. H. Norcross, "Water-Works Extensions and Improvements," *American City* 23 (Aug. 1920): 209; Mary McWilliams, *Seattle Water Department History, 1854–1954* (Seattle: City of Seattle, 1955), 103.

35. "Assessing Cost of Extensions in a Municipally-Owned Water-Works Plant," *American City* 12 (June 1915): 491–92.

36. Winfred D. Hubbard and Wynkoop Kiersted, *Water-Works Management and Maintenance* (New York: John Wiley and Sons, 1907), 366–76; E. Kuichling, "The Financial Management of Water-Works," *Transactions of the ASCE* 38 (Dec. 1897): 1–40; "Assessing Cost of Extensions," 491–92.

37. R. E. McDonnell, "The Value of Pure Water Supply," *American Municipalities* 27 (June 1914): 70.

38. George C. Whipple, "Clean Water as a Municipal Asset," *American City* 4 (Apr. 1911): 162.

39. Gilbert H. Grosvenor, "The New Method of Purifying Water," *Century Magazine* 69 (Dec. 1904): 208.

40. Turneaure and Russell, *Public Water-Supplies* (1911), 122–25; Whipple, "Clean Water as a Municipal Asset," 161.

41. Turneaure and Russell, *Public Water-Supplies* (1911), 125–31.

42. George C. Whipple, "Municipal Water-Works Laboratories," *Popular Science Monthly* 58 (Dec. 1900): 172–75.

43. AWWA, *Water Quality and Treatment*, 2nd ed. (New York: AWWA, 1951), 30–31; Whipple, "Municipal Water-Works Laboratories," 174–82.

44. "Typhoid Fever and Water Supply in 66 American and Foreign Cities," *Engineering News* 35 (May 21, 1896): 336.

45. Rosenkrantz, *Public Health and the State*, 104–5.

46. Allen Hazen, *The Filtration of Public Water-Supplies* (New York: John Wiley and Sons, 1905); Allen Hazen, *Clean Water and How to Get It* (New York: John Wiley and Sons, 1914), 73–88; John W. Hill, *The Purification of Public Water Supplies* (New York: D. Van Nostrand, 1898), 255–66; George C. Whipple, "History of Water Purification," *Transactions of the ASCE* 85 (1922): 476–81; Eddy, "Water Purification," 83–84; Rudolph Hering, "Water Purification," *Journal of the Franklin Institute* 139 (Feb. 1895): 135–44, and 140 (Mar. 1895): 215–24.

47. Baker, "Sketch of the History of Water Treatment," 902–38.

48. See Symons, "History of Water Supply," 191–94.

49. Baker, "Sketch of the History of Water Treatment," 911.

50. Baker, *Quest for Pure Water*, 179; Baker, "Sketch of the History of Water Treatment," 916; Turneaure and Russell, *Public Water-Supplies* (1911), 432–34,

502–3; Fuller, "Progress in Water Purification," 1568; Hazen, "Public Water Supplies," 696.

51. Symons, "History of Water Supply," 191–94; Baker, "Sketch of the History of Water Treatment," 916–17; "Present Tendencies in Water-Works Practice," 233; Alvord, "Recent Progress and Tendencies," 284–85; Edwin O. Jordan, "The Purification of Water Supplies," *Scientific American Supplement*, June 24, 1916, 406–7.

52. Turneaure and Russell, *Public Water-Supplies* (1948), 124–26; Hazen, *Filtration of Public Water-Supplies*, 1–2; Floyd Davis, "Impure Water and Public Health," *Engineering Magazine* 2 (Dec. 1891): 362; William T. Sedgwick, "Water Supply Sanitation in the Nineteenth Century and in the Twentieth," *Journal of the New England Water Works Association* 30 (June 1916): 185–86.

53. Galishoff, "Triumph and Failure," 44; Baker, *Quest for Pure Water*, 139–40; Baker, "Sketch of the History of Water Treatment," 915; Fuller, "Progress in Water Purification," 1569, 1570.

54. Turneaure and Russell, *Public Water-Supplies* (1911), 506–11; James H. Fuertes, *Water Filtration Works* (New York: John Wiley and Sons, 1904), 246–55.

55. "Golden Decade for Philadelphia Water," 37.

56. William P. Mason, "Sanitary Problems Connected with Municipal Water Supply," *Journal of the Franklin Institute* 143 (May 1897): 354; Waterman, *Elements of Water Supply Engineering*, 38–39; McDonnell, "Value of Pure Water Supply," 70; Whipple, "Clean Water as a Municipal Asset," 161; Davis, "Impure Water and Public Health," 362; George A. Johnson, "The High Cost of Sanitary Ignorance," *American City* 14 (June 1916): 586.

57. AWWA, *Water Chlorination* (New York: AWWA, 1973), 3–4; N. J. Howard, "Twenty Years of Chlorination of Public Water-Supplies," *American City* 36 (June 1927): 791–94; Morris M. Cohn, "Chlorination of Water," *Municipal Sanitation* 2 (July 1931): 333–34; "Is the Chlorination of Water-Supplies Worth While?" *American City* 20 (June 1919): 524–25; George W. Fuller, "The Influence of Sanitary Engineering on Public Health," *AJPH* 12 (Jan. 1922): 16; Fuller, "Water-Works," 1601.

58. "Water-Supply Statistics of Metered Cities," (Dec. 1920): 613–20.

59. Symons, "History of Water Supply," 191–94; Baker, "Sketch of the History of Water Treatment," 919–21; Baker, *Quest for Pure Water*, 346–49; Fuller, "Progress in Water Purification," 1572; Alvord, "Recent Progress and Tendencies," 287–88; Jennings, "Uses and Accomplishments of Chlorine Compounds," 296–97; Turneaure and Russell, *Public Water-Supplies* (1948), 127; Francis E. Longley, "Present Status of Disinfection of Water Supplies," *JAWWA* 2 (Dec. 1915): 680–86; Joseph W. Ellms, "Disinfection of Public Water Supplies," *American City* 9 (Dec. 1913): 564–68.

60. Alvord, "Recent Progress and Tendencies," 283–84; Symons, "History of Water Supply," 191–94; Baker, "Sketch of the History of Water Treatment," 922–24; Baker, *Quest for Pure Water*, 253–64.

61. Death rates from typhoid in cities with over 100,000 in population were approximately half the rate for rural and urban populations combined in 1920. See Turneaure and Russell, *Public Water-Supplies* (1948), 132.

62. Alvord, "Recent Progress and Tendencies," 283; Turneaure and Russell, *Public Water-Supplies* (1948), 423; Waterman, *Elements of Water Supply Engineering*, 32–33.

63. Marshall O. Leighton, "Industrial Wastes and Their Sanitary Significance," *Public Health: Papers and Reports* 31 (1906): 29; "Progress Report of Committee on Industrial Wastes in Relation to Water Supply," *JAWWA* 10 (May 1923): 415; Melosi, "Hazardous Waste and Environmental Liability," 753; Joel A. Tarr, "Historical Perspectives on Hazardous Wastes in the United States," *Waste Management and Resources* 3 (1985): 91; L. F. Warrick, "Relative Importance of Industrial Wastes in Stream Pollution," *Civil Engineering* 3 (Sept. 1933): 495.

64. Biological oxygen demand (BOD) measures the oxygen used by microorganisms to decompose organic waste. If there is a large amount of organic waste in the water supply, there will also be considerable bacteria present to decompose the waste. In this case, the demand for oxygen will be high, and the BOD level will be high. As the waste is consumed or dispersed in the water, BOD levels will decline. See Bess Furman, *A Profile of the United States Public Health Service, 1798–1948* (Bethesda, Md.: National Institute of Health, 1973), 295; Joel A. Tarr, "Industrial Wastes and Public Health" *AJPH* 75 (Sept. 1985): 1060.

65. Tarr, "Industrial Wastes and Public Health," 1060.

66. W. P. Mason, "Dangers of Sanitary Neglect at the Watersheds from Which Come Supplies of City Water," *Sanitarian* 38 (May 1897): 385–93; Philip P. Micklin, "Water Quality," in *Congress and the Environment,* ed. Richard A. Cooley and Geoffrey Wandesforde-Smith (Seattle: University of Washington Press, 1970), 131; Tarr, "Industrial Wastes and Public Health," 1059–61; George W. Rafter, "Epidemic Water Pollution," *Engineering Magazine* 1 (Apr. 1891): 156–67; E. B. Besselievre, "Statutory Regulation of Stream Pollution and the Common Law," *Transactions of the American Institute of Chemical Engineers* 16 (1924): 225; Joel A. Tarr and Charles Jacobson, "Environmental Risk in Historical Perspective," in *The Social and Cultural Construction of Risk,* ed. Branden B. Johnson and Vincent T. Covello (Boston: D. Reidel, 1987), 318; X. H. Goodnough, "Some Results of the Systematic Examination of the Water of Public Water Supplies," *Journal of the New England Water Works Association* 4 (Sept. 1899): 66; Edwin B. Goodell, *A Review of the Laws Forbidding Pollution of Inland Waters in the United States* (Washington, D.C.: GPO, 1905), 7–47; John Emerson Monger, "Administrative Phases of Stream Pollution Control," *Journal of the APHA* 16 (Aug. 1926): 788; James A. Tobey, "Legal Aspects of the Industrial Wastes Problem," *Industrial and Engineering Chemistry* 31 (Nov. 10, 1939): 1322; Warrick, "Relative Importance of Industrial Wastes," 496.

Chapter 8. Battles at Both Ends of the Pipe

1. James B. Crooks, *Politics and Progress* (Baton Rouge: Louisiana State University Press, 1968), 132–33.

2. Joel A. Tarr, "Sewerage and the Development of the Networked City in the United States, 1850–1930," in Tarr and Dupuy, *Technology and the Rise of the Networked City,* 169.

3. Carol Hoffecker, "Water and Sewage Works in Wilmington, Delaware, 1810–1910," *Essays in Public Works History* 12 (Chicago: Public Works Historical Society, 1981), 9; Henry W. Taylor, "A Privately Financed System of Sewers," *Engineering Record* 71 (Jan. 16, 1915): 79–80.

4. Frederick Moore, "The New Drainage and Sewerage System of New Orleans," *Scientific American*, Dec. 7, 1901, 564; "New Orleans: Drainage and Sewerage," *Sanitarian* 43 (Oct. 1899): 299–314; Advisory Board on Drainage, *Report on the Drainage of the City of New Orleans* (New Orleans, 1895); "The New Orleans Sewerage System," *American City* 19 (July 1918): 26–28.

5. Samuel W. Abbott, "The Past and Present Condition of Public Hygiene and State Medicine in the United States," *Monographs on American Social Economics* 19 (1900): 40–43; Pearse, *Modern Sewage Disposal*, 13.

6. Teaford, *Unheralded Triumph*, 219–20; Schultz, *Constructing Urban Culture*, 174; Tarr et al., "Development and Impact," 66–67.

7. John H. Ellis, "Memphis' Sanitary Revolution, 1880–1890," *Tennessee Historical Quarterly* 23 (Mar. 1964): 59–61; Lynette B. Wrenn, "The Memphis Sewer Experiment," *Tennessee Historical Quarterly* 44 (Fall 1985): 340; "American Sewerage Practice," in *Sewering the Cities*, ed. Barbara Gutmann Rosenkrantz (New York: Arno Press, 1977), 24.

8. Wrenn, "Memphis Sewer Experiment," 340; Ellis, "Memphis' Sanitary Revolution," 60; Sorrels, *Memphis' Greatest Debate*, 42.

9. G. B. Thornton, *The Death-Rate of Memphis* (Memphis, 1882), 4–5; Wrenn, "Memphis Sewer Experiment," 340–41.

10. Ellis, "Memphis' Sanitary Revolution," 62–65; U.S. Department of the Interior, Census Office, *Report on the Social Statistics of Cities, Tenth Census, 1880*, comp. George E. Waring Jr. (Washington, D.C., 1886), 144–45.

11. U.S. Department of the Interior, *Report on the Social Statistics of Cities*, 144–45.

12. Ibid., 145; Wrenn, "Memphis Sewer Experiment," 343–44; Ellis, "Memphis' Sanitary Revolution," 65–66.

13. Wrenn, "Memphis Sewer Experiment," 342–45; U.S. Department of the Interior, *Report on the Social Statistics of Cities*, 146.

14. George Preston Brown, *Sewer-Gas and Its Dangers* (Chicago, 1881), 17.

15. Tarr and McMichael, "Decisions about Wastewater Technology," 52; Tarr, "Separate vs. Combined Sewer Problem," 315–18; Schultz and McShane, "To Engineer the Metropolis," 394; Schultz, *Constructing Urban Culture*, 167–69.

16. Cited in Sorrels, *Memphis' Greatest Debate*, 44–45.

17. Frederick S. Odell, "The Sewerage of Memphis," *Transactions of the ASCE* 216 (Feb. 1881): 26.

18. Ellis, "Memphis' Sanitary Revolution," 66; Wrenn, "Memphis Sewer Experiment," 345–46; John Lundie, *Report on the Water Works System of Memphis, Tenn.* (Memphis, 1898), 4.

19. Melosi, *Garbage in the Cities*, 44–48.

20. "American Sewerage Practice," 25.

21. Melosi, *Garbage in the Cities*, 47–49.

22. Metcalf and Eddy, *American Sewerage Practice*, vol. 1, 16.

23. Tarr, "Separate vs. Combined Sewer Problem," 317–30; Metcalf and Eddy, *American Sewerage Practice*, vol. 1, 26.

24. Henry N. Ogden, *Sewer Design* (New York: Wiley, 1913), 2–6, 8; Harrison P. Eddy, "Use and Abuse of Systems of Separate Sewers and Storm Drains," *Proceed-

ings of the *ASMI* 28 (1922): 131; Staley and Pierson, *Separate System of Sewerage*, 38.

25. Tarr and McMichael, "Historic Turning Points," 84; Tarr, "Sewerage and the Development of the Networked City in the United States," 167.

26. Rudolph Hering, "Sewerage Systems," *Transactions of the ASCE* 230 (Nov. 1881): 362–63.

27. Sam Bass Warner Jr., *Streetcar Suburbs* (1962; reprint, Cambridge, Mass.: Harvard University Press, 1978), 30.

28. Charles C. Euchner, "The Politics of Urban Expansion," *Maryland Historical Magazine* 86 (Fall 1991): 270–87.

29. Harris, *Political Power in Birmingham*, 178–79; Zunz, *Changing Face of Inequality*, 114–16; Doyle, *Nashville in the New South*, 83, 86; Miller, *Memphis during the Progressive Era*, 67–68; Zane L. Miller, *Boss Cox's Cincinnati* (New York: Oxford University Press, 1968), 67.

30. Morris Knowles, "Keeping Boundary Waters from Pollution," *Survey* 33 (Dec. 19, 1914): 313.

31. Metcalf and Eddy, *American Sewerage Practice*, vol. 1, 32.

32. Rudolph Hering, "Sewers and Sewage Disposal," *Engineering Magazine* 8 (Mar. 1895): 1013.

33. Christopher Hamlin, *What Becomes of Pollution?* (New York: Garland, 1987), 1.

34. Ibid., 2–3; Tarr, "Industrial Wastes and Public Health," 1060.

35. Whipple, "Principles of Sewage Disposal," 20.

36. "Sewage Treatment in Great Britain and Some Comparisons with Practice in the United States," *Engineering News* 52 (Oct. 6, 1904): 310–11; "British Practice in Sewage Disposal," *Engineering Record* 66 (Nov. 2, 1912): 496.

37. Hyde, "Review of Progress in Sewage Treatment," 3.

38. Pearse, *Modern Sewage Disposal*, 13.

39. Langdon Pearse, "The Dilution Factor," *Transactions of the ASCE* 85 (1922): 451; Metcalf and Eddy, *American Sewerage Practices*, vol. 1, 30–31; H. B. Hommon et al., "Treatment and Disposal of Sewage," *Public Health Reports* 35 (Jan. 16, 1920): 102–3; Hyde, "Review of Progress in Sewage Treatment," 3; Kinnicutt et al., *Sewage Disposal*, 60–61.

40. George W. Fuller, *Sewage Disposal* (New York: McGraw-Hill, 1912), 204–5, 225; E. Sherman Chase, "Progress in Sanitary Engineering in the United States," *Transactions of the ASCE* CT (1953): 562; Tarr, "Sewerage and the Development of the Networked City," 1701.

41. H. W. Streeter, "Disposal of Sewage in Inland Waterways," in Pearse, *Modern Sewage Disposal*, 192; Metcalf and Eddy, *American Sewerage Practice*, vol. 1, 30–31; Charles F. Mebus, "Sanitary Sewerage and Sewage Disposal," *American City* 3 (Oct. 1910): 167–71.

42. Leonard Metcalf and Harrison P. Eddy, *American Sewerage Practice*, vol. 3 (New York: McGraw-Hill, 1915), 40–42; Fuller and McClintock, *Solving Sewage Problems*, 19–21.

43. James E. Foster, "Water-Borne Disease and the Law," *Hygeia* 6 (June 1928): 319–21; John Wilson, "Legal Responsibility for a Pure Water-Supply" *American*

City 21 (Sept. 1919): 237; Milton P. Adams, "River Pollution Relieved and Sewer System Expanded," *Civil Engineering* 1 (Dec. 1931): 1370; Metcalf and Eddy, *American Sewerage Practice,* vol. 2, 31; Tarr et al., "Development and Impact," 70–71.

44. Tarr et al., "Development and Impact," 72; Tarr, "Industrial Wastes and Public Health," 1061; "The Pollution of Streams," *Engineering Record* 60 (Aug. 7, 1909): 157–59; Monger, "Administrative Phases of Stream Pollution Control," 788; Tobey, "Legal Aspects of the Industrial Wastes Problem," 1322; Warrick, "Relative Importance of Industrial Wastes," 496.

45. Tarr and McMichael, "Decisions about Wastewater Technology," 55; John E. Allen, "Sewage Treatment for Philadelphia," *Proceedings of the ASMI* 35 (1929): 221; George Peter Gregory, "A Study in Local Decision Making," *Western Pennsylvania Historical Magazine* 57 (Jan. 1974): 33, 42.

46. "Chicago Drainage Canal and the City of St. Louis," *Scientific American,* June 20, 1903, 464; O'Connell, "Chicago's Quest for Pure Water," 1, 5–15, 17, 69.

47. Hommon et al., "Treatment and Disposal of Sewage," 120–21; Joel A. Tarr, "From City to Farm: Urban Wastes and the American Farmer," *Agricultural History* 49 (Oct. 1975): 599, 607–10; Kinnicutt et al., *Sewage Disposal,* 205–7; Eddy, "Sewerage and Sewage Disposal," 694; Mary Taylor Bissell, *A Manual of Hygiene* (New York, 1894), 214; Hollis Godfrey, "City Water and City Waste," *Atlantic Monthly,* Sept. 1906, 380; C.-E. A. Winslow, "The Scientific Disposal of City Sewage," *Technology Quarterly* 17 (Dec. 1905): 320.

48. Hommon et al., "Treatment and Disposal of Sewage," 117–18; Winslow, "Scientific Disposal of City Sewage," 321–23; Rudolph Hering, "New Method of Sewage Sludge Treatment," *AJPH* 2 (Feb. 1912): 113.

49. Morris M. Cohn, "Present Status of Sewage Treatment Reviewed by APHA," *Municipal Sanitation* 6 (Nov. 1935): 341; Bissell, *Manual of Hygiene,* 210; Hommon et al., "Treatment and Disposal of Sewage," 108.

50. Hamlin, "William Dibdin," 190.

51. Ibid.

52. Winslow, "Scientific Disposal of City Sewage," 324–25; Kinnicutt et al., *Sewage Disposal,* 270–73; Metcalf and Eddy, *American Sewerage Practice,* vol. 2, 10–11; Hommon et al., "Treatment and Disposal of Sewage," 114–15.

53. Hommon et al., "Treatment and Disposal of Sewage," 115–16.

54. Metcalf and Eddy, *American Sewerage Practice,* vol. 2, 17–19; Fuller, *Sewage Disposal,* 690–91; Kinnicutt et al., *Sewage Disposal,* 316; Eddy, "Sewerage and Sewage Disposal," 694; A. Marston, "Present Status of Sewage Disposal in the United States," *City Hall* 10 (Nov. 1908): 159; Winslow, "Scientific Disposal of City Sewage," 325–26; Chase, "Progress in Sanitary Engineering," 563.

55. Easby, "Beginnings of Sanitary Science," 104–5; Godfrey, "City Water and City Waste," 382; Winslow, "Scientific Disposal of City Sewage," 323; E. Sherman Chase, "Modern Methods of Sewage Disposal," *American City* 22 (Apr. 1920): 394; Metcalf and Eddy, *American Sewerage Practice,* vol. 2, 12; Marston, "Present Status of Sewage Disposal," 157.

56. Metcalf and Eddy, *American Sewerage Practice,* vol. 2, 20–22; Hommon et al., "Treatment and Disposal of Sewage," 108–9.

57. Metcalf and Eddy, *American Sewerage Practice,* vol. 2, 23–24; Hommon et

al., "Treatment and Disposal of Sewage," 109; Kinnicutt et al., *Sewage Disposal,* 175; Chase, "Progress in Sanitary Engineering," 563.

58. Hering, "Sewerage Systems," 366–68.

59. Hommon et al., "Treatment and Disposal of Sewage," 119–20; Eddy, "Sewerage and Sewage Disposal," 695; Earle P. Phelps, "The Chemical Disinfection of Water and Sewage," *Journal of the APHA* 1 (Sept. 1911): 618–19; Earle B. Phelps, "Stream Pollution by Industrial Wastes and Its Control," in Ravenel, *Half Century of Public Health,* 73–76.

60. Bissell, *Manual of Hygiene,* 214; J. D. Glasgow, "Sewage Disposal," *American Municipalities* 26 (Jan. 1914): 120; Hommon et al., "Treatment and Disposal of Sewage," 125–28.

61. Hommon et al., "Treatment and Disposal of Sewage," 121–22.

62. Kinnicutt et al., *Sewage Disposal,* 381–86; Chase, "Progress in Sanitary Engineering," 564; Edward Barow, "The Development of Sewage Treatment by the Activated Sludge Process," *American City* 32 (Mar. 1925): 296–97; William B. Fuller, "Sewage Disposal by the Activated Sludge Process," *American City* 14 (Jan. 1916): 78–79; T. Chalkley Hatton, "Activated-Sludge Process of Sewage Disposal Firmly Established," *Engineering Record* 75 (Jan. 6, 1917): 16–17; Walter C. Roberts, "Activated Sludge Processes," *Public Works* 57 (Nov. 1926): 378; M. N. Baker, "Activated Sludge in America," *Engineering News* 74 (July 22, 1915): 164–71; Anthony M. Rud, "Activated Sludge—A Modern Miracle," *Illustrated World* 25 (Mar. 1916): 91–92.

63. Tarr et al., "Development and Impact," 74–75; Tarr, "Sewerage and the Development of the Networked City," 170; Tarr and McMichael, "Decisions About Wastewater Technology," 84.

64. Tarr, "Separate vs. Combined Sewer Problem," 330–32; Tarr and McMichael, "Historic Turning Points," 85–86; Joel A. Tarr, Terry Yosie, and James McCurley III, "Disputes Over Water Quality Policy: Professional Cultures in Conflict, 1900–1917," *AJPH* 70 (Apr. 1980): 429, 433.

65. Mebus, "Sanitary Sewerage and Sewage Disposal," 168.

Chapter 9. The Third Pillar of Sanitary Services

1. Portions of this chapter were derived from Martin V. Melosi, "Refuse Pollution and Municipal Reform," in Melosi, *Pollution and Reform in American Cities,* 105–33; Melosi, *Garbage in the Cities.*

2. U.S. Bureau of the Census, *Historical Statistics of the United States* (Washington, D.C.: Department of Commerce, 1975), 224–25, 320–21, 328–32.

3. "Disposal of Refuse in American Cities," *Scientific American,* Aug. 29, 1891, 136; John McGaw Woodbury, "The Wastes of a Great City," *Scribner's Magazine,* Oct. 1903, 392; Henry Smith Williams, "How New York Is Kept Partially Clean," *Harper's Weekly,* Oct. 13, 1894, 973; Rudolph Hering and Samuel A. Greeley, *Collection and Disposal of Municipal Refuse* (New York: McGraw-Hill, 1921), 40.

4. Susan Strasser, *Satisfaction Guaranteed* (New York: Pantheon Books, 1989), 6–7, 101; Grant D. McCracken, *Culture and Consumption* (Bloomington: Indiana University Press, 1988), 22–27.

5. Andrew R. Heinze, *Adapting to Abundance* (New York: Columbia University Press, 1990), 12.

6. Melosi, *Garbage in the Cities*, 146–49; U.S. Bureau of the Census, *Statistics of Cities Having a Population of Over 30,000: 1907* (Washington, D.C.: Department of Commerce, 1910), 452–57.

7. Franz Schneider Jr., "The Disposal of a City's Waste," *Scientific American*, July 13, 1912, 24.

8. Hering and Greeley, *Collection and Disposal of Municipal Refuse*, 13, 28.

9. Tarr, "Urban Pollution," 65–66; "Clean Streets and Motor Traffic," *Literary Digest* 49 (Sept. 5, 1914): 413; "Disposal of Refuse in American Cities," 52; Armstrong et al., *History of Public Works*, 127. See also Clay McShane and Joel A. Tarr, *The Horse in the City* (Baltimore: Johns Hopkins University Press, 2007).

10. Williams, "How New York Is Kept Partially Clean," 974.

11. City of Newton, Mass., *Report of the Board of Health (1895)*, 5; Philadelphia, Bureau of Health, *Annual Report* (1892), 18–19; Chicago, Department of Public Works, *Annual Report* (1899), 23–24; Detroit, Bureau of Health, *Annual Report* (1882), 115.

12. U.S. Department of the Interior, *Report on the Social Statistics of Cities*; Wilson G. Smillie, *Public Health* (New York: Macmillan, 1955), 352.

13. Ravenel, *Half Century of Public Health*, 190–91; Hering and Greeley, *Collection and Disposal of Municipal Refuse*, 2.

14. G. T. Ferris, "Cleansing of Great Cities," *Harper's Weekly*, Jan. 10, 1891, 33.

15. Mary E. Trautmann, "Women's Health Protective Association," *Municipal Affairs* 2 (Sept. 1898): 439–43; Duffy, *History of Public Health in New York City*, 124, 130, 132; New York Ladies' Health Protective Association, *Memorial to Abram S. Hewitt on the Subject of Street Cleaning* (New York, 1887), 4–5.

16. Philadelphia Department of Public Works, Bureau of Street Cleaning, *Annual Report* (1893), 58; Mrs. C. G. Wagner, "What the Women Are Doing for Civic Cleanliness," *Municipal Journal and Engineer* 11 (July 1901): 35.

17. League of American Municipalities, *Proceedings, 1899*, 13–27; *City Government* 7 (Sept. 1899): 49ff.; Carol Aronovici, "Municipal Street Cleaning and Its Problems," *National Municipal Review* 1 (Apr. 1912): 218–25.

18. Mildred Chadsey, "Municipal Housekeeping," *Journal of Home Economics* 7 (Feb. 1915): 53–59.

19. Samuel Greeley, "The Work of Women in City Cleansing," *American City* 6 (June 1912): 873–75.

20. Regina Markell Morantz, "Making Women Modern: Middle-Class Women and Health Reform in Nineteenth-Century America," in *Women and Health in America*, ed. Judith Walzer Leavitt (Madison: University of Wisconsin Press, 1984), 349.

21. Maureen A. Flanagan, "The City Profitable, the City Livable," *Journal of Urban History* 22 (Jan. 1996): 173–74; Suellen Hoy, *Chasing Dirt* (New York: Oxford University Press, 1995), 72–73; Suellen Hoy, "Municipal Housekeeping," in Melosi, *Pollution and Reform*, 173–98.

22. "A Street-Cleaning Nurse," *Literary Digest* 52 (Mar. 18, 1916): 709–10; "Street Cleaning Brigade of Women," *Municipal Journal and Engineer* 9 (Dec. 1900): 152; Mrs. Lee Bernheim, "A Campaign for Sanitary Collection and Disposal

of Garbage," *American City* 15 (Aug. 1916): 134–36; Greeley, "Work of Women in City Cleansing," 73–75; Hester M. McClung, "Women's Work in Indianapolis," *Municipal Affairs* 2 (Sept. 1898): 523; Ewing Galloway, "How Sherman Cleans Up," *American City* 9 (July 1913): 40.

23. Maureen A. Flanagan, "Gender and Urban Political Reform," *American Historical Review* 95 (Oct. 1990): 1036–39, 1044–50.

24. William Parr Capes and Jeanne Daniels Carpenter, *Municipal Housecleaning* (New York: Dutton, 1918), 6–9, 213–32; Gustavus A. Weber, "A 'Clean-up' Campaign Which Resulted in a 'Keep Clean' Ordinance," *American City* 10 (Mar. 1914): 231–34; "Philadelphia's Second Annual Clean-up Week," *Municipal Journal and Engineer* 37 (Sept. 10, 1914): 348–49; Galloway, "How Sherman Cleans Up," 40–41.

25. U.S. Department of the Interior, *Report on the Social Statistics of Cities.*

26. Ibid.

27. Melosi, *Garbage in the Cities*, 21–22, 61–63, 96–97, 111–24.

28. Chicago, Department of Health, *Annual Report* (1892), 15–16.

29. "Garbage Contracts," *City and State* 4 (Jan. 20, 1898): 259.

30. Hering and Greeley, *Collection and Disposal of Municipal Refuse*, 155–56; U.S. Department of Interior, *Report on the Social Statistics of Cities;* "Garbage Collection and Disposal," *City Manager Magazine* (July 1924): 12–14.

31. P. M. Hall, "The Collection and Disposal of City Waste and the Public Health," *AJPH* 3 (Apr. 1913): 315–17; C. E. Terry "The Public Dump and the Public Health," *AJPH* 3 (Apr. 1913): 338–41; R. H. Bishop Jr., "Infantile Paralysis and Cleanable Streets," *American City* 15 (Sept. 1916): 313.

32. George E. Waring Jr., "The Cleaning of the Streets of New York," *Harper's Weekly*, Oct. 29, 1895, 1022.

33. George E. Waring Jr., "The Disposal of a City's Waste," *North American Review* 161 (July 1895): 4, 49–54; George E. Waring Jr., *A Report on the Final Disposition of the Wastes of New York by the Dept. of Street Cleaning* (New York, 1896), 3–6; George E. Waring Jr., "The Cleaning of a Great City," *McClure's Magazine*, Sept. 1897, 917–19; Woodbury, "Wastes of a Great City," 388–90; Hering and Greeley, *Collection and Disposal of Municipal Refuse*, 299.

34. Waring, "Cleaning of a Great City," 917–21; E. Burgoyne Baker, "The Refuse of a Great City," *Munsey's Magazine* 23 (Apr. 1900): 83–84; "Street Cleaning," *Outlook*, Oct. 20, 1900, 427; George E. Waring Jr., "The Garbage Question in the Department of Street Cleaning of New York," *Municipal Affairs* 1 (Sept. 1897): 515–24.

35. Baker, "Refuse of a Great City," 90; Waring, "Cleaning of a Great City," 921–23; David Willard, "The Juvenile Street-Cleaning Leagues," in George E. Waring Jr., *Street-Cleaning and the Disposal of a City's Wastes* (New York: Doubleday and McClure, 1898), 177–86.

36. "The Delehanty Dumping-Scow," *Harper's Weekly*, Oct. 24, 1896, 1051; George E. Waring Jr., "The Fouling of the Beaches," *Harper's Weekly*, July 2, 1898, 663; Waring, "Cleaning of a Great City," 917–19; Waring, *Street Cleaning and the Disposal of a City's Wastes*, 47; Baker, "Refuse of a Great City," 89.

37. George E. Waring Jr., "The Utilization of a City's Garbage," *Cosmopolitan* 24 (Feb. 1898): 406–10; Waring, "Cleaning of a Great City" 919–21; Waring, *Street*

Cleaning and the Disposal of a City's Wastes, 47–49; Baker, "Refuse of a Great City," 87.

38. Charles Zueblin, *American Municipal Progress*, rev. ed. (New York: Macmillan, 1916), 75–76, 82; Delos F. Wilcox, *The American City* (New York: Macmillan, 1906), 118, 224; George A. Soper, *Modern Methods of Street Cleaning* (New York: Engineering News, 1907), 165; John A. Fairlie, *Municipal Administration* (New York: Macmillan, 1906), 258–59; "Tammany and the Streets," *Outlook*, Oct. 20, 1906, 427–28; "The Disposal of New York's Refuse," *Scientific American*, Oct. 24, 1903, 292–94.

39. Helen Gray Cone, "Waring," *Century*, Feb. 1900, 547; Albert Shaw, *Life of Col. Geo. E. Waring, Jr.: The Greatest Apostle of Cleanliness* (New York, 1899), preface, 10–11, 14, 30–34.

40. Ellen H. Richards, *Conservation by Sanitation* (New York: John Wiley and Sons, 1911), 216ff.; Gerhard, *Sanitation and Sanitary Engineering*, 59ff.; H. de B. Parsons, *The Disposal of Municipal Refuse* (New York: John Wiley and Sons, 1906), 8; M. N. Baker, *Municipal Engineering and Sanitation* (New York: Macmillan, 1902), 164; Carl S. Dow, "Sanitary Engineering," *Chautauquan* 66 (Mar. 1912): 80–98; R. Winthrop Pratt, "Sanitary Engineering," *Scientific American*, Supplement, Mar. 7, 1914, 150; "Sanitary Engineer," 286–87.

41. Hering and Greeley, *Collection and Disposal of Municipal Refuse*, 4.

42. "Report of the Committee on the Disposal of Garbage and Refuse," in APHA, *Public Health: Papers and Reports* 23 (1897): 206; ASMI, *Proceedings of the ASMI* (1918), 296; *Municipal Journal and Engineer* 41 (Nov. 23, 1916): 646.

43. Hering and Greeley, *Collection and Disposal of Municipal Refuse*, 12–20.

44. "Report of the Committee on Refuse Collection and Disposal," *AJPH* 5 (Sept. 1915): 933–34; "Refuse Disposal in America," *Engineering Record* 58 (July 25, 1908): 85.

45. "Report of the Committee on the Disposal of Garbage and Refuse," 207; Harry R. Crohurst, "Municipal Wastes," *USPHS Bulletin* 107 (Oct. 1920): 79ff.; ASMI, *Proceedings of the ASMI* (1916), 244–45; M. N. Baker, "Condition of Garbage Disposal in United States," *Municipal Journal and Engineer* 11 (Oct. 1901): 147–48; "Report of the Committee on Street Cleaning," *AJPH* (Mar. 1915): 255–59.

46. Hering and Greeley, *Collection and Disposal of Municipal Refuse*, 104–5; Parsons, *Disposal of Municipal Refuse*, 43–44; William F. Morse, *The Collection and Disposal of Municipal Waste* (New York: Municipal Journal and Engineer, 1908), 36; John H. Gregory, "Collection of Municipal Refuse," *AJPH* 2 (Dec. 1912): 919; Robert H. Wild, "Modern Methods of Municipal Refuse Disposal," *American City* 5 (Oct. 1911): 205–7.

47. Melosi, *Garbage in the Cities*, 144–46.

48. Martin V. Melosi, "Technology Diffusion and Refuse Disposal," in Tarr and Dupuy, *Technology and the Rise of the Networked City*, 207–13.

49. Ibid., 214–22.

50. William F. Morse, "Disposal of the City's Waste," *American City* 2 (May 1910): 271.

51. "Report of the Committee on the Disposal of Garbage and Refuse," 215ff.

52. "Disposal of Garbage," *City Government* 5 (Aug. 1898): 67.

53. "Recent Refuse Disposal Practice," *Municipal Journal and Engineer* 37 (Dec. 10, 1914): 848–49.

54. "Report of Committee on Refuse Disposal and Street Cleaning," in ASMI, *Proceedings of the ASMI* (1916), 245.

55. Capes and Carpenter, *Municipal Housecleaning*, 194–99.

56. Frederick L. Stearns, *The Work of the Department of Street Cleaning* (New York, 1913), 210.

57. Crohurst, "Municipal Wastes," 42–43; Parsons, *Disposal of Municipal Refuse*, 93; William P. Munn, "Collection and Disposal of Garbage," *City Government* 2 (Jan. 1897): 6–7; Capes and Carpenter, *Municipal Housecleaning*, 175.

58. Terry, "Public Dump and the Public Health," 338–39.

59. Cleveland, Chamber of Commerce, Committee on Housing and Sanitation, *Report on Collection and Disposal of Cleveland's Waste* (1917), 7.

60. Hering and Greeley, *Collection and Disposal of Municipal Refuse*, 257; Crohurst, "Municipal Wastes," 43–45; Parsons, *Disposal of Municipal Refuse*, 78–80; D. C. Faber, "Collection and Disposal of Refuse," *American Municipalities* 30 (Feb. 1916): 185–86; Wild, "Modern Methods of Municipal Refuse Disposal," 207–8; A. M. Compton, "The Disposal of Municipal Waste by the Burial Method," *AJPH* 2 (Dec. 1912): 925–29; Charles A. Meade, "City Cleansing in New York City," *Municipal Affairs* 4 (Dec. 1900): 735–36; New York City, Department of Street Cleaning, *Annual Report* (1902–5), 74; "Waste-Material Disposal of New York," *Engineering News* 77 (Jan. 18, 1917): 119.

61. George H. Norton, "Recoverable Values of Municipal Refuse," *Municipal Engineering* 45 (Dec. 1913): 550–52; "Revenue from Municipal Waste," *Municipal Journal and Engineer* 32 (June 6, 1912): 868; William F. Morse, *The Disposal of Refuse and Garbage* (New York: J. J. O'Brien and Sons, 1899), 3–7.

62. "Effect of the War on Garbage Disposal," *Municipal Engineering* 53 (Sept. 1917): 110–11; "Save the Garbage Waste," *American Municipalities* 33 (July 1917): 105; Irwin S. Osborn, "Effect of the War on the Production of Garbage and Methods of Disposal," *AJPH* 7 (May 1918): 368–72; E. G. Ashbrook and A. Wilson, "Feeding Garbage to Hogs," *Farmer's Bulletin*, No. 1133 (Washington, D.C., 1921), 3–26; Charles V. Chapin, "Disposal of Garbage by Hog Feeding," *AJPH* 7 (Mar. 1918): 234–35; U.S. Food Administration, *Garbage Utilization, with Particular Reference to Utilization by Feeding* (Washington, D.C., 1918), 3–11.

63. "Clean Streets and Motor Traffic," 413–14; Woodbury, "Wastes of a Great City," 396–98.

Chapter 10. The Great Depression, World War II, and Public Works

1. Chudacoff and Smith, *Evolution of American Urban Society*, 207.

2. Carl Abbott, *Urban America in the Modern Age* (Arlington Heights, Ill.: Harlan Davidson, 1987), 2.

3. Chudacoff and Smith, *Evolution of American Urban Society*, 216–17; Abbott, *Urban America in the Modern Age*, 4–5; Joseph Interrante, "The Road to Autopia," *Michigan Quarterly Review* 19–20 (Fall/Winter 1980–81): 502–17; Amos H. Hawley, *The Changing Shape of Metropolitan America* (Glencoe, Ill.: Free Press,

1956), 2; Barry Edmonston, *Population Distribution in American Cities* (Lexington, Mass.: D. C. Heath, 1975), 68.

4. Kenneth T. Jackson, *Crabgrass Frontier* (New York: Oxford University Press, 1985), 190.

5. Chudacoff and Smith, *Evolution of American Urban Society*, 211; Goldfield and Brownell, *Urban America*, 289; Abbott, *Urban America in the Modern Age*, 7.

6. Interrante, "Road to Autopia," 502–17; Jackson, *Crabgrass Frontier*, 162–63; Jon C. Teaford, *The Twentieth-Century American City* (Baltimore: Johns Hopkins University Press, 1986), 63; Chudacoff and Smith, *Evolution of American Urban Society*, 212; Abbott, *Urban America in the Modern Age*, 36, 43.

7. Miller and Melvin, *Urbanization of Modern America*, 143–47; Teaford, *Twentieth-Century American City*, 67–72; Chudacoff and Smith, *Evolution of American Urban Society*, 214–15; Goldfield and Brownell, *Urban America*, 302–6.

8. Jackson, *Crabgrass Frontier*, 203–17.

9. Abbott, *Urban America in the Modern Age*, 42.

10. G. E. Gordon, "Water Works Financing," *JAWWA* 26 (Apr. 1934): 519.

11. Abbott, *Urban America in the Modern Age*, 15, 47–48; Goldfield and Brownell, *Urban America*, 323–24; Teaford, *Twentieth-Century American City*, 74–80; Chudacoff and Smith, *Evolution of American Urban Society*, 233–34.

12. Goldfield and Brownell, *Urban America*, 325–26.

13. John H. Mollenkopf, *The Contested City* (Princeton, N.J.: Princeton University Press, 1983), 47.

14. Roger Daniels, "Public Works in the 1930s," in *The Relevancy of Public Works History* (Washington, D.C.: Public Works Historical Society, 1975), 3.

15. Ibid., 3–4; L. Evans Walker, comp., *Preliminary Inventory of the Records of the PWA* (Washington, D.C.: National Archives, 1960), 1; J. Kerwin Williams, *Grants-in-Aid under the PWA* (New York: AMS Press, 1939), 22–28; Charles Trout, "The New Deal and the Cities," in *Fifty Years Later*, ed. Harvard Sitkoff (New York: Knopf, 1985), 134; Goldfield and Brownell, *Urban America*, 326.

16. Mollenkopf, *Contested City*, 55.

17. Teaford, *Twentieth-Century American City*, 82–83; Daniels, "Public Works," 1; Bonnie Fox Schwartz, *The Civil Works Administration* (Princeton, N.J.: Princeton University Press, 1984), 45.

18. Daniels, "Public Works," 7; Mollenkopf, *Contested City*, 65–66; John J. Gunther, *Federal-City Relations in the United States* (Newark: University of Delaware Press, 1990), 78–79.

19. Trout, "New Deal and the Cities," 137, 139, 141, 144–45; Mark Gelfand, *A Nation of Cities* (New York: Oxford University Press, 1975), 48–49; Mollenkopf, *Contested City*, 71–72; Monkkonen, *America Becomes Urban*, 134–35.

20. Teaford, *Twentieth-Century American City*, 90–91; Goldfield and Brownell, *Urban America*, 336–41.

21. Gelfand, *Nation of Cities*, 148–51, 242–45; Walker, *Preliminary Inventory*, 3.

Chapter 11. Water Supply as a National Issue

1. Fuller, "Water-Works," 1588; Turneaure and Russell, *Public Water-Supplies* (1948), 9.

2. "Water-Supply Statistics for Municipalities of Less than 5,000 Population," *American City* 32 (Feb. 1925): 185–91, 33 (Mar. 1925): 309–23, 34 (Apr. 1925): 435–45, 35 (May 1925): 555–65, 36 (June 1925): 665–77, and 37 (July 1925): 47–59.

3. Calvin V. Davis, "Water Conservation—The Key to National Development," *Scientific American*, Feb. 1933, 92.

4. C. B. Hoover, "As Cheap as Water," *Civil Engineering* 1 (Aug. 1931): 1027; George W. Biggs Jr., "Distribution System Practices of a Large Group of Water Companies," *Engineering News-Record* 104 (May 22, 1930): 851.

5. "A Survey of Public Water Supplies," *American City* 50 (June 1935): 63; Turneaure and Russell, *Public Water-Supplies* (1911), 21.

6. Leonard Metcalf, "Effect of Water Rates and Growth in Population upon Per Capita Consumption," *JAWWA* 15 (Jan. 1926): 2, 4–5, 12–17, 19–20.

7. Louis Brownlow, "The Water-Supply and the City Limits," *American City* 37 (June 1927): 27.

8. V. Bernard Siems, "The Advantages of Metropolitan Water-Supply Districts," *American City* 32 (June 1925): 644–45.

9. Sarah S. Elkind, *Bay Cities and Water Politics* (Lawrence: University Press of Kansas, 1998).

10. John Bauer, "How to Set Up Utility Districts," *National Municipal Review* 33 (Oct. 1944): 462–68.

11. George H. Fenkell, "The Management of Water-Works Business from the Executive Standpoint," *American City* 39 (Nov. 1928): 115; "Should Water and Sewerage Systems Be Managed Jointly?" *American City* 60 (Feb. 1945): 11.

12. Armstrong et al., *History of Public Works*, 228, 231–32; Daniels, "Public Works," 9; PWA, *America Builds* (Washington, D.C.: PWA, 1939), 170, 173–78.

13. Abel Wolman, "Some Recent Federal Activities in the Conservation of Water Resources," *JAWWA* 28 (Sept. 1936): 1252–56.

14. Williams, *Grants-in-Aid*, 34; Daniels, "Public Works," 8.

15. U.S. Department of Labor, Bureau of Labor Statistics, *Public Works Administration and Industry* (Washington, D.C.: U.S. Department of Labor, 1938), 9; Mollenkopf, *Contested City*, 66–67.

16. "Water Supplies Will Be Widely Extended after the War," *Scientific American*, July 1944, 18.

17. Armstrong et al., *History of Public Works*, 229–30.

18. "Cities Cooperate to Meet Water Crisis," *American City* 58 (Aug. 1943): 53; L. A. Smith, "Operating the Water Department during Wartime," *American City* 58 (June 1943): 56–57; "Water Conservation," *American City* 58 (Nov. 1943): 51; "Water Supply," *AJPH* 35 (July 1945): 743; Becker, *Century of Milwaukee Water*, 225; "Opportunities for Improving Water-Works Economy," *American City* 43 (Dec. 1930): 103–5; "A National Project for Water Works Betterment," *Civil Engineering* 2 (Apr. 1932): 268.

19. "Important Events, Developments, and Trends in Water Supply Engineer-

ing during the Decade Ending with the Year 1939," *Transactions of the ASCE* 105 (1940): 1740, 1744–54, 1756–61, 1765–69.

20. Kahrl, "Politics of California Water," 106, 109, 111–15.

21. Abel Wolman, *Water, Health, and Society* (Bloomington: Indiana University Press, 1969), 96.

22. Fuller, "Progress in Water Purification," 1574–75; Harry E. Jordan, "Water Supply and Treatment," *Transactions of the ASCE CT* (1953): 573; Nicholas S. Hill Jr., "Twenty-one Years of Progress in Water-Supply and Purification Practice," *American City* 43 (Sept. 1930): 88–89; Eskel Nordell, "Water Treatment Today— And What of the Future?" *American City* 46 (June 1932): 71–73; AWWA, *Water Quality and Treatment,* 255.

23. "Filtration versus Chlorination," *American City* 47 (July 1932): 7; Paul Hansen, "Some Relations between Sewage Treatment and Water Purification," *American City* 36 (June 1927): 765–67; "The Unsolved Problems of Water Supply," *American City* 52 (Feb. 1937): 9.

24. Waterman, *Elements of Water Supply Engineering,* 6; "The Present Status of Public Water Supply," *American City* 53 (Oct. 1938): 9; Baker, "Sketch of the History of Water Treatment," 922–26; "Report of the Committee on Water Supply Engineering of the Sanitary Engineering Division," *Transactions of the ASCE* 105 (1940): 1777–78; "Inventory of Water Supply Facilities," *Engineering News-Record* 123 (Sept. 28, 1939): 60–62.

25. "Advantages of Chloramine Treatment of Water," *American City* 43 (July 1930): 7; "Chlorine and Ammonia in Water Purification," *American City* 53 (Jan. 1938): 9; Charles E. Dalton, "Offensive Tastes in Public Water Supplies," *AJPH* 14 (Oct. 1924): 845–46; LaNier, "Municipal Water Systems," 177; Hansen, "Developments in Water-Purificatiori Practice," 842; Baker, *Quest for Pure Water,* 453–57; "Inventory of Water Supply Facilities," 60–62; Wellington Donaldson, "Water Purification—A Retrospect," *JAWWA* 26 (Aug. 1934): 1058–59; "Water Supply and Purification," *AJPH* 17 (July 1927): 684–85; H. W. Streeter, "Chlorination—A Reserve Protection? Or an Integral Part of Purification?" *American City* 35 (Dec. 1926): 788–91; AWWA, *Water Chlorination,* 3–5; Howard, "Twenty Years of Chlorination," 793; Cohn, "Chlorination of Water," 386–90; M. N. Baker, *Quest for Pure Water,* 2; Turneaure and Russell, *Public Water-Supplies* (1948), 135; Baker, "Sketch of the History of Water Treatment," 931–32; "Chlorination—Five Years' Experience," *American City* 59 (Dec. 1944): 9; Symons, "History of Water Supply," 194.

26. James W. Armstrong, "History of Water Supply with Local Reference to Baltimore," *JAWWA* 24 (Apr. 1932): 539; Charles B. Burdick, "Developments in Water Works Construction," *Civil Engineering* 6 (July 1936): 452; "Improved Water Softening with Precipitators," *American City* 53 (Mar. 1938): 55–56.

27. Abel Wolman, "Pollution Control—Where Does It Stand?" *Municipal Sanitation* 11 (Feb. 1940): 64.

28. Seth G. Hess, "Pollution—And the Pocketbook," *Municipal Sanitation* 10 (July 1939): 356–58; Cornelius W. Kruse, "Our Nation's Water," in *Advances in Environmental Sciences,* ed. James N. Pitts Jr. and Robert C. Metcalf, vol. 1 (New York: Wiley-Interscience, 1969), 44–45; Advisory Committee on Water Pollution, U.S. National Resources Committee, *Water Pollution in the United States* (Wash-

ington, D.C.: GPO, 1939), 38; David M. Neuberger, "The Disastrous Results of Pollution of Our Waters," *Outlook*, May 23, 1923, 8; Scotland G. Highland, "Stream Pollution an Indictable Offense," *American City* 41 (Dec. 1929): 117.

29. Advisory Committee on Water Pollution, *Water Pollution*, 4.

30. "Typhoid Fever in the Large Cities of the United States in 1933," *JAWWA* 26 (July 1934): 947–48; Baker, "Sketch of the History of Water Treatment," 932; "Typhoid Fever in the Large Cities of the United States in 1926," *JAWWA* 17 (June 1927): 769.

31. J. Frederick Jackson, "Stream Pollution by Industrial Wastes, and Its Control," *American City* 31 (July 1924): 23; Sheppard T. Powell, "Industrial-Waste Problems and Their Correction," *Mechanical Engineering* 61 (May 1939): 364; Ernest W. Steel, "By-products from Industrial Wastes," *Scientific American*, Nov. 1930, 379; "Progress Report of Committee on Industrial Wastes," 415; Warrick, "Relative Importance of Industrial Wastes," 495; E. F. Eldridge, *Industrial Waste Treatment Practice* (New York: McGraw-Hill, 1942), 1–4; Wellington Donaldson, "Industries and Water Supplies," *JAWWA* 22 (Feb. 1930): 203; Phelps, "Stream Pollution by Industrial Wastes," 201; L. M. Fisher, "Pollution Kills Fish," *Scientific American*, Mar. 1939, 144–46.

32. Pitts and Metcalf, *Advances in Environmental Sciences*, vol. 1, 43, 46, 48–50; Almon L. Fales, "Effects of Industrial Wastes on Sewage Treatment," *Sewage Works Journal* 9 (Nov. 1937): 970–71; "New Sewage Plants Check Stream Pollution," *American City* 54 (May 1939): 15; E. B. Besselievre, "The Disposal of Industrial Chemical Waste," *Chemical Age* 25 (Dec. 12, 1931): 517; N. T. Veatch Jr., "Stream Pollution and Its Effects," *JAWWA* 17 (Jan. 1927): 62.

33. Veatch, "Stream Pollution and Its Effects," 62; Almon L. Fales, "The Problem of Stream Cleansing," *Civil Engineering* 3 (Sept. 1933): 493; Pearse, "Dilution Factor," 451; Jackson, "Stream Pollution by Industrial Wastes," 24.

34. Craig E. Colten, "Industrial Wastes before 1940," paper presented at Forests, Habitats, and Resources: A Conference in World Environmental History, Durham, N.C., Apr. 1987, 4.

35. Robert Sperr Weston, "Water Pollution," *Industrial and Engineering Chemistry* 31 (Nov. 1939): 1314; Leighton, "Industrial Wastes," 29; "Progress Report of Committee on Industrial Wastes," 415–16; Warrick, "Relative Importance of Industrial Wastes," 496; Fales, "Effects of Industrial Wastes," 971–72; Tarr, "Industrial Wastes and Public Health," 1062.

36. Colten, "Industrial Wastes before 1940," 11-1–2, 11-4–8; Fales, "Effects of Industrial Wastes," 973–74; Leighton, "Industrial Wastes," 32–33; Hervey J. Skinner, "Waste Problems in the Pulp and Paper Industry," *Industrial and Engineering Chemistry* 31 (Nov. 1939): 1331–32.

37. AWWA, *Water Quality and Treatment*, 34.

38. Tarr, "Industrial Wastes and Public Health," 1060, 1062, 1066.

39. Duffy, *Sanitarians*, 218, 256–58, 261–63, 269; H. A. Kroeze, "The Expanded Role of the Sanitarian," *AJPH* 32 (June 1942): 613–14.

40. Edward G. Sheibley, "The Sanitary Engineer—His Value in Health Administration," *American City* 25 (Nov. 1921): 365.

41. Abel Wolman, "The Training for the Sanitarian of Environment," *AJPH* 14 (June 1924): 472–73.

---

42. Bizzell, "Sanitary Engineering as a Career," 509–11; S. C. Prescott, "Training for the Public Health Engineer," *AJPH* 21 (Oct. 1931): 1092–97; Whipple, "Sanitation," 97.

43. Arthur Richards, "Status of Employment among Municipal Engineers," *American City* 49 (Dec. 1934): 58; Reynolds, "Engineer in Twentieth-Century America," 179.

44. Jackson, "Stream Pollution by Industrial Wastes," 23; "Industrial Wastes in City Sewers—I," *American City* 52 (May 1937): 86.

45. H. R. Crohurst, "Water Pollution Abatement in the United States," *AJPH* 36 (Feb. 1936): 177.

46. P. Aarne Vesilind, "Hazardous Waste," in *Hazardous Waste Management*, ed. J. Jeffrey Peirce and P. Aarne Vesilind (Ann Arbor, Mich.: Ann Arbor Science, 1981), 26; Micklin, "Water Quality," 131; Tarr, "Industrial Wastes and Public Health," 1059, 1061, 1064; Tarr, "Historical Perspectives on Hazardous Wastes," 96; Lawrence M. Freidman, *A History of American Law* (New York: Simon and Schuster, 1973), 162–63; Warrick, "Relative Importance of Industrial Wastes," 496; Monger, "Administrative Phases of Stream Pollution Control," 790; James A. Tobey, "Legal Aspects of the Industrial Wastes Problem," *Industrial and Engineering Chemistry* 31 (Nov. 10, 1939): 1322; Hopkins, *Elements of Sanitation*, 183.

47. Donaldson, "Industrial Wastes in Relation to Water Supplies," 198; Edmund B. Besselievre, *Industrial Waste Treatment* (New York: McGraw-Hill, 1952), 325–44; Fales, "Progress in the Control of Pollution by Industrial Wastes," 715–17; Skinner, "Waste Problems," 1332.

48. Baity, "Aspects of Governmental Policy," 1302–3; Donaldson, "Industries and Water Supplies," 207; "State Laws Governing Pollution," 506; John D. Rue, "Disposal of Industrial Wastes," *Sewage Works Journal* 1 (Apr. 1929): 365–69; "Industrial Wastes in City Sewers—I," 87; Besselievre, "Statutory Regulation of Stream Pollution," 217ff.; Warrick, "Relative Importance of Industrial Wastes," 496; John H. Fertig, "The Legal Aspects of the Stream Pollution Problem," *AJPH* 16 (Aug. 1926): 786; Tobey, "Legal Aspects of the Industrial Wastes Problem," 1322; Tarr, "Industrial Wastes and Public Health," 1060–61.

49. M. C. Hinderlider and R. I. Meeker, "Interstate Water Problems and Their Solution," *Proceedings of the ASCE* 52 (Apr. 1926): 606–8.

50. Hopkins, *Elements of Sanitation*, 185–88; "Interstate Sanitation Commission Starts Work in Pollution Abatement," *American City* 52 (May 1937): 93; Kruse, "Our Nation's Water," 50.

51. *Water Supply and Sewage Disposal*, 12; Cooley and Wandesforde-Smith, *Congress and the Environment*, 131; Warrick, "Relative Importance of Industrial Wastes," 496; Hopkins, *Elements of Sanitation*, 183.

52. Albert E. Cowdrey, "Pioneering Environmental Law," *Pacific Historical Review* 44 (Aug. 1975): 331–49; William H. Rodgers Jr., "Industrial Water Pollution and the Refuse Act," *University of Pennsylvania Law Review* 119 (1971): 322–35; Lettie McSpaden Wenner, "Federal Water Pollution Control Statutes in Theory and Practice," *Environmental Law* 4 (Winter 1974): 252–63; "The Refuse Act of 1899," *Ecology Law Quarterly* 1 (1971): 173–202.

53. Martin V. Melosi, *Coping with Abundance* (New York: Knopf, 1985), 151–

52; Joseph A. Pratt, "The Corps of Engineers and the Oil Pollution Act of 1924," unpublished manuscript; Colten, "Industrial Wastes before 1940," 13–14.

Chapter 12. Sewerage, Treatment, and the "Broadening Viewpoint"

1. Pearse, *Modern Sewage Disposal*, 13.

2. John R. Thoman and Kenneth H. Jenkins, *Statistical Summary of Sewage Works in the United States* (Washington, D.C.: USPHS, HEW, 1958), 27.

3. John R. Thoman, *Statistical Summary of Sewage Works in the United States* (Washington, D.C.: Federal Security Agency, 1946), 7.

4. Alden Wells, "Investigating Inadequate Sewers—I," *American City* 35 (Oct. 1926): 481–82.

5. Thoman, *Statistical Summary of Sewage Works*, 7.

6. Martin V. Melosi, "Sanitary Services and Decision Making in Houston," *Journal of Urban History* 20 (May 1994): 367, 376–83, 393–95.

7. Willis T. Knowlton, "The Sewage Disposal Problem of Los Angeles, California," *Transactions of the ASCE* 92 (1928): 985; Los Angeles, Special Sewage Disposal Commission, *Report*, pt. 1 (1921), 3–6; Franklin Thomas, "The Sewage Situation of the City of Los Angeles," *Sewage Works Journal* 12 (Sept. 1940): 879–80; B. D. Phelps and R. C. Stockman, "New Sanitary Sewage Facilities for San Diego," *Civil Engineering* 12 (Jan. 1942): 17.

8. "New Occasions—New Duties," *AJPH* 22 (Nov. 1932): 1169.

9. Williams, *Grants-in-Aid*, 35–37; "Selling Sewage Treatment on Facts—Not Fancy," *AJPH* 22 (Oct. 1932): 1069; Harrison P. Eddy, "Developments in Sewerage and Sewage Treatment during 1933," *Water Works and Sewerage* 81 (Feb. 1934): 39; Daniels, "Public Works," 9; PWA, *America Builds*, 279, 291.

10. "WPA Sewer Work in New York City," *American City* 51 (Dec. 1936): 11; H. E. Hargis, "Sewer Connections—A Health and Financial Problem," *American City* 52 (June 1937): 119; E. J. Cleary, "Sanitation Stirs the South," *Engineering News-Record* 118 (June 10, 1937): 872–74; "Large Cities Make Progress in Sewage Treatment," *AJPH* 27 (May 1937): 272; L. G. Pearce, "Atlanta Builds Modern Disposal Plants," *Municipal Sanitation* 10 (Jan. 1939): 27; Dana E. Kepner, "Status of Sewage Disposal in Western States," *Civil Engineering* 8 (Sept. 1938): 608.

11. Williams, *Grants-in-Aid*, 256–58.

12. Douglas L. Smith, *The New Deal in the Urban South* (Baton Rouge: Louisiana State University Press, 1988), 106–11. Race remained an issue in the allocation of project funds in the South. Blacks played no role in administering projects, black workers hired for the projects were limited in number, and black neighborhoods did not receive comparable treatment to white neighborhoods. See Smith, *New Deal*, 234–35.

13. "Cost of Cities' Sanitation Service," *Engineering News-Record* 118 (Jan. 1937): 57; "Financial Statistics of Cities: 1926–1934 Sanitation Service," *American City* 51 (Dec. 1936): 11–19.

14. Samuel A. Greeley, "Organizing and Financing Sewage Treatment Projects," *Transactions of the ASCE* 109 (1944): 256–57.

15. Frank A. Marston, "Why Charge for Sewerage Service?—Why Not?" *Ameri-

can City 42 (Feb. 1936): 145–47; Samuel A. Greeley, "Charges for Sewerage Service," *American City* 48 (Jan. 1933): 65–66; "What Charges for Sewerage Service?" *American City* 48 (June 1933): 68–69; Wagner, "Making Sewage Disposal Pay Its Way," 158; Ernest Boyce, "Service Charges for Sewers and Sewage Treatment Plants," *American City* 40 (Feb. 1929): 106–7; E. E. Smith, "What Does Sewage Disposal Cost?" *American City* 44 (Apr. 1930): 120–22; George J. Schroepfer, "Economics of Sewage Treatment," *Transactions of the ASCE* 104 (1939): 1210–38.

16. *Municipal Yearbook, 1945* (1946), 350; Donald C. Stone, *The Management of Municipal Public Works* (Chicago: Public Administration Service, 1939), 241; "Cities Charge Sewer Rentals to Meet Sewerage Costs," *Texas Municipalities* 26 (Apr. 1939): 108; "Status of Sewer Rental Laws in the United States," *American City* 48 (Nov. 1933): 13.

17. "Sewerage and Water Systems under Joint Management," *American City* 60 (Apr. 1945): 78.

18. "California Cities Plan Joint Sewage Facilities," *National Municipal Review* 30 (Dec. 1941): 716; "Single Trunk Sewer Proposed to Serve Seventeen Massachusetts Communities," *National Municipal Review* 13 (Dec. 1924): 715; Edward S. Rankin, "A Notable Example of Municipal Cooperation," *American City* 36 (Feb. 1927): 215; "California Cities Cooperate for Sewage Disposal," *National Municipal Review* 29 (1940): 127; C. A. Holmquist, "Essential Features of an Efficient Municipal Sewerage System," *American City* 37 (Nov. 1927): 610; "Passaic Valley Trunk Sewer Completed," *American City* 31 (Nov. 1924): 315; "Formula for Sewerage Excellence," *American City* 60 (1945): 92.

19. Greeley, "Organizing and Financing Sewage Treatment Projects," 255; Charles Haydock, "Municipal Authorities," *American City* 59 (Mar. 1944): 85–86.

20. E. French Chase, "Regional Planning and Sewage Disposal," *American City* 49 (Aug. 1934): 39; C. A. Holmquist, "Interstate Sanitation Compact and Its Implications," *AJPH* 26 (Oct. 1936): 989–95.

21. Hansen, "Some Relations between Sewage Treatment and Water Purification," 766–67.

22. George B. Gascoigne, "Sewers, Sewerage and Sewage Disposal," *American City* 43 (Sept. 1930): 97–98.

23. Willem Rudolfs, "Needed Research in Sewage Disposal," *AJPH* 17 (Jan. 1927): 24–26.

24. H. W. Clark, "Past and Present Developments in Sewage Disposal and Purification," *Sewage Works Journal* 2 (Oct. 1930): 563.

25. Charles Gilman Hyde, "Decade of Sewage Treatment: 1928–1938," *Municipal Sanitation* 9 (Oct. 1938): 482.

26. "New Sewage Plants Check Stream Pollution," 15.

27. Thoman and Jenkins, *Statistical Summary of Sewage Works*, 21; Pearse, *Modern Sewage Disposal*, 6, 8, 13; E. K. Gubin, "Sewage Treatment and National Defense," *Hygeia* 19 (Dec. 1941): 987; Tarr, "Sewerage and the Development of the Networked City in the United States," 171; "Sewage Facilities in the United States," *American City* 53 (June 1938): 11; Thoman, *Statistical Summary of Sewage Works*, 9.

28. Linn H. Enslow, "Chemical Precipitation Processes," *Civil Engineering* 5 (Apr. 1935): 235; Wellington Donaldson, "Outlook for Chemical Sewage Treat-

ment," *Civil Engineering* 5 (Apr. 1935): 245; Niles, "Early Environmental Engineering," 10.

29. Charles Gilman Hyde, "Recent Trends in Sewerage and Sewage Treatment," *Municipal Sanitation* 7 (Feb. 1936): 48–49.

30. Willem Rudolfs, "Developments in Sewage Treatment—1940," *Sewage Works Engineering* 12 (Feb. 1941): 65–71.

31. Ibid., 68; Carpenter, "Progress in Sewerage and Sewage Treatment," 49–50; Harry A. Mount, "Sewage: The Price of Civilization," *Scientific American,* Feb. 1922, 125–26; "Activated Sludge Process Grows in Favor," *American City* 28 (Feb. 1923): 122; Edward Bartow, "The Development of Sewage Treatment by the Activated Sludge Process," *American City* 32 (Mar. 1925): 296–98; Roberts, "Activated Sludge Processes," 378–81; Arthur J. Martin, *The Activated Sludge Process* (London: Macdonald and Evans, 1927), 107–27, 355–60; Robert T. Regester, "Problems and Trends in Activated Sludge Practice," *Transactions of the ASCE* 106 (1941): 158–79; William H. Trinkaus, "Chicago's New Activated Sludge Plant Is Largest in the World," *Civil Engineering* 9 (May 1939): 285–88.

32. Paul Hansen, "Trends in Sewage Treatment," *Sewage Works Engineering* 15 (Mar. 1944): 134.

33. Harold W. Streeter, "Surveys for Stream Pollution Control," *Proceedings of the ASCE* 64 (Jan. 1938): 6.

34. Advisory Committee on Water Pollution, *Water Pollution,* 4.

35. Streeter, "Surveys for Stream Pollution Control," 6, 46.

36. The interest in sewage farming in the West exhibited elements of conservation of natural resources. See "Sewage and Conservation," *AJPH* 20 (May 1930): 520–21; E. B. Black, "Sewage Disposal at Denver, Colo.," *Civil Engineering* 8 (Sept. 1938): 578.

37. Tarr et al., "Development and Impact," 75.

38. "Shall Waterways Be Sewers Forever?" *American City* 35 (Aug. 1926): 197.

39. "Three Groups Consider Costs of Sewage Treatment vs. Stream Pollution," *American City* 47 (Aug. 1932): 13.

40. Hommon, "Brief History," 175; Gustav H. Radebaugh, "Selling Sewage Treatment to the Public," *Municipal Sanitation* 7 (Oct. 1936): 340–41; "Izaak Walton League Urges Sewage Treatment as No. 1 Postwar Job," *Sewage Works Engineering* 15 (Oct. 1944): 507.

41. A.L.H. Street, "The City's Legal Rights and Duties," *American City* 35 (Dec. 1926): 875; "Liability for Pollution of Waters by Sewage," *National Municipal Review* (Oct. 1928): 605.

42. Leo T. Parker, "Right of City to Pollute Water," *Municipal Sanitation* 2 (Oct. 1931): 489; Leo T. Parker, "Review of Important 1943 Sewage Suits," *Sewage Works Engineering* 15 (Mar. 1944): 143.

43. Quoted in Melosi, "Hazardous Waste and Environmental Liability," 755.

44. Ibid., 757.

45. "Laxity in the Operation of Sewage Disposal Plants," *National Municipal Review* 14 (1925): 577.

46. Louis P. Cain, "Unfouling the Public's Nest," *Technology and Culture* 15 (Oct. 1974): 605–6, 608–10.

Chapter 13. The "Orphan Child of Sanitary Engineering"

1. George W. Fuller, "The Place of Sanitary Engineering in Public Health Activities," *AJPH* 15 (Dec. 1925): 1072.

2. Samuel A. Greeley, "An Analysis of Garbage Disposal," *American City* 31 (Aug. 1924): 104.

3. Melosi, *Coping with Abundance,* 103–5.

4. APWA, *Solid Waste Collection Practice,* 4th ed. (Chicago: APWA, 1975), 22.

5. Lent D. Upson, *Practice of Municipal Administration* (New York: Century, 1926), 449–57.

6. Samuel A. Greeley, "Administrative and Engineering Work in the Collection and Disposal of Garbage," *Transactions of the ASCE* 89 (1926): 800; "Financial Statistics of Cities: 1926–1934 Sanitation Service," *American City* 51 (Dec. 1936): 11, 13, 15, 17, 19.

7. Harrison P. Eddy, "Why Not Make Garbage Collection and Disposal Self-Sustaining?" *American City* 47 (Oct. 1932): 52–53; Stone, *Management of Municipal Public Works,* 241.

8. Trout, "New Deal and the Cities," 136, 144; "Sanitary Landfill and the Decline of Recycling as a Solid Waste Management Strategy in American Cities," unpublished paper, 15, in author's possession; PWA, *America Builds,* 279; Harrison P. Eddy, "Refuse Disposal—A Review," *Municipal Sanitation* 8 (Jan. 1937): 86.

9. Samuel A. Greeley, "Modern Methods of Disposal of Garbage, and Some of the Troubles Experienced in Their Use," *American City* 28 (Jan. 1923): 15.

10. Ibid., 17; George B. Gascoigne, "A Year's Progress in Refuse Disposal and Street Cleaning," American Society of Municipal Engineers, *Official Proceedings of the 38th Annual Convention* 38 (Jan. 1933): 191–92.

11. C. G. Gillespie and E. A. Reinke, "Municipal Refuse Problems and Procedures," *Civil Engineering* 4 (Sept. 1934): 487–88; Stone, *Management of Municipal Public Works,* 241.

12. E. S. Savas, ed., *The Organization and Efficiency of Solid Waste Collection* (Lexington, Mass.: D. C. Heath, 1977), 35–37, 43.

13. Upson, *Practice of Municipal Administration,* 459; Greeley, "Street Cleaning," 1245.

14. Upson, *Practice of Municipal Administration,* 458; *Municipal Index, 1926,* 162–83; *Municipal Index and Atlas, 1930,* 618–35; Armstrong et al., *History of Public Works,* 442.

15. Armstrong et al., *History of Public Works,* 441–42; Gascoigne, "Year's Progress in Refuse Disposal," 188–89; Upson, *Practice of Municipal Administration,* 459–60; J. E. Doran, "The Economical Collection of Municipal Wastes," *American City* 39 (Oct. 1928): 98; Greeley, "Street Cleaning," 1245.

16. Roger J. Bounds, "Refuse Disposal in American Cities," *Municipal Sanitation* 2 (Sept. 1931): 431–32.

17. George W. Schusler, "The Disposal of Municipal Wastes," *American City* 51 (Aug. 1936): 86; Stone, *Management of Municipal Public Works,* 259–60.

18. Martin V. Melosi, "Waste Management," *Environment* 23 (Oct. 1981): 12; Eddy, "Refuse Disposal," 79; Bounds, "Refuse Disposal," 431; Gillespie and

Reinke, "Municipal Refuse Problems," 432, 490; "Sanitary Landfill and the Decline of Recycling," 18; Upson, *Practice of Municipal Administration,* 462.

19. Rachel Maines and Joel Tarr, "Municipal Sanitation," case study, Carnegie-Mellon University, Dec. 1980, 16; Bounds, "Refuse Disposal," 431–32.

20. Melosi, *Garbage in the Cities,* 182.

21. Martin V. Melosi, "Historic Development of Sanitary Landfills and Subtitle D," *Energy Laboratory Newsletter* 31 (1994): 20.

22. "Sanitary Landfill and the Decline of Recycling," 19–21; "An Interview with Jean Vincenz," *Public Works Historical Society Oral History Interview* 1 (1980); Jean L. Vincenz, "Sanitary Fill at Fresno," *Engineering News-Record* 123 (Oct. 26, 1939): 539–40; Jean L. Vincenz, "The Sanitary Fill Method of Refuse Disposal," *Public Works Engineers' Yearbook* (1940): 187–201; Jean L. Vincenz, "Refuse Disposal by the Sanitary Fill Method," *Public Works Engineers' Yearbook* (1944): 88–96; "The Sanitary Fill as Used in Fresno," *American City* 55 (Feb. 1940): 42–43.

23. "Sanitary-Fill Refuse Disposal at San Francisco," *Engineering News-Record* 116 (Feb. 27, 1936): 314–17; "Fill Disposal of Refuse Successful in San Francisco," *Engineering News-Record* 116 (July 6, 1939): 27–28; J. C. Geiger, "Sanitary Fill Method," *Civil Engineering* 10 (Jan. 1940): 42; John J. Casey, "Disposal of Mixed Refuse by Sanitary Fill Method at San Francisco," *Civil Engineering* 9 (Oct. 1939): 590–92.

24. "Sanitary Landfill and the Decline of Recycling," 22–25.

25. "Interview with Jean Vincenz," 17–19; Vincenz, "Refuse Disposal by the Sanitary Fill Method," 88–89; Melosi, "Historic Development of Sanitary Landfills," 20; APWA, *Municipal Refuse Disposal,* 3rd ed. (Chicago: APWA, 1970), 91–92.

26. Armstrong et al., *History of Public Works,* 448; Eddy, "Refuse Disposal," 80; Bounds, "Refuse Disposal," 433–34.

27. Armstrong et al., *History of Public Works,* 448; Melosi, "Waste Management," 12.

28. Eddy, "Refuse Disposal," 79; Bounds, "Refuse Disposal," 433–34; Greelcy, "Street Cleaning," 1246; Upson, *Practice of Municipal Administration,* 463–64; Harry A. Mount, "A Garbage Crisis," *Scientific American,* Jan. 1922, 38.

29. Mount, "Garbage Crisis," 38.

30. "Trends in Refuse Disposal," *American City* 51 (May 1939): 13.

31. Upson, *Practice of Municipal Administration,* 465–66; Maines and Tarr, "Municipal Sanitation," 6; APWA, *Municipal Refuse Disposal,* 337; Nathan B. Jacobs, "What Future for Municipal Refuse Disposal?" *Municipal Sanitation* 1 (July 1930): 384.

32. Suellen M. Hoy and Michael C. Robinson, *Recovering the Past* (Chicago: Public Works Historical Society, 1979), 20–22.

33. Cyril E. Marshall, "Incinerator Knocks Out Garbage Dump in Long Island Town," *American City* 40 (June 1929): 129; Hering and Greeley, *Collection and Disposal of Municipal Refuse,* 313; *Municipal Index, 1924,* 68; "Garbage Collection and Disposal," 12–13.

34. Crohurst, "Municipal Wastes," 48–49.

35. Michael R. Greenberg et al., *Solid Waste Planning in Metropolitan Regions*

(New Brunswick, N.J.: Center for Urban Policy Research, Rutgers University, 1976), 8; G. C. Holbrook, "The Modern Refuse Incinerator—A Sanitary Municipal Utility," *American City* 51 (Dec. 1936): 59.

36. Greenberg et al., *Solid Waste Planning in Metropolitan Regions*, 8.

37. Henry W. Taylor, "Incineration of Municipal Refuse," *Municipal Sanitation* 6 (May 1935): 142; Henry W. Taylor, "Incineration of Municipal Refuse," *Municipal Sanitation* 6 (Oct. 1935): 300; Henry W. Taylor, "Incineration of Municipal Refuse," *Municipal Sanitation* 6 (Aug. 1935): 239; George L. Watson, "What Constitutes a Low Bid on an Incinerator?" *American City* 49 (Oct. 1934): 66.

38. "Public Still Wants No Incinerator as a Next Door Neighbor," *Municipal Sanitation* 8 (Nov. 1937): 585.

39. Rolf Eliassen, "Incinerator Mechanization Wins Increasing Favor," *Civil Engineering* 19 (Apr. 1949): 17–19; Morris M. Cohn, "Highlights of Incinerator Construction—1941," *Sewage Works Engineering* 13 (Feb. 1942): 87.

40. "Combined Treatment of Sewage and Garbage," *National Municipal Review* 13 (Aug. 1924): 450–51; C. E. Keefer, "The Disposal of Garbage with Sewage," *Civil Engineering* 6 (Mar. 1936): 18–80; "Disposal of Ground Garbage into Sewers Arouses Interest," *Municipal Sanitation* 7 (Mar. 1936): 94; Hyde, "Recent Trends in Sewerage and Sewage Treatment," 46–47; "Send Out the Garbage with the Sewage from the Home," *American City* 50 (Sept. 1935): 13.

41. Suellen Hoy, "The Garbage Disposer," *Technology and Culture* 26 (Oct. 1985): 761.

42. Susan Strasser, "Leftovers and Litter," paper presented at the Organization of American Historians meeting, Atlanta, Apr. 1994, 6–8.

43. Jacobs, "What Future for Municipal Refuse Disposal?" 384–85.

44. Committee on Refuse Collection and Disposal, *Refuse Collection Practice* (Chicago: APWA, 1941), 350–70.

Chapter 14. The Challenge of Suburban Sprawl and the "Urban Crisis" in the Age of Ecology

1. Dennis R. Judd, *The Politics of American Cities*, 3rd ed. (Glenview, Ill.: Scott, Foresman, 1988), 192.

2. U.S. Bureau of the Census, *Historical Statistics of the United States* (1975), 8, 11, 39; Miller and Melvin, *Urbanization of Modern America*, 184–85; John C. Bollens and Henry J. Schmandt, *The Metropolis*, 2nd ed. (New York: Harper and Row, 1970), 17, 19; Alfred H. Katz and Jean Spencer Felton, eds., *Health and the Community* (New York: Free Press, 1965), 25.

3. Jackson, *Crabgrass Frontier*, 139–40.

4. Mollenkopf, *Contested City*, 244–45; Bollens and Schmandt, *Metropolis*, 284–86.

5. Gelfand, *Nation of Cities*, 158; Miller and Melvin, *Urbanization of Modern America*, 213–14; Chudacoff and Smith, *Evolution of American Urban Society*, 261; Mollenkopf, *Contested City*, 214; Jackson, *Crabgrass Frontier*, 138–39; Teaford, *Twentieth-Century American City*, 98, 109.

6. Abbott, *Urban America in the Modern Age*, 100, 110; Miller and Melvin, *Ur-*

banization of Modern America, 180–81; Richard M. Bernard and Bradley R. Rice, eds., *Sunbelt Cities* (Austin: University of Texas Press, 1983), 1–26.

7. Goldfield and Brownell, *Urban America*, 345; Chudacoff and Smith, *Evolution of American Urban Society*, 259–60; Teaford, *Twentieth-Century American City*, 100–102, 216–22; Abbott, *Urban America in the Modern Age*, 64–66; Gelfand, *Nation of Cities*, 158.

8. Jackson, *Crabgrass Frontier*, 215–16.

9. Kenneth Fox, *Metropolitan America* (New Brunswick, N.J.: Rutgers University Press, 1985), 171, 174, 185–86, 189; Judd, *Politics of American Cities*, 179, 183; Jackson, *Crabgrass Frontier*, 242; Christopher Silver, "Housing Policy and Suburbanization," in *Race, Ethnicity, and Minority Housing in the United States*, ed. Jamshid A. Momeni (New York: Greenwood Press, 1986), 73, 76; Jamshid A. Momeni, "Su casa no es mi casa," in Momeni, *Race, Ethnicity, and Minority Housing*, 141; W. Dennis Keating, *The Suburban Racial Dilemma* (Philadelphia: Temple University Press, 1994), 11–12.

10. Abbott, *Urban America in the Modern Age*, 113; Chudacoff and Smith, *Evolution of American Urban Society*, 261: Teaford, *Twentieth-Century American City*, 105, 107.

11. Teaford, *Twentieth-Century American City*, 98–99; Jackson, *Crabgrass Frontier*, 162–63; Abbott, *Urban America in the Modern Age*, 86.

12. Melosi, *Coping with Abundance*, 270–71; Abbott, *Urban America in the Modern Age*, 86.

13. Judd, *Politics of American Cities*, 175.

14. Chudacoff and Smith, *Evolution of American Urban Society*, 262, 271–77; Goldfield and Brownell, *Urban America*, 360–63; Jackson, *Crabgrass Frontier*, 244; Teaford, *Twentieth-Century American City*, 115–18.

15. Teaford, *Twentieth-Century American City*, 111, 113; Goldfield and Brownell, *Urban America*, 354.

16. Abbott, *Urban America in the Modern Age*, 76, 82, 84–85; Goldfield and Brownell, *Urban America*, 348–54; Teaford, *Twentieth-Century American City*, 114, 118–26; Jon C. Teaford, *The Rough Road to Renaissance* (Baltimore: Johns Hopkins University Press, 1990), 105–7; Miller and Melvin, *Urbanization of Modern America*, 207, 247; Gelfand, *Nation of Cities*, 160, 168, 205–6; Chudacoff and Smith, *Evolution of American Urban Society*, 270.

17. Gelfand, *Nation of Cities*, 349; Teaford, *Twentieth-Century American City*, 127–36; Abbott, *Urban America in the Modern Age*, 91–95, 117–25; Miller and Melvin, *Urbanization of Modern America*, 201.

18. Miller and Melvin, *Urbanization of Modern America*, 177–78.

19. Teaford, *Twentieth-Century American City*, 107.

20. Bollens and Schmandt, *Metropolis*, 103; Harrigan, *Political Change in the Metropolis*, 248.

21. Gelfand, *Nation of Cities*, 164–65, 196.

22. Teaford, *Twentieth-Century American City*, 136–40; Chudacoff and Smith, *Evolution of American Urban Society*, 278; Miller and Melvin, *Urbanization of Modern America*, 205–6; Goldfield and Brownell, *Urban America*, 363–67.

23. Bollens and Schmandt, *Metropolis*, 252–58; Robert L. Lineberry and Ira

Sharkansky, *Urban Politics and Public Policy* (New York: Harper and Row, 1971), 207–8.

24. Institute for Training in Municipal Administration, *Municipal Public Works Administration*, 5th ed. (Chicago: International City Managers' Association, 1957), 51; Bollens and Schmandt, *Metropolis*, 260–63; Lineberry and Sharkansky, *Urban Politics and Public Polity*, 206, 208–9.

25. Bollens and Schmandt, *Metropolis*, 268, 271–73.

26. Clifford B. Knight, *Basic Concepts of Ecology* (New York: Macmillan, 1965), 2.

27. Donald Worster, *Nature's Economy* (Cambridge: Cambridge University Press, 1977), 289, 378; Carolyn Merchant, ed., *Major Problems in American Environmental History* (Lexington, Mass.: D. C. Heath, 1993), 444.

28. Worster, *Nature's Economy*, 339–40; Robert Gottlieb, *Forcing the Spring* (Washington, D.C.: Island Press, 1993), 36; Eugene Odum, "Ecology as a Science," in *The Encyclopedia of the Environment*, ed. Ruth A. Eblen and William R. Eblen (Boston: Houghton Mifflin, 1994), 171.

29. Victor B. Scheffer, *The Shaping of Environmentalism in America* (Seattle: University of Washington Press, 1991), 4.

30. Daniel Faber and James O'Connor, "Environmental Politics," in Merchant, *Major Problems in American Environmental History*, 553.

31. Scheffer, *Shaping of Environmentalism*, 113; Melosi, *Coping with Abundance*, 296–97.

32. Gottlieb, *Forcing the Spring*, 46, 81, 87–98.

33. Environmental Pollution Panel, President's Science Advisory Committee, *Restoring the Quality of Our Environment* (1965), quoted in *American Environmentalism*, 3rd ed., ed. Roderick Nash (New York: McGraw-Hill, 1990), 201; Eric A. Walker, "Technology Can Become More Human," in *Americans and Environment*, ed. John Opie (Lexington, Mass.: D. C. Heath, 1971), 179.

34. Abel Wolman, "The Civil Engineer's Role in Environmental Development," *Civil Engineering—ASCE* 40 (Oct. 1970): 42.

35. Franklin Thomas, "Sanitary Engineers Face Problems Incident to Rapid Expansion in Field of Sanitation," *Civil Engineering* 20 (Feb. 1950): 38; Mark D. Hollis, "Our Rapidly Changing Technology—Its Impact on Sanitary Engineering," *Civil Engineering* 24 (May 1954): 54–55; Mark D. Hollis, "Role of the Sanitary Engineer in Public Health," in *Centennial of Engineering, 1852–1952*, ed. Lenox R. Lohr (Chicago: Centennial of Engineering, 1953), 989–94.

36. Abel Wolman, "The Sanitary Engineer Looks Forward," *AJPH* 36 (Nov. 1946): 1278; Hollis, "Our Rapidly Changing Technology," 54–55.

37. Duffy, *Sanitarians*, 273–74, 280–82, 285–86.

38. Gary Cross and Rick Szostak, *Technology and American Society* (Englewood Cliffs, N.J.: Prentice Hall, 1995), 306.

39. Joseph A. Salvato Jr., *Environmental Sanitation* (New York: Wiley, 1958), 15.

40. Ibid., 574, 577.

41. Victor M. Ehlers and Ernest W. Steel, *Municipal and Rural Sanitation* (New York: McGraw-Hill, 1965), 587–90, 599; David Keith Todd, ed., *The Water Ency-*

clopedia (Port Washington, N.Y.: Water Information Center, 1990), 491; Hanlon, *Principles of Public Health Administration*, 64.

Chapter 15. A Time of Unease

1. George P. Hanna Jr., "Domestic Use and Reuse of Water Supply," *Journal of Geography* 60 (Jan. 1961): 22.

2. Ibid.

3. "The Water Picture in the United States," *American City* 76 (Sept. 1961): 181; "Water-Works Men Want Faster Progress," *American City* 81 (July 1965): 105; Edward A. Ackerman and George O. G. Lof, *Technology in American Water Development* (Baltimore: Johns Hopkins University Press, 1959), 7.

4. Bollens and Schmandt, *Metropolis*, 176; Edward T. Thompson, "The Worst Public-Works Problem," *Fortune*, Dec. 1958, 102.

5. Babbitt and Doland, *Water Supply Engineering*, 40; G. M. Fair, J. L. Geyer, and Daniel Alexander Okun, *Elements of Water Supply and Wastewater Disposal*, 2nd ed. (New York: Wiley, 1977), 14.

6. Joseph A. Salvato Jr., *Environmental Engineering and Sanitation*, 2nd ed. (New York: Wiley-Interscience, 1972), 103–4.

7. Todd, *Water Encyclopedia*, 351.

8. Fair et al., *Elements of Water Supply*, 14.

9. John D. Wright and Don R. Hassall, "Trends in Water Financing," *American City* 86 (Dec. 1971): 61; Ernest W. Steel, *Water Supply and Sewerage*, 4th ed. (New York: McGraw-Hill, 1960), 617; "Water-Works Men Want Faster Progress," 106.

10. Water Resources Council, *The Nation's Water Resources* (Washington, D.C.: Water Resources Council, 1968), 4-1-1, 4-1-2; Murray Stein, "Problems and Programs in Water Pollution," *Natural Resources Journal* 2 (Dec. 1962): 395; Jack Hirshleifer, James C. DeHaven, and Jerome W. Milliman, *Water Supply* (Chicago: University of Chicago Press, 1960), 2, 26; Fair et al., *Elements of Water Supply*, 27–28.

11. Bollens and Schmandt, *Metropolis*, 176, 178.

12. Todd, *Water Encyclopedia*, 226–27, 345–49.

13. Bollens and Schmandt, *Metropolis*, 176.

14. Water Resources Council, *Nation's Water Resources*, 5-1-3.

15. Robert L. Lineberry, *Equality and Urban Policy* (Beverly Hills, Calif.: Sage, 1977), 130.

16. Rodney R. Fleming, "The Big Questions," *American City* 82 (June 1967): 94–95; Charles M. Bolton, "A Metropolitan Water Works Is Best," *American City* 74 (Jan. 1959): 67–68; Wright and Hassall, "Trends in Water Financing," 61; Kruse, "Our Nation's Water," 54; Martin V. Melosi, "Community and the Growth of Houston," *Houston Review* 11 (1989): 110–12.

17. Bollens and Schmandt, *Metropolis*, 177; *Municipal Year Book, 1947*, 295, 297; Michael N. Danielson, *The Politics of Exclusion* (New York: Columbia University Press, 1976), 223, 233.

18. T. E. Larson, "Deterioration of Water Quality in Distribution Systems," *JAWWA* 58 (Oct. 1968): 1316; Steel, *Water Supply and Sewerage*, 618.

19. Donald E. Stearns, "Expanding and Improving Water Distribution Sys-

tems," *Water and Sewage Works* 104 (June 1957): 256; William D. Hudson, "Studies of Distribution System Capacity in Seven Cities," *JAWWA* 58 (Feb. 1966): 157, 159, 161–63.

20. Lineberry, *Equality and Urban Policy*, 130–31.

21. Kenneth J. Ives, "Progress in Filtration," *JAWWA* 56 (Sept. 1964): 1225, 1231; J. T. Ling, "Progress in Technology of Water Filtration," *Water and Sewage Works* 109 (Aug. 1962): 315–16.

22. Ling, "Progress in Technology," 317–19.

23. J. Carrell Morris, "Future of Chloridation," *JAWWA* 58 (Nov. 1968): 1475, 1481; AWWA, *Water Chlorination*, 5.

24. Larry E. Jordan, "Outstanding Achievements in Water Supply and Treatment," *Civil Engineering* 22 (Sept. 1952): 137; "Water Supply," *AJPH* 37 (May 1947): 556; Armstrong et al., *History of Public Works*, 240.

25. Jordan, "Outstanding Achievements," 137; "Water Supply" (May 1947), 556; Herman E. Hilleboe, "Public Health Aspects of Water Fluoridation," *AJPH* 41 (Nov. 1951): 1370–71; "Fluoridation OK," *Newsweek*, Dec. 10, 1951, 46; "Fluoridation of Public Water Supplies," *AJPH* 42 (Mar. 1952): 339; "Fluoridation," *Bulletin of Atomic Scientists* 20 (Sept. 1964): 30.

26. J. C. Furnas, "The Fight Over Fluoridation," *Saturday Evening Post*, May 19, 1956, 37, 142–44; Fred Merryfield, "Water Supply Progress in 1956," *Water and Sewage Works* 104 (Jan. 1957): 12–13.

27. AWWA, *Water Quality and Treatment*, 406.

28. Furnas, "Fight Over Fluoridation," 143–44.

29. AWWA, *Water Quality and Treatment*, 409.

30. "Fluoridation Decade," *Scientific American*, Feb. 1956, 58.

31. "Water Getting Scarce," *Business Week*, Feb. 28, 1948, 24–25; "Water: New York Feels the Pinch," *Business Week*, Dec. 3, 1949, 31–33; Arthur H. Carhart, "Turn Off That Faucet!" *Atlantic Monthly*, Feb. 1950, 39–42; Sherwood D. Ross, "Water Pollution: A National Disgrace," *Progressive*, Aug. 1960, 18.

32. Hugh Hammond Bennett, "Warning: The Water Problem Is National," *Saturday Evening Post*, May 13, 1950, 32–33.

33. "The People-Water Crisis," *Newsweek*, Aug. 23, 1965, 48; "Plenty of Water—But Not to Waste," *Business Week*, Sept. 9, 1950, 82, 84, 86; "Where Is Water Short?" *Chemical Industries* 66 (Apr. 1950): 515; Fairfield Osborn, "Water, Water Everywhere?" *Today's Health* 28 (July 1950): 18–19; Francis Bello, "How Are We Fixed for Water?" *Fortune*, Mar. 1954, 120–23; "Water Crisis Still a Reality," *American City* 71 (Nov. 1956): 23; "Will Water Become Scarce?" *U.S. News and World Report*, Apr. 27, 1956, 84–93; John Robbins, "Water: How Fast Can We Waste It?" *Atlantic Monthly*, July 1957, 31, 33; "Great U.S. Water Shortage," *Newsweek*, Nov. 11, 1957, 45–46; "National Water Shortage," *Science* 127 (Mar. 21, 1958): 634; "Year of the Great Thirst," *Newsweek*, June 28, 1965, 56–57; "Water Crisis—Why?" *U.S. News and World Report*, Aug. 2, 1965, 39–41; "Water Problem in U.S.—What Can Be Done about It," *U.S. News and World Report*, Oct. 25, 1965, 66–68, 73–75; "Water—Too Little and Too Much," *Nation*, Aug. 1965, 91.

34. "Water Scheme Assures City's Growth," *Engineering News-Record* 160 (Apr. 17, 1958): 33–34, 36, 39–40.

35. "People-Water Crisis," 52; "Plenty of Water," 98; "Water-Supply Trends and

Responsibilities," *American City* 76 (June 1961): 7; "Key West Gets Largest De-
salting Plant," *American City* 81 (July 1966): 22.

36. Wallace Stegner, "Myths of the Western Dam," *Saturday Review*, Oct. 23, 1965, 29.

37. Robbins, "Water," 32; "People-Water Crisis," 49, 52.

38. Hopkins et al., *Practice of Sanitation*, 131–32, 137; Water Resources Council, *Nation's Water Resources*, 5-4-1.

39. Samuel P. Hays, *Beauty, Health, and Permanence* (New York: Cambridge
University Press, 1987), 77.

40. Scheffer, *Shaping of Environmentalism*, 50; Tarr and Jacobson, "Environ-
mental Risk in Historical Perspective," 328; Stein, "Problems and Programs in
Water Pollution," 401; Richard J. Frankel, "Water Quality Management," *Water
Resource Research* 1 (June 1965): 173; Kruse, "Our Nation's Water," 62–67, 151–54;
Bello, "How Are We Fixed for Water?" 124; AWWA, *Water Quality and Treatment*, 441; "Nation's Water Crisis," 83; "Plenty of Water," 88; Carhart, "Turn Off
That Faucet!" 42; Robbins, "Water," 34; Water Resources Council, *Nation's Water Resources*, 5-4-2; "Water Supply," *APHA Yearbook, 1949–1950* 40 (May 1950):
118.

41. Ross, "Water Pollution," 17–18.

42. Scheffer, *Shaping of Environmentalism*, 50–53; "Pollution Kills 7,800,000
Fish," *American City* 80 (Mar. 1965): 133.

43. AWWA, *Water Quality and Treatment*, 20–31, 69; Water Resources Council, *Nation's Water Resources*, 5-4-2; Armstrong et al., *History of Public Works*,
244.

44. P. H. McGauhey, "Folklore in Water Quality Standards," *Civil Engineering—ASCE* 35 (June 1965): 71; "Water Supply" (May 1947), 558.

45. Tarr and Jacobson, "Environmental Risk in Historical Perspective," 329;
John W. Clark, Warren Viessman Jr., and Mark J. Hammer, *Water Supply and Pollution Control*, 2nd ed. (Scranton, Pa.: International Textbook, 1971), 228.

46. Ross, "Water Pollution," 18; Mark D. Hollis, "Water Pollution Abatement in
the United States," *Sewage and Industrial Waste* 23 (Jan. 1951): 89; "What Stream
Pollution Means Nationally," *American City* 67 (Jan. 1952): 139; M. D. Hollis and
G. E. McCallum, "Federal Water Pollution Control Legislation," *Sewage and In-
dustrial Waste* 28 (Mar. 1956): 308; David H. Howells, "We Need More Municipal
Waste Treatment Works," *Civil Engineering* 33 (Sept. 1963): 54.

47. Samuel A. Greeley, "Water Resource and Pollution Control Legislation,"
Civil Engineering 31 (Dec. 1961): 62–63; "Where We Stand on Pollution Control,"
Engineering News-Record 137 (Dec. 26, 1946): 78; Warren J. Scott, "Federal and
State Legislation for Stream Pollution Control," *Sewage Works Journal* 19 (Sept.
1947): 884; Hollis, "Water Pollution Abatement," 91–92; W. B. Hart, "Antipollu-
tion Legislation and Technical Problems in Water Pollution Abatement," in *Wa-
ter for Industry*, ed. Jack B. Graham and Meredith F. Burrill (Washington, D.C.:
AAAS, 1956), 79–81; Allen V. Kneese, "Scope and Challenge of the Water Pollution
Situation," in *Water Pollution*, ed. Ted L. Willrich and N. William Hines (Ames:
Iowa State University Press, 1965), 56, 60; Greeley, "Water Resource and Pollution
Control Legislation," 62–63.

48. Henry J. Graeser, "America's Drinking Water Is/Is Not Safe," *American City* 85 (June 1970): 79; "Where We Stand on Pollution Control," 78.

49. Scott, "Federal and State Legislation," 886–88; "Where We Stand on Pollution Control," 78–79.

50. J. Clarence Davies III, *The Politics of Pollution* (New York: Pegasus, 1970), 40–41; Camp, "Pollution Abatement Policy," 252–53; Federal Security Agency, USPHS, *Water Pollution in the United States* (Washington, D.C.: GPO, 1951), 36; Hollis, "Water Pollution Abatement," 89; Hollis and McCallum, "Federal Water Pollution Control Legislation," 307; Micklin, "Water Quality," 131.

51. Davies, *Politics of Pollution*, 40–41; Hollis, "Water Pollution Abatement," 91; Kruse, "Our Nation's Water," 52; Micklin, "Water Quality," 131–32; Ross, "Water Pollution," 18; Hollis and McCallum, "Federal Water Pollution Control Legislation," 308; Murray Stein, "Legal Aspects Stimulate Pollution Control Program," *Civil Engineering* 32 (July 1962): 50; Martin Reuss, "The Management of Stormwater Systems," in *Water and the City*, ed. Howard Rosen and Ann Durkin Keating (Chicago: Public Works Historical Society, 1991), 327.

52. Davies, *Politics of Pollution*, 40; Micklin, "Water Quality," 132; Stein, "Legal Aspects Stimulate Pollution Control Program," 50.

53. Howells, "We Need More Municipal Waste Treatment Works," 54; Stein, "Legal Aspects Stimulate Pollution Control Program," 51; Greeley "Water Resource and Pollution Control Legislation," 61; Water Resources Council, *Nation's Water Resources*, 4-1-2; Armstrong et al., *History of Public Works*, 232.

54. Graeser, "America's Drinking Water Is/Is Not Safe," 79; Stein, "Problems and Programs in Water Pollution," 403, 411, 413.

55. Micklin, "Water Quality," 133–34.

56. Ibid., 136–41; Kneese, "Scope and Challenge of the Water Pollution Situation," 3; John F. Timmons, "Economics of Water Quality," in Willrich and Hines, *Water Pollution*, 3–4.

57. "Water Pollution," *Science* 150 (Oct. 8, 1965): 198.

58. Ibid.; Micklin, "Water Quality," 141; Timmons, "Economics of Water Quality," 63.

59. "Water Pollution: Federal Role," *Science* 150 (Oct. 8, 1965): 199; Micklin, "Water Quality," 142–44.

60. "Vietnam Peace Could Bring Surge in Water Pollution Control," *American City* 83 (Sept. 1968): 124.

Chapter 16. Beyond Their Limits

1. "Nationwide Sewage Treatment—How It Looks Today," *American City* 65 (June 1950): 137; Fair et al., *Elements of Water Supply and Wastewater Disposal*, 14.

2. "Sewage Works Show Growth in Small Communities," *American City* 71 (Mar. 1956): 148; "More Sewers Than Ever," *American City* 74 (Feb. 1959): 14; *Water Supply and Sewage Disposal* (Paris: Organization of European Economic Cooperation, 1953), 69.

3. Abel Wolman, "The Metabolism of Cities," *Scientific American*, Sept. 1965, 184.

4. Kruse, "Our Nation's Water," 23–24, 53; Thompson, "Worst Public-Works Problem," 6.

5. Institute for Training in Municipal Administration, *Municipal Public Works Administration,* 52, 62; "How Air Conditioning Affects Water Supply" *American City* 63 (July 1948): 9; "273 Cities Charge Sewer Rentals," *American City* 65 (Aug. 1950): 21.

6. "User Charges, Not Grants, Should Pay Utility Costs," *American City* 91 (Nov. 1976): 72; Robie L. Mitchell, "Sewer Revenue Financing," *American City* 61 (Nov. 1946): 121–22.

7. *Water Supply and Sewage Disposal,* 70.

8. Teaford, *Rough Road to Renaissance,* 91; William Edwin Ross and George Erganian, "Sewer Extension Promotes Municipal Growth," *Civil Engineering* 33 (Mar. 1963): 48–49.

9. Joseph A. Salvato, "Problems of Wastewater Disposal in Suburbia," *Public Works* 95 (Mar. 1964): 120.

10. Kruse, "Our Nation's Water," 55; V. G. MacKenzie, "Research Studies on Individual Sewage Disposal Systems," *AJPH* 42 (May 1952): 411; Salvato, "Problems of Wastewater Disposal," 120.

11. Lewis Herber, *Crisis in Our Cities* (Englewood Cliffs, N.J.: Prentice-Hall, 1965), 16.

12. John E. Kiker, "Developments in Septic Tank Systems," *Transactions of the ASCE* 123 (1958): 77, 83.

13. Teaford, *Rough Road to Renaissance,* 90; "Cities Increase Service Charges to Suburbs," *National Municipal Review* 35 (July 1946): 379.

14. David J. Galligan, "Townships Consolidate Sewerage Systems," *American City* 69 (Oct. 1954): 88; Phil Holley, "Inter-City Water-Sewerage Agreements," *American City* 79 (Dec. 1964): 73–75; Arthur D. Caster, "County-Owned, City-Managed Sewerage System," *American City* 84 (Aug. 1969): 117–18.

15. William S. Foster, "Metropolitan Sewerage Pacts," *American City* 75 (Oct. 1960): 87–89.

16. William S. Foster, "Metropolitan Sewerage Pacts," *American City* 75 (Nov. 1960): 1769–76; William S. Foster, "Metropolitan Sewerage Pacts," *American City* 75 (Dec. 1960): 143–47; "Sewerage Authorities—Their Short-comings and Advantages," *American City* 68 (Aug. 1953): 15; "Sewage Problems Can Be Solved Without Separate Districts," *American City* 71 (Feb. 1956): 138–41.

17. Martin Lang et al., "Sewering the City of New York," *Civil Engineering—ASCE* 46 (Jan. 1976): 55.

18. "80-Year-Old Sewer Collapses," *American City* 70 (Mar. 1955): 122.

19. "Better Sewer Pipe Needed," *American City* 69 (Mar. 1954): 27.

20. Thoman and Jenkins, *Statistical Summary of Sewage Works,* 5, 8–9, 20, 29.

21. Allen Cynwin and William A. Rosenkranz, "Advances in Storm and Combined Sewer Pollution Abatement Technology," paper presented at the Water Pollution Control Federation meeting, San Francisco, Oct. 3–8, 1971, 18.

22. Sullivan, "Assessment of Combined Sewer Problems," 108.

23. Ibid., 109–11, 115; APWA, *Report on Problems of Combined Sewer Facilities,* xiv–xv, 2–4; Edward Scott Hopkins, W. McLean Bingley, and George Wayne

Schucker, *The Practice of Sanitation in Its Relation to the Environment,* 4th ed. (Baltimore: Williams and Wilkins, 1970), 324.

24. Sullivan, "Assessment of Combined Sewer Problems," 109; "No More Combined Sewers," *American City* 81 (May 1966): 30; J. A. Bronow, "Separate Those Sewers," *American City* 82 (Feb. 1967): 94–96; Cynwin and Rosenkranz, "Advances in Storm and Combined Sewer Pollution Abatement Technology," 1; APWA, *Report on Problems of Combined Sewer Facilities,* xvii; Vinton W. Bacon, "Separate Storm and Sanitary Sewers Not the Answer in Chicago," *American City* 82 (Jan. 1967): 67.

25. Cynwin and Rosenkranz, "Advances in Storm and Combined Sewer Pollution Abatement Technology," 5.

26. "Whose Responsibility Is Sewage Treatment?" *American City* 61 (Mar. 1946): 15.

27. "Nationwide Sewage Treatment," 137; Thoman and Jenkins, *Statistical Summary of Sewage Works,* 10, 22, 25; Stein, "Problems and Programs in Water Pollution," 397.

28. "Sewerage and Sewage Treatment Planning," *American City* 61 (Apr. 1946): 119; Federal Security Agency, *Water Pollution in the United States,* 22; Morris M. Cohn, "Over 1,100 Treatment Plants under Construction in 1950," *Wastes Engineering* 22 (Mar. 1951): 125–35; John R. Thoman and Kenneth H. Jenkins, *Municipal Sewage Treatment Needs* (Washington, D.C.: USPHS, HEW, 1958), 5–6, 30.

29. Geoffrey A. Parkes, "Los Angeles Aims at Perfection," *American City* 66 (June 1951): 79–81, 169.

30. W. W. Eckenfelder Jr., "Theory and Practice of Activated Sludge Process Modifications," *Water and Sewage Works* 108 (Apr. 1961): 145–50; "More Sewers Than Ever," 15.

31. "The Treatment of Industrial Wastes," *AJPH* 36 (Mar. 1946): 281.

32. "Eight-State Drive to Clean Up Rivers," *Business Week,* July 31, 1948, 26.

33. Federal Security Agency, *Water Pollution in the United States,* 10.

34. E. Weisberg, R. A. Phillips, and T. Helfgott, "New Aspects of Waste-Water Reclamation," *American City* 79 (Aug. 1964): 91.

35. Ibid.

36. W. Wesley Eckenfelder and John W. Hood, "Detergents and Foaming in Sewage Treatment," *American City* 65 (May 1950): 132.

37. Detergents—Are They Really to Blame?" *American City* 68 (Jan. 1953): 105; "Detergents—Not Necessarily Guilty," *American City* 71 (Mar. 1956): 120; Eckenfelder and Hood, "Detergents and Foaming," 133; "The Problem of Detergents in Sanitary Engineering," *American City* 66 (Sept. 1950): 115.

38. "Soapless Opera," *Scientific American,* July 1953, 48.

39. "Uncle Sam Steps Up Drive to Halt Pollution by Cities, Industries," *Wall Street Journal,* Nov. 14, 1960; Rolf Eliassen, "Stream Pollution," in Katz and Felton, *Health and the Community,* 94; "Federal Aid for Sewage Treatment," *American City* 71 (Sept. 1956): 5.

40. Samuel I. Zack, "The Case for Federal Aid for Sewage Works," *American City* 71 (Aug. 1956): 165.

41. Howells, "We Need More Municipal Waste Treatment Works," 54; "446

Cities Share in Aid," *Texas Municipalities* 44 (Oct. 1957): 285; "Sewage Works Get Federal Aid," *American City* 71 (Aug. 1956): 15.

42. "Federal Grants for Sewage Works—A Sugar-Coated Warning," *American City* 71 (Aug. 1956): 18.

43. "Legal War on Water Pollution," *Business Week*, July 16, 1960, 132, 134; Howells, "We Need More Municipal Waste Treatment Works," 54; Hopkins et al., *Practice of Sanitation*, 341–42.

44. Hopkins et al., *Practice of Sanitation*, 321, 348; Cynwin and Rosenkranz, "Advances in Storm and Combined Sewer Pollution Abatement Technology," 1.

45. "Collision Course on Pollution," *Business Week*, Nov. 29, 1969, 44.

Chapter 17. Solid Waste as "Third Pollution"

1. William E. Small, *Third Pollution* (New York: Praeger, 1970), 7.

2. Kenneth A. Hammond, George Micinko, and Wilma B. Fairchild, eds., *Sourcebook on the Environment* (Chicago: University of Chicago Press, 1978), 327.

3. William D. Ruckelshaus, "Solid Waste Management," *Public Management* (Oct. 1972): 2–4.

4. Melosi, "Waste Management," 9.

5. Bernard Baum et al., *Solid Waste Disposal*, vol. 1 (Ann Arbor, Mich.: Science, 1974), 3–4; Brian J. L. Berry and Frank E. Horton, *Urban Environmental Management* (Englewood Cliffs, N.J.: Prentice Hall, 1974), 259; Alfred J. Van Tassel, ed., *Our Environment* (Lexington, Mass.: Lexington Books, 1973), 460; Melosi, *Garbage in the Cities*, 207–8; EPA, *Characterization of Municipal Solid Waste in the United States: 1990 Update; Executive Summary* (Washington, D.C.: EPA, June 13, 1990), 3; William Rathje and Cullen Murphy, *Rubbish!* (New York: Harper-Collins, 1992), 101; George Tchobanoglous, Hilary Theisen, and Rolf Eliassen, *Solid Wastes* (New York: McGraw-Hill, 1977), 7.

6. EPA, *The Solid Waste Dilemma: An Agenda for Action: Background Documents* (Washington, D.C.: EPA, Sept. 1988), 1-18, 1-19.

7. John A. Burns and Michael J. Seaman, "Some Aspects of Solid Waste Disposal," in Van Tassel, *Our Environment*, 457–58; Baum et al., *Solid Waste Disposal*, i–vi; Laurent Hodges, *Environment Pollution* (New York: Holt, Rinehart, and Winston, 1973), 219; C. L. Mantell, ed., *Solid Wastes* (New York: Wiley, 1975), 32, 35.

8. EPA, *Solid Waste Dilemma: An Agenda for Action*, 1–9; Keep America Beautiful, "An Introduction to Municipal Solid Waste Management," *Focus* 1 (1991): 1; U.S. HEW, USPHS, Environmental Health Service, Bureau of Solid Waste Management, *Solid Waste Management, Abstracts and Excerpts from the Literature*, Pub. No. 2038, 2 vols (Washington, D.C.: GPO, 1970), 35.

9. Berry and Horton, *Urban Environmental Management*, 260; Amos Turk, Jonathan Turk, and Janet T. Wittes, *Ecology, Pollution, Environment* (Philadelphia: Saunders, 1972), 138.

10. Baum et al., *Solid Waste Disposal*, i–vi; D. Joseph Hagerty, Joseph L. Pavoni, and John E. Heer Jr., *Solid Waste Management* (New York: Van Nostrand Rein-

hold, 1973), 13–14; APWA, *Municipal Refuse Disposal*, viii–ix; National League of Cities and U.S. Conference of Mayors, Solid Waste Management Task Force, *Cities and the Nation's Disposal Crisis* (Washington, D.C.: National League of Cities and U.S. Conference of Mayors, Mar. 1973), 3, 32; Melosi, "Waste Management," 9; Werner Z. Hirsch, "Cost Functions of an Urban Government Service," *Review of Economics and Statistics* 47 (Feb. 1965): 92.

11. Hopkins et al., *Practice of Sanitation*, 226.

12. Peter Kemper and John M. Quigley, *Economics of Refuse Collection* (Cambridge, Mass.: Ballinger, 1976), 109–11; Hagerty et al., *Solid Waste Management*, 13.

13. APWA, *Solid Waste Collection Practice*, 36–39; Hagerty et al., *Solid Waste Management*, 20; William E. Korbitz, ed., *Urban Public Works Administration* (Washington, D.C.: International City Management Association, 1976), 439.

14. Homer A. Neal and J. R. Korbitz, *Solid Waste Management and the Environment* (Englewood Cliffs, N.J.: Prentice-Hall, 1987), 29.

15. Armstrong et al., *History of Public Works*, 442, 444–46.

16. Hoy, "Garbage Disposer," 758; Strasser, "Leftovers and Litter," 7; Melosi, *Garbage in the Cities*, 180–81.

17. Dennis Young, *How Shall We Collect the Garbage?* (Washington, D.C.: Urban Institute, 1972), 12.

18. Savas, *Organization and Efficiency of Solid Waste Collection*, 37–38.

19. Baum et al., *Solid Waste Disposal*, 1:5.

20. E. S. Savas, "Intracity Competition between Public and Private Service Delivery," *Public Administration Review* 41 (Jan./Feb. 1981): 47–48; Young, *How Shall We Collect the Garbage?* 8–10.

21. Armstrong et al., *History of Public Works*, 446–47; Charles G. Burck, "There's Big Business in All That Garbage," *Fortune*, Apr. 7, 1980, 106–7.

22. Peter Reuter, "Regulating Rackets," *Regulation* 8 (Sept.–Dec. 1984): 29–30, 33–34.

23. Esher Shaheen, *Environmental Pollution* (Mahomet, Ill.: Engineering Technology, 1974), 261.

24. "Sanitary Landfill or Incineration?" *American City* 66 (Mar. 1951): 99; Tarr, "Historical Perspectives on Hazardous Wastes," 99.

25. Thomas J. Sorg and Thomas W. Bendixen, "Sanitary Landfill," in Mantell, *Solid Wastes*, 71–72; Armstrong et al., *History of Public Works*, 450.

26. Melosi, *Garbage in the Cities*, 184–85.

27. Melosi, "Waste Management," 13.

28. Joel A. Tarr, "Risk Perception in Waste Disposal," unpublished paper, 20–22.

29. Eliassen, "Incinerator Mechanization Wins Increasing Favor," 17–19.

30. Ibid.; Cohn, "Highlights of Incinerator Construction," 87; E. R. Bowerman, "What Cities Use Incinerators—And Why?" *American City* 67 (Mar. 1952): 100.

31. Casimir A. Rogus, "Refuse Incineration—Trends and Developments," *American City* 74 (July 1959): 94–97.

32. "The Incinerator—'A Machine of Beauty,'" *American City* 69 (Aug. 1954): 85; "Sanitary Landfill or Incineration?" 98–99.

33. Mantell, *Solid Wastes*, 21.

34. "Incinerator—Residue Study Under Way," *American City* 80 (Mar. 1965): 20; Junius W. Stephenson, "Planning for Incineration," *Civil Engineering* 34 (Sept. 1964): 38, 40; APWA, *Municipal Refuse Disposal*, viii.

35. Melosi, *Garbage in the Cities*, 187–88.

36. Melosi, "Waste Management," 12; Fenton, "Current Trends in Municipal Solid Waste Disposal," 170; Hodges, *Environment Pollution*, 260; Shaheen, *Environmental Pollution*, 260.

37. Shaheen, *Environmental Pollution*, 271; Hopkins et al., *Practice of Sanitation*, 247; Max L. Panzer and Harvey F. Ludwig, "Should We Reconsider Composting of Organic Refuse?" *Civil Engineering* 21 (Feb. 1951): 40–41; APWA, *Municipal Refuse Disposal*, 296–98.

38. EPA, *Legal Compilation: Statutes and Legislative History, Executive Orders, Regulations, Guidelines and Reports*, Suppl. 2, vol. 1, *Solid Waste* (Washington, D.C.: EPA, Jan. 1974), 45–50; J. Rodney Edwards, "Recycling Waste Paper," in Mantell, *Solid Wastes*, 883–90.

39. Frank P. Grad, "The Role of the Federal and State Governments," in Savas, *Organization and Efficiency of Solid Waste Collection*, 169.

40. Stanley D. Degler, *Federal Pollution Control Programs*, rev. ed. (Washington, D.C.: Bureau of National Affairs, 1971), 36.

41. APWA, *Municipal Refuse Disposal*, x.

42. Melosi, *Garbage in the Cities*, 200–201; Grad, "Role of the Federal and State Governments," 169–70.

43. Mantell, *Solid Wastes*, 3–7; Hagerty et al., *Solid Waste Management*, 268–69; APWA, *Municipal Refuse Disposal*, i–z; Shaheen, *Environmental Pollution*, 9.

44. Ernest Flack and Margaret C. Shipley, *Man and the Quality of His Environment* (Boulder: University of Colorado Press, 1968), 117–19.

45. Tchobanoglous et al., *Solid Wastes*, 40; Mantell, *Solid Wastes*, 11–12; Kemper and Quigley, *Economic of Refuse Collection*, 5.

46. Armstrong et al., *History of Public Works*, 453; Tchobanoglous et al., *Solid Wastes*, 40–43; Hagerty et al., *Solid Waste Management*, 269, 283–91; Douglas B. Cargo, *Solid Wastes* (Chicago: University of Chicago, Department of Geography, 1978), 74.

47. Melosi, "Waste Management," 7; Degler, *Federal Pollution Control Programs*, 37–38.

48. Van Tassel, *Our Environment*, 468; Grad, "Role of the Federal and State Governments," 169–83; Armstrong et al., *History of Public Works*, 453; Peter S. Menell, "Beyond the Throwaway Society," *Ecology Law Quarterly* 7 (1990): 671; William L. Kovacs, "Legislation and Involved Agencies," in *The Solid Waste Handbook*, ed. William D. Robinson (New York: Wiley, 1986), 9; Berry and Horton, *Urban Environmental Management*, 361–62.

49. Craig E. Colten, "Chicago's Waste Lands," *Journal of Historical Geography* 20 (1994): 124.

Chapter 18. From Earth Day to Infrastructure Crisis

1. Goldfield and Brownell, *Urban America*, 375.

2. Jon C. Teaford, *The Metropolitan Revolution* (New York: Columbia University Press, 2006), 240.

3. U.S. Council on Environmental Quality, *Environmental Quality: Twenty-fourth Annual Report of the Council on Environmental Quality* (Washington, D.C.: GPO, 1993), 385; U.S. Bureau of the Census, *Statistical Abstract of the United States* (Washington, D.C.: Department of Commerce, 1995), 43.

4. U.S. Bureau of the Census, *Statistical Abstract of the United States*, 39; U.S. Bureau of the Census, *1990 Census of Population and Housing, Supplemental Reports: Urbanized Areas of the United States and Puerto Rico*, sec. 1 (Washington, D.C.: Department of Commerce, 1993), II-a; U.S. Census Bureau, *Population Change and Distribution, 1990–2000* (Apr. 2001), 5, http://www.census.gov/prod/2001pubs/c2kbr01-2.pdf.

5. Chudacoff and Smith, *Evolution of American Urban Society*, 289–90; Abbott, *Urban America in the Modern Age*, 136; Goldfield and Brownell, *Urban America*, 380–81, 414–23.

6. Silver, "Housing Policy and Suburbanization," 71; Bernard H. Ross and Myron A. Levine, *Urban Politics*, 5th ed. (Itasca, Ill.: F. E. Peacock, 1996), 59, 285–89.

7. Wendell Cox, "The Role of Urban Planning in the Decline of Central Cities," *Demographia* (June 2005), 16, http://demographia.com/db-xplannerscities.pdf.

8. Teaford, *Twentieth-Century American City*, 153.

9. Jackson, *Crabgrass Frontier*, 284.

10. Teaford, *Twentieth-Century American City*, 153–54; Teaford, *Metropolitan Revolution*, 242–43; Chudacoff and Smith, *Evolution of American Urban Society*, 288–89, 301; Abbott, *Urban America in the Modern Age*, 111, 113–15, 132; Miller and Melvin, *Urbanization of Modern America*, 213; Goldfield and Brownell, *Urban America*, 435–48.

11. Chudacoff and Smith, *Evolution of American Urban Society*, 294.

12. Howard Chernick and Andrew Reschovsky, "Urban Fiscal Problems," in *The Urban Crisis*, ed. Burton A. Weisbrod and James C. Worthy (Evanston, Ill.: Northwestern University Press, 1997), 132, 135–36; Teaford, *Metropolitan Revolution*, 167–84.

13. Daniel T. Lichter and Martha L. Crowley, "Poverty Rates Vary Widely across the United States," Population Reference Bureau (2007), http://www.prb.org/Articles/2002/PovertyRatesVaryWidelyAcrosstheUnitedStates.aspx.

14. Chernick and Reschovsky, "Urban Fiscal Problems," 138–41.

15. Teaford, *Twentieth-Century American City*, 142; Teaford, *Rough Road to Renaissance*, 218, 225, 262, 265; Chudacoff and Smith, *Evolution of American Urban Society*, 294–95; Lawrence J. R. Herson and John M. Bolland, *The Urban Web* (Chicago: Nelson-Hall, 1990), 347.

16. Goldfield and Brownell, *Urban America*, 385–87; Teaford, *Rough Road to Renaissance*, 227–30; Teaford, *Twentieth-Century American City*, 143–46; Abbott, *Urban America in the Modern Age*, 130; Chudacoff and Smith, *Evolution of American Urban Society*, 295–96.

17. Marian Lief Palley and Howard A. Palley, *Urban America and Public Poli-*

cies, 2nd ed. (Lexington, Mass.: D. C. Heath, 1981), 24, 59; Teaford, *Twentieth-Century American City*, 140–41; Chudacoff and Smith, *Evolution of American Urban Society*, 293; Herson and Bolland, *Urban Web*, 335–36.

18. Benjamin Kleinberg, *Urban America in Transformation* (Thousand Oaks, Calif.: Sage, 1995), 187–93; Miller and Melvin, *Urbanization of Modern America*, 210–11, 236–38; Teaford, *Twentieth-Century American City*, 140.

19. Herson and Bolland, *Urban Web*, 306; Palley and Palley, *Urban America and Public Policies*, 92; Goldfield and Brownell, *Urban America*, 388–90; Abbott, *Urban America in the Modern Age*, 130, 131; Chudacoff and Smith, *Evolution of American Urban Society*, 292–93; Miller and Melvin, *Urbanization of Modern America*, 211.

20. Kleinberg, *Urban America in Transformation*, 210–13; Goldfield and Brownell, *Urban America*, 392–94.

21. George E. Peterson and Carol W. Lewis, eds., *Reagan and the Cities* (Washington, D.C.: Urban Institute Press, 1986), 1; Chernick and Reschovsky, "Urban Fiscal Problems," 146. An enterprise zone was an area in a low-income neighborhood in which the federal government would offer tax concessions and other inducements to businesses in exchange for the promise of economic development. Some were established in the 1980s, but few were successful.

22. Chernick and Reschovsky, "Urban Fiscal Problems," 146.

23. Peterson and Lewis, *Reagan and the Cities*, 1–10; Kleinberg, *Urban America in Transformation*, 226–36, 242–44; Miller and Melvin, *Urbanization of Modern America*, 228–39; Abbott, *Urban America in the Modern Age*, 131; Chudacoff and Smith, *Evolution of American Urban Society*, 294, 296–97; Goldfield and Brownell, *Urban America*, 433–35.

24. U.S. Census Bureau (2007), http://www.census.gov/govs/estimate/0400ussl_1.htm.

25. Committee on Infrastructure Innovation, National Research Council, *Infrastructure for the 21st Century* (Washington, D.C.: National Academy Press, 1987), 9; NCPWI, *The Nation's Public Works: Defining the Issues* (Washington, D.C.: NCPWI, Sept. 1986), 57; NCPWI, *Fragile Foundations* (Washington, D.C.: GPO, Feb. 1988), 12–13.

26. Pat Choate and Susan Walter, *America in Ruins* (Durham, N.C.: Duke Press Paperbacks, 1981), xi.

27. Ibid., xi–xii, 1–4.

28. George E. Peterson and Mary John Miller, *Financing Public Infrastructure* (Washington, D.C.: Community and Economic Development Task Force, HUD, 1982), 1.

29. Government Finance Research Center, Municipal Finance Officers Association, *Building Prosperity* (Washington, D.C.: Municipal Finance Officers Association, Oct. 1983), 2–3, 72; U.S. Congress, Congressional Budget Office, *Public Works Infrastructure* (Washington, D.C.: GPO, Apr. 1983), 1, 6–7, 14; Touche Ross and Co., *The Infrastructure Crisis* (New York: Touche Ross, 1983), 1.

30. John P. Eberhard and Abram B. Bernstein, eds., *Technological Alternatives for Urban Infrastructure* (Washington, D.C.: National Research Council and Urban Land Institute, Dec. 1985), 1–2, 40, 46, 51–52.

31. NCPWI, *Defining the Issues*, 1, 6.

32. NCPWI, *Fragile Foundations*, 1, 6.

33. Ibid., 7–8, 10, 43, 45.

34. Roger W. Caves, *Exploring Urban America* (Thousand Oaks, Calif.: Sage, 1995), 247–55; Ralph Gakenheimer, "Infrastructure Shortfall," *American Planning Association Journal* 55 (Winter 1987): 22; Bruce Seely, "A Republic Bound Together," *Wilson Quarterly* 17 (Winter 1993): 19–20, 38; Michael Pagano, "Local Infrastructure," *Public Works Management and Policy* 1 (July 1996): 19–22; Carol T. Everett, "So Is There an Infrastructure Crisis or What?" *Public Works Management and Policy* (July 1996): 88–95.

35. American Society for Civil Engineers, "Report Card for America's Infrastructure, 2005," http://www.asce.org/reportcard/2005/index.cfm.

36. Melosi, *Coping with Abundance*, 297; Gottlieb, *Forcing the Spring*, 105–14.

37. Melosi, *Coping with Abundance*, 297–98; Gottlieb, *Forcing the Spring*, 109–10.

38. Wallis E. McClain Jr., ed., *U.S. Environmental Laws: 1994 Edition* (Washington, D.C.: Bureau of National Affairs, 1994), 9-1; Gottlieb, *Forcing the Spring*, 124–25; Melosi, *Coping with Abundance*, 298.

39. Melosi, *Coping with Abundance*, 298; McClain, *U.S. Environmental Laws*, 9-1; Gottlieb, *Forcing the Spring*, 128–29.

40. Richard N. L. Andrews, "Environmental Protection Agency," in *Conservation and Environmentalism*, ed. Robert Paehlke (New York: Garland, 1995), 256; Gottlieb, *Forcing the Spring*, 129; Joseph Petulla, *Environmental Protection in the United States* (San Francisco: San Francisco Study Center, 1987), 48–49.

41. Gottlieb, *Forcing the Spring*, 126, 129.

42. Daniel H. Henning and William R. Mangun, *Managing the Environmental Crisis* (Durham, N.C.: Duke University Press, 1989), 20–21, 27.

43. Petulla, *Environmental Protection*, 56–57.

44. Terry Davies, "Environmental Protection Agency," in Eblen and Eblen, *Encyclopedia of the Environment*, 223; Henning and Mangun, *Managing the Environmental Crisis*, 27–29.

45. Sir Shridath Ramphal, "Sustainable Development," in Eblen and Eblen, *Encyclopedia of the Environment*, 680–83; Lester W. Milbrath, "Sustainability," in Paehlke, *Conservation and Environmentalism*, 612–13.

46. Eileen Maura McGurty, "From NIMBY to Civil Rights," *Environmental History* 2 (July 1997): 305–14.

47. Andrew Szasz, *Ecopopulism* (Minneapolis: University of Minnesota Press, 1994), 5.

48. Lois Marie Gibbs, "Celebrating Ten Years of Triumph," *Everyone's Backyard* 11 (Feb. 1993): 2.

49. Szasz, *Ecopopulism*, 6, 69–72.

50. Cynthia Hamilton, "Coping with Industrial Exploitation," in *Confronting Environmental Racism*, ed. Robert D. Bullard (Boston: South End Press, 1993), 63.

51. Robert D. Bullard, *Dumping in Dixie*, 2nd ed. (Boulder, Colo.: Westview Press, 1994), xiii.

52. Quoted in Karl Grossman, "The People of Color Environmental Summit,"

in *Unequal Protection,* ed. Robert D. Bullard (San Francisco: Sierra Club Books, 1994), 272.

53. Bunyan Bryant and Paul Mohai, eds., *Race and the Incidence of Environmental Hazards* (Boulder, Colo.: Westview Press, 1992), 1–2; Dana A. Alston, ed., *We Speak for Ourselves* (Washington, D.C.: Panos Institute, 1990), 3; "From the Front Lines of the Movement for Environmental Justice," *Social Policy* 22 (Spring 1992): 12; Robert D. Bullard, "Anatomy of Environmental Racism and the Environmental Justice Movement," in Bullard, *Confronting Environmental Racism,* 22–23.

54. Charles Lee, "Toxic Waste and Race in the United States," in Bryant and Mohai, *Race and the Incidence of Environmental Hazards,* 10–16, 22–27; Karl Grossman, "Environmental Racism," *Crisis* 98 (Apr. 1991): 16–17; Karl Grossman, "From Toxic Racism to Environmental Justice," *E: The Environmental Magazine* 3 (May/June 1992): 30–32; Dick Russell, "Environmental Racism," *Amicus Journal* 11 (Spring 1989): 22–25.

55. Daniel Kevin, "'Environmental Racism' and Locally Undesirable Land Uses," *Villanova Environmental Law Journal* 8 (1997): 122; Rachel D. Godsil, "Remedying Environmental Racism," *Michigan Law Review* 90 (Nov. 1991): 397–98; Bullard, *Confronting Environmental Racism;* Vicki Been, "What's Fairness Got to Do with It?" *Cornell Law Review* 78 (Sept. 1993): 1014–15, 1018–24; Vicki Been, "Locally Undesirable Land Uses in Minority Neighborhoods," *Yale Law Journal* (Apr. 1994): 1386, 1406.

56. Bullard, *Dumping on Dixie,* xv.

57. Bullard, *Unequal Protection,* xvi.

58. President William Clinton, "Executive Order on Federal Actions to Address Environmental Justice in Minority Populations and Low-Income Populations," Washington, D.C., Feb. 11, 1994; "Not in My Backyard," *Human Rights* 20 (Fall 1993): 27–28; Bryant and Mohai, *Race and the Incidence of Environmental Hazards,* 5; Grossman, "People of Color Environmental Summit," 287.

59. Gottlieb, *Forcing the Spring,* 235–69; Mark Dowie, *Losing Ground* (Cambridge, Mass.: MIT Press, 1997), 125–35, 170–72.

60. Gottlieb, *Forcing the Spring,* 207, 227–30, 233–34; Carolyn Merchant, *Radical Ecology* (New York: Routledge, 1992), 183–209.

61. Richard N. L. Andrews, *Managing the Environment, Managing Ourselves* (New Haven, Conn.: Yale University Press, 2006), 351.

62. Ibid., 350–95.

63. Ibid., 351.

64. Odum, "Ecology as a Science," 171; Herman Koren, ed., *Handbook of Environmental Health and Safety,* vol. 1 (Boca Raton, Fla.: Lewis, 1991), 81.

65. Daniel Woltering and Talbot Page, "Ecological Risk," in Eblen and Eblen, *Encyclopedia of the Environment,* 163.

66. Hays, *Beauty, Health, and Permanence,* 338; Walter A. Rosenbaum, *Environmental Politics and Policy,* 3rd ed. (Washington, D.C.: Congressional Quarterly Press, 1995), 9–11, 75–78, 175.

67. P. Aarne Vesilind, J. Jeffrey Peirce, and Ruth F. Weiner, *Environmental Engineering,* 3rd ed. (Boston: Butterworth-Heinemann, 1994), 11; Robert A. Corbitt, ed., *Standard Handbook of Environmental Engineering* (New York: McGraw-Hill,

1990), 1.2–1.6; C. Maxwell Stanley, "The Engineer and the Environment," *Civil Engineering—ASCE* 42 (July 1972): 79.

68. Arcadio P. Sincero and Gregoria A. Sincero, *Environmental Engineering* (Upper Saddle River, N.J.: Prentice-Hall, 1996), xv, 1–2.

69. Jeffrey K. Stine, "Engineering a Better Environment," paper presented at the SHOT/HSS Critical Problems and Research Frontiers Conference, Madison, Wis., 1991, 11.

70. Vesilind et al., *Environmental Engineering*, 11.

71. Paul N. Cheremisinoff and Richard A. Young, eds., *Pollution Engineering Practice Handbook* (Ann Arbor, Mich.: Ann Arbor Science, 1975), v.

72. William J. Mitsch and Sven Erik Jorgensen, eds., *Ecological Engineering* (New York: Wiley, 1989), 4–5.

73. Peter C. Schulze, ed., *Engineering within Ecological Constraints* (Washington, D.C.: National Academy Press, 1996), 1, 3, 6, 31, 33, 66–68, 79, 112–14, 131.

Chapter 19. Beyond Broken Pipes and Tired Treatment Plants

1. NCPWI, *The Nation's Public Works: Executive Summaries of Nine Studies* (Washington, D.C.: NCPWI, May 1987), 37–38.

2. Neil S. Grigg, *Urban Water Infrastructure* (New York: Wiley, 1986), 7–8.

3. Everett, "So Is There an Infrastructure Crisis or What?" 91; Jesse H. Ausubel and Robert Herman, eds., *Cities and Their Vital Systems* (Washington, D.C.: National Academy Press, 1988), 265.

4. Grigg, *Urban Water Infrastructure*, 17; NCPWI, *The Nation's Public Works: Report on Water Supply* (Washington, D.C.: NCPWI, 1987), 14; David Holtz and Scott Sebastian, eds., *Municipal Water Systems* (Bloomington: Indiana University Press, 1978), 71.

5. Sam M. Cristofano and William S. Foster, eds., *Management of Local Public Works* (Washington, D.C.: International City Management Association, 1986), 280; Duane D. Baumann and Daniel Dworkin, *Water Resources for Our Cities* (Carbondale.: Southern Illinois University Press, 1978), 8; U.S. Water Resources Council, *The Nation's Water Resources, 1975–2000*, vol. 1 (Washington, D.C.: U.S. Water Resources Council, Dec. 1978), 2.

6. CEQ, *Environmental Quality: Twenty-fourth Annual Report of the Council on Environmental Quality* (Washington, D.C.: GPO, 1993), 55–57; U.S. Bureau of the Census, *Statistical Abstract of the United States* (Washington, D.C.: Department of Commerce, 1995), 232; Conservation Foundation, *State of the Environment: A View from the Nineties* (Washington, D.C.: Conservation Foundation, 1987), 225–26, 232.

7. Susan S. Hutson et al., "Estimated Use of Water in the United States in 2000," *U.S. Geological Survey Circular 1268* (Mar. 2004; revised Feb. 2005), http://pubs.usgs.gov/circ/2004/circ1268.

8. NCPWI, *Report on Water Supply*, 7, 17–18, 91; Hutson et al., "Estimated Use of Water in the United States in 2000."

9. Janet Werkman and David L. Weterling, "Privatizing Municipal Water and Wastewater Systems," *Public Works Management and Policy* 5 (July 2000): 52.

10. Anthony Lenze, "Liquid Assets," *Pittsburgh Post-Gazette*, Sept. 16, 2003;

Christopher D. Cook, "Drilling for Water in the Mojave," *Progressive*, Oct. 2002, 19–20; Lolis Eric Elie, "Privatization Argument Has Its Leaks," *New Orleans Times-Picayune*, Mar. 31, 2003.

11. Norris Hundley Jr., *The Great Thirst* (Berkeley: University of California Press, 1992), 332–47.

12. NCPWI, *Report on Water Supply*, 16; Holtz and Sebastian, *Municipal Water Systems*, 71; Grigg, *Urban Water Infrastructure*, 1, 85.

13. NCPWI, *The Nation's Public Works: Report on Wastewater Management* (Washington, D.C.: NCPWI, 1987), 1, 35; NCPWI, *Executive Summaries of Nine Studies*, 41; NCPWI, *Defining the Issues*, 13; NCPWI, *Fragile Foundations*, 54, 158, 164; Robert B. Williams and Gordon L. Culp, eds., *Handbook of Public Water Systems* (New York: Van Nostrand Reinhold, 1986), 801; U.S. Census Bureau, *Statistical Abstract of the United States*, 2000, http://www.census.gov/prod/www/statistical-abstract-1995_2000.html.

14. Government Finance Research Center, Municipal Finance Officers Association, *Building Prosperity* (Washington, D.C.: Government Finance Research Center, Municipal Finance Officers Association, Oct. 1983), 4; NCPWI, *Report on Wastewater Management*, 86–87; SCS Engineers, *Sewer Moratoria* (Washington, D.C.: Office of Policy Development and Research, HUD, July 1977), 1, 4, 9–15, 17–22.

15. NCPWI, *Fragile Foundations*, 54; George Tchobanoglous and Franklin L. Burton, eds., *Wastewater Engineering* (New York: McGraw-Hill, 1991), 4; U.S. Census Bureau, *Statistical Abstract of the United States*, 2000, http://www.census.gov/prod/www/statistical-abstract-1995_2000.html.

16. H. E. Hudson Jr., "Water Treatment—Present, Near Future, Futuristic," *JAWWA* 68 (June 1976): 275–76; Graham Walton, "Developments in Water Clarification in the U.S.A.," in Society for Water Treatment and Examination and the Water Research Association, "Water Treatment in the Seventies," proceedings of a symposium, Reading, Pa., Jan. 1970, 69; EPA, Office of Water Programs Operations, *Primer for Wastewater Treatment* (Washington, D.C.: EPA, July 1980), 4; Koren, *Handbook of Environmental Health and Safety*, 513–14; William S. Foster, "Wastewater Plants Using Computers," *American City* 90 (Dec. 1975): 66; George A. Sawyer, "New Trends in Wastewater Treatment and Recycle," *Chemical Engineering* 79 (July 24, 1972): 120–28; Russell L. Culp, "No Innovation in Wastewater Treatment?" *Civil Engineering—ASCE* 42 (July 1972): 46–48; Conservation Foundation, *State of the Environment*, 446.

17. ACIR, *Financing Public Physical Infrastructure* (Washington, D.C.: ACIR, June 1984), 12; APWA, *Proceedings of the National Water Symposium* (Washington, D.C.: APWA, Nov. 1982), 11; Peterson and Miller, *Financing Public Infrastructure*, 31–32; Betsy A. Cody et al., *Federally Supported Water Supply and Wastewater Treatment Programs* (Washington, D.C.: Congressional Research Service, Library of Congress, Mar. 25, 2003), 1, 3, http://weller.house.gov.

18. NCPWI, *Report on Wastewater Management*, 14–15, 17; Richard Pinkham and Scott Chaplin, *Water 2010* (Denver: Rocky Mountain Institute, 1996), 4.

19. Steven J. Burian et al., "Urban Wastewater Management in the United States," *Journal of Urban Technology* 7 (2000): 54.

20. CEQ, *Environmental Quality*, 83, 86–87.

21. J. Carrell Morris, "Chlorination and Disinfection—State of the Art," *JAWWA* 63 (Dec. 1971): 769, 772–73; William Whipple Jr., *New Perspectives in Water Supply* (Boca Raton, Fla.: Lewis, 1994), 15–21; Charles D. Larson, O. Thomas Love, and James M. Symons, "Recent Developments in Chlorination Practice," *Journal of the New England Water Works Association* 91 (Sept. 1977): 279; George E. Symons and Kenneth W. Henderson, "Disinfection—Where Are We?" *JAWWA* 69 (Mar. 1977): 148–54.

22. J. Carrell Morris, "Chlorination and Practice," *Proceedings of the Annual Public Water Supply Engineers' Conference* (1978): 31; Joseph T. Ling, "Research—Key to Quality Water Supply in the 1980s," *JAWWA* 68 (Dec. 1976): 659; NCPWI, *Report on Water Supply*, 3, 5.

23. John Cary Stewart, *Drinking Water Hazards* (Hiram, Ohio: Envirographics, 1990), 121–26; "The Fluoridation Controversy," *Health Matrix* 2 (Summer 1984): 66–76.

24. NCPWI, *Report on Wastewater Management*, 10.

25. Williams and Culp, *Handbook of Public Water Systems*, 552; William H. Rodgers Jr., *Environmental Law: Air and Water*, vol. 1 (St. Paul, Minn.: West, 1986), 230–37.

26. Conservation Foundation, *State of the Environment*, 103; Council on Environmental Quality, *Environmental Quality*, 66–69; Koren, *Handbook of Environmental Health and Safety*, 472.

27. EPA, "Policy and Guidance: Fact Sheet," (Apr. 1998), http://www.epa.gov/waterscience/standards/planfs.html.

28. NCPWI, *Report on Water Supply*, 18; Sarah E. Lewis, "The 1986 Amendments to the Safe Drinking Water Act and Their Effect on Groundwater," *Syracuse Law Review* 40 (1989): 894; U.S. Geological Survey, "Ground Water Use in the United States," http://ga.water.usgs.gov/edu/wugw.html.

29. CEQ, *Environmental Quality*, 63; Conservation Foundation, *State of the Environment*, 231; Sally Benjamin and David Belluck, *State Groundwater Regulation* (Washington, D.C.: Bureau of National Affairs, 1994), 9–10; GAO, *Water Supply for Urban Areas* (Washington, D.C.: GPO, June 15, 1979), 10; U.S. Geological Survey, "Ground Water Use in the United States"; "Long Range Planning for Drought Management—The Groundwater Component," U.S. Department of Agriculture, Natural Resources Conservation Service, http://wmc.ar.nrcs.usda.gov/technical/GW/Drought.html.

30. Melosi, "Sanitary Services and Decision Making in Houston," 393.

31. Tames J. Geraghty and David W. Miller, "Status of Groundwater Contamination in the U.S.," *JAWWA* 70 (Mar. 1978): 162.

32. Conservation Foundation, *State of the Environment*, xlii, 96; Lewis, "1986 Amendments to the Safe Drinking Water Act," 897; Benjamin and Belluck, *State Groundwater Regulation*, 3, 7, 10; Carol Wekesser, ed., *Water* (San Diego, Calif.: Greenhaven Press, 1994), 81; Geraghty and Miller, "Status of Groundwater Contamination," 162, 166; Joan Goldstein, *Demanding Clean Food and Water* (New York: Plenum Press, 1990), 113, 116; David E. Lindorff, "Ground-Water Pollution—A Status Report," *Ground Water*, vol. 1 of *Proceedings of the Fourth NWWA-EPA National Ground Water Quality Symposium* (Jan./Feb. 1979), 9–12; U.S. Congress, Office of Technology Assessment, *Protecting the Nation's Groundwater from*

Contamination (Washington, D.C.: GPO, 1984), 7; EPA Region 5 and Agricultural and Biological Engineering, Purdue University, *Ground Water Primer* (May 8, 1998), http://www.purdue.edu/envirosoft/groundwater/src/title.htm.

33. NCPWI, *Report on Water Supply*, 6; Stewart, *Drinking Water Hazards*, 225–26; International Bottled Water Association Web site, http://www.bottled water.org.

34. Laurel Berman et al., *Urban Runoff Water Quality Solutions* (Chicago: APWA Research Foundation, May 1991), 9–10; ACIR, *Sourcebook of Working Documents to Accompany High Performance Public Works* (Washington, D.C.: ACIR, Sept. 1994), 453; Vladimir Novotny and Gordon Chesters, *Handbook of Nonpoint Pollution* (New York: Van Nostrand Reinhold, 1981), 2–3, 7–9, 11.

35. NCPWI, *Report on Wastewater Management*, 18.

36. EPA, *Nonpoint Source Pollution: The Nation's Largest Water Quality Problem*, Pointer No. 1, EPA841-F-96-004A (updated 2006), http://www.epa.gov/nps/facts/point1.htm.

37. NCPWI, *Report on Wastewater Management*, 11; Richard Field and John A. Lager, "Urban Runoff Pollution Control—State-of-the-Art," *Journal of the Environmental Engineering Division—ASCE* 101 (Feb. 1975): 107.

38. Richard Field and Robert Turkeltaub, "Don't Underestimate Urban Runoff Problems," *Water and Wastes Engineering* 17 (Oct. 1980): 48.

39. Ibid., 50–51; Vladimir Novotny and Harvey Olem, *Water Quality* (New York: Van Nostrand Reinhold, 1994), 7–8; Kevin B. Smith, "Combined Sewer Overflows and Sanitary Sewer Overflows," *Environmental Law Reporter* 26 (June 1996): 26–27; EPA, "National Pollutant Discharge Elimination System (NPDES)," (Sept. 12, 2002), http://cfpub.epa.gov/npdes/cso/cpolicy.cfm?program_id=5.

40. Ralph A. Luken and Edward H. Pechan, *Water Pollution Control* (New York: Praeger, 1977), 4; NCPWI, *Report on Wastewater Management*, 5.

41. NCPWI, *Report on Wastewater Management*, 6, 13, 56–57; Russell V. Randle and Suzanne R. Shaeffer, "Water Pollution," in *Environmental Law Handbook*, ed. Timothy A. Vanderver Jr. (Washington, D.C.: Bureau of National Affairs, 1994), 148–49; Koren, *Handbook of Environmental Health and Safety*, 535; Grigg, *Urban Water Infrastructure*, 87.

42. Paul B. Downing, *Environmental Economics and Policy* (Boston: Little, Brown, 1984), 5; George S. Tolley, Philip E. Graves, and Glenn C. Blomquist, *Environmental Policy*, vol. 1 (Cambridge, Mass.: Ballinger, 1981), 182; Luken and Pechan, *Water Pollution Control*, 3; Koren, *Handbook of Environmental Health and Safety*, 458–59, 536–39.

43. NCPWI, *Report on Wastewater Management*, 13, 57–58.

44. Rodgers, *Environmental Law*, 16.

45. Downing, *Environmental Economic and Policy*, 7–8.

46. "The Push to Ease Water Rules," *Business Week*, Mar. 21, 1977, 69.

47. Palley and Palley, *Urban America and Public Policies*, 291–92.

48. Rodgers, *Environmental Law*, 19–20.

49. Grigg, *Urban Water Infrastructure*, 86.

50. NCPWI, *Report on Water Supply*, 8–9.

51. Petulla, *Environmental Protection*, 54–55.

52. J. Clarence Davies III and Barbara S. Davies, *The Politics of Pollution*, 2nd

ed. (Indianapolis: Bobbs-Merrill, 1975), 184, 187, 194–96; "Water Quality: Problems," *Water and Sewage Works* 123 (Dec. 1976): 41; Hays, *Beauty, Health, and Permanence*, 78–79; Downing, *Environmental Economics and Policy*, 5; Wekesser, *Water*, 63–66.

53. G. E. Eden and M.D.F. Haigh, *Water and Environmental Management in Europe and North America* (New York: Ellis Horwood, 1994), 33–35. One of several interesting Web sites that monitors antipollution legislation, including the Clean Water Act, is Scorecard: The Pollution Information Site, http://www.score card.org/.

54. Lewis, "1986 Amendments to the Safe Drinking Water Act," 898–99.

55. Benjamin and Belluck, *State Groundwater Regulation*, 6–7, 11.

56. Environment and Natural Resources Policy Division, Congressional Research Service, *Nonpoint Pollution and the Area-Wide Waste Treatment Management Program under the Federal Water Pollution Control Act* (Washington, D.C.: GPO, 1980), 14.

57. McClain, *U.S. Environmental Laws*, 2-1-1; Koren, *Handbook of Environmental Health*, 534, 540–41.

58. Randle and Shaeffer, "Water Pollution," 192; Smith, "Combined Sewer Overflows," 31; Berman et al., *Urban Runoff*, 7; Burian et al., "Urban Wastewater Management," 53–58.

59. Williams and Culp, *Handbook of Public Water Systems*, 10; Whipple, *New Perspectives in Water Supply*, 73; Goldstein, *Demanding Clean Food and Water*, 133; CONSAD Research Corporation, *Study of Public Works Investment in the United States*, vol. 4 (Pittsburgh: CONSAD, Mar. 1980), E27; Grigg, *Urban Water Infrastructure*, 55–56; Alan Levin, "Safe Drinking Water Act and Its Implications," *Proceedings of the Third Domestic Water Quality Symposium*, St. Louis, Feb. 27–Mar. 1, 1979, 1020–21.

60. Daniel A. Okun, "Drinking Water for the Future," *AJPH* 66 (July 1976): 639.

61. Ibid.; McClain, *U.S. Environmental Laws*, 5-1.

62. Randle and Shaeffer, "Water Pollution," 220–21.

63. Kenneth F. Gray, "Drinking-Water Act Amendments Will Tap New Sources of Strength," *National Law Journal* 8 (Sept. 1, 1986): 16.

64. Randle and Shaeffer, "Water Pollution," 235; Ling, "Research," 661; Palley and Palley, *Urban America and Public Policies*, 293–94; Okun, "Drinking Water for the Future," 639.

65. "The Clean Water Act Turns 30," *Newspaper in Education*, Oct. 22, 2002, http://www.cincinnati.com/nie/archive/10-22-02.

Chapter 20. Out of State, Out of Mind

Note: Major portions of this chapter are based on Martin V. Melosi, "Down in the Dumps," in *Urban Public Policy*, ed. Martin V. Melosi (University Park: Pennsylvania State University Press, 1993), 100–127; Melosi, *Garbage in the Cities*, chapter 8; and other of my publications listed in the notes.

1. See *Houston Chronicle*, May 18, 1987.

2. Mount, "Garbage Crisis," 38.

3. National League of Cities and U.S. Conference of Mayors, *Cities and the Nation's Disposal Crisis*, 1; Neal and Korbitz, *Solid Waste Management and the Environment*, 5; Robert Emmet Long, ed., *The Problem of Waste Disposal* (New York: H. W. Wilson, 1989), 9; "An Interview with Sylvia Lowrance," *EPA Journal* 15 (Mar./Apr. 1989): 10.

4. William L. Rathje, "Rubbish!" *Atlantic Monthly*, Dec. 1989, 99.

5. "Waste and the Environment," *Economist* 327 (May 29, 1993): 3.

6. U.S. Bureau of the Census, *Statistical Abstract of the United States* (Washington, D.C.: Department of Commerce, 1995), 236; "Comparative Data on National Solid Waste Generation and Economic Output," in *Recycling and Incineration*, ed. Richard A. Denison and John Ruston (Washington, D.C.: Island Press, 1990), 34–35; Franklin Associates, Ltd., *Analysis of Trends in Municipal Solid Waste Generation, 1972 to 1987* (Proctor and Gamble, Browning-Ferris Industries, General Mills, and Sears, Jan. 1992), ES-1–2, 1–4; Richard Stren, Rodney White, and Joseph Whitney, eds., *Sustainable Cities* (Boulder, Colo.: Westview Press, 1992), 184; EPA, *Municipal Solid Waste: Basic Facts* (Mar. 2, 2007), http://www.epa.gov/msw/facts.htm; EPA, *Municipal Solid Waste in the United States, 2005: Facts and Figures*, 1, http://www.epa.gov/msw/facts.htm.

7. U.S. Bureau of the Census, *Statistical Abstract of the United States, 1996* (Washington, D.C.: Department of Commerce, 1995), 236; Stratford P. Sherman, "Trashing a $150 Billion Business," *Fortune*, Aug. 28, 1989, 90; Philip O'Leary and Patrick Walsh, "Introduction to Solid Waste Landfills," *Waste Age* 22 (Jan. 1991): 44; EPA, *Municipal Solid Waste: Basic Facts* (Mar. 2, 2007).

8. EPA, *Solid Waste Dilemma: An Agenda for Action*, 1-18, 1-19; EPA, *Municipal Solid Waste in the United States*, 33.

9. Lewis Erwin and L. Hall Healy Jr., *Packaging and Solid Waste* (Washington, D.C.: AMA Membership Publications Division, American Management Association, 1990), 19; Melosi, *Garbage in the Cities*, 207–8; EPA, *Municipal Solid Waste in the United States*, 40.

10. EPA, *Solid Waste Dilemma: An Agenda for Action*, 1-19, 1-20, Appendixes A-B-C, A.A-1-48; EPA, *Municipal Solid Waste in the United States*, 49–50.

11. EPA, *Solid Waste Dilemma: An Agenda for Action*, Appendixes A-B-C, A.A-1-48; Franklin Associates, *Analysis of Trends in Municipal Solid Waste Generation*, 1-10–11, 4-1–6, 4-14, 5-3–4, 5-11, 5-14, 6-1–2, 6-10, 7-3; EPA, *Municipal Solid Waste in the United States*, 37.

12. APWA, *Solid Waste Collection Practice*, 1; E. S. Savas, "How Much Do Government Services Really Cost?" *Urban Affairs Quarterly* 15 (Sept. 1979): 23–42.

13. Neal and Korbitz, *Solid Waste Management and the Environment*, 29.

14. Karen Tumulty, "No Dumping (There's No More Dump)," *Los Angeles Times*, Sept. 2, 1988.

15. Matthew Gandy, *Concrete and Clay* (Cambridge, Mass.: MIT Press, 2002), 192–93; Hans Tammemagi, *The Waste Crisis* (New York: Oxford University Press, 1999), 194–95; "Waste Disposal in New York City," *Waste Age* 12 (Dec. 1981): 45; Bill Breen, "Landfills Are #1," *Garbage* 2 (Sept./Oct. 1990): 43; "What to Do with

Our Waste," *Newsweek*, July 27, 1987, 51; J. Tevere MacFadyen, "Where Will All the Garbage Go?" *Atlantic*, Mar. 1985, 29.

16. "Waste Disposal in New York City," 45; Tammemagi, *Waste Crisis*, 194–95.

17. Jim Johnson, "New York City 'Nightmare' Ends," *Waste News* 6 (Mar. 26, 2001): 1, 35.

18. "Attack Resurrects NYC's Fresh Kills," *Waste News* 6 (Nov. 12, 2001): 13; "Debris Gone: Memories Remain," *Waste News* 7 (Sept. 2, 2002): 1.

19. Casey Bukro, "Eastern Trash Being Dumped in America's Heartland," *Houston Chronicle*, Nov. 24, 1989.

20. Edward W. Repa, "Interstate Movement: 1995 Update," *Waste Age* 28 (June 1997): 41–44, 48, 50.

21. Susanna Duff, "Interstate Waste Keeps Crossing the Lines," *Waste News* 6 (Aug. 6, 2001): 4.

22. Repa, "Interstate Movement," 52, 54, 56.

23. Deb Starkey and Kelly Hill, *A Legislator's Guide to Municipal Solid Waste Management* (Washington, D.C.: National Conference of State Legislatures, Aug. 1996), 20–21; EPA, Solid Waste and Emergency Response, Office of Solid Waste, *Environmental Fact Sheet: Report to Congress on Flow Control and Municipal Solid Waste* (EPA530-F-95-008, Mar. 1995), www.epa.gov/fedrgstr/EPA-WASTE/1995/March/Day-21/Pr-177/html; H. Lanier Hickman Jr., *Principles of Integrated Solid Waste Management* (New York: American Academy of Environmental Engineers, 1999), 2.6.3–2.6.7; Herbert F. Lund, *The McGraw-Hill Recycling Handbook* (New York: McGraw-Hill, 1998), 2.3–2.4; Larry S. Luton, *The Politics of Garbage* (Pittsburgh: University of Pittsburgh Press, 1996), 29, 107–8, 117–18, 133–34; John Aquino, "The Tie That Binds?" *Waste Age* 27 (Sept. 1996): 90; Deanna L. Ruffer, "Life after Flow Control," *Waste Age* 28 (Jan. 1997): 73.

24. J. J. Dunn Jr. and Penelope Hong, "Landfill Siting—An Old Skill in a New Setting," *APWA Reporter* 46 (June 1979): 12.

25. Quoted in Peter Steinhart, "Down in the Dumps," *Audubon*, May 19, 1986, 106.

26. "Solid Waste Organizing Project," *Everyone's Backyard* 11 (Feb. 1993): 8.

27. Martin V. Melosi, "Equity, Eco-racism and Environmental History," *Environmental History Review* 19 (Fall 1995): 1–16.

28. Neal and Korbitz, *Solid Waste Management and the Environment*, 116; Sue Darcey, "Landfill Crisis Prompts Action," *World Wastes* 32 (May 1989): 28; Joanna D. Underwood, Allen Hershkowitz, and Maarten de Kadt, *Garbage* (New York: INFORM, 1988), 8–12; Denison and Ruston, *Recycling and Incineration*, 4–5.

29. "Municipal Solid Waste Management," *State Factor* 15 (June 1989): 2; Edward W. Repa and Allen Blakey, "Municipal Solid Waste Disposal Trends: 1996 Update," *Waste Age* 27 (Jan. 1996): 43; NSWMA, *Landfill Capacity in the Year 2000* (Washington, D.C., 1989), 1–3; Edward W. Repa, "Landfill Capacity: How Much Really Remains," *Waste Alternatives* 1 (Dec. 1988): 32; Ishwar P. Murarka, *Solid Waste Disposal and Reuse in the United States*, vol. 1 (Boca Raton, Fla.: CRC Press, 1987), 5; "Land Disposal Survey," *Waste Age* 12 (Jan. 1981): 65; NCPWI, *Fragile Foundations*, 193; Conservation Foundation, *State of the Envi-*

ronment, 107; "The State of Garbage in America," *BioCycle* 41 (Mar. 2000): 30, www.jgpress.com/BCArticles/2000/040032.html; Chaz Miller, "Garbage by the Numbers," *NSWMA Research Bulletin* (July 2002): 2; "The State of Garbage in America," *BioCycle* 31 (Mar. 1990): 49; "The State of Garbage in America," *Bio-Cycle* 32 (Apr. 1991): 34–36; "The State of Garbage in America," *BioCycle* 41 (Mar. 2000): 30; EPA, *Municipal Solid Waste in the United States,* 140.

30. EPA, *Municipal Solid Waste in the United States,* 140; EPA, *Solid Waste Dilemma: An Agenda for Action,* 2.E-1; O. P. Kharbanda and E. A. Stallworthy, *Waste Management* (New York: Auburn House, 1990), 67.

31. Melosi, "Waste Management," 12; Kharbanda and Stallworthy, *Waste Management,* 67; Greenberg et al., *Solid Waste Planning in Metropolitan Regions,* 8.

32. Eileen B. Berenyi and Robert N. Gould, "Municipal Waste Combustion in 1993," *Waste Age* (Nov. 1993): 51.

33. EPA, *Solid Waste Dilemma: An Agenda for Action,* 2.D-1–3, 2.D-5.

34. "Hard Road Ahead for City Incinerators," in *Solid Wastes—II,* ed. Stanton S. Miller (Washington, D.C.: American Chemical Society, 1973), 110–11; Van Tassel, *Our Environment,* 464–65; "Moving to Garbage Power," *Time,* Jan. 9, 1978, 46.

35. Neil Seldman, "Mass Burn Is Dying," *Environment* 31 (Sept. 1989): 42; U.S. Congress, Office of Technology Assessment, *Facing America's Trash* (Washington, D.C.: Office of Technology Assessment, 1989), 222.

36. David Tillman et al., *Incineration of Municipal and Hazardous Solid Wastes* (San Diego, Calif.: Academic Press, 1989), 59, 113; Institute for Local Self-Reliance, *An Environmental Review of Incineration Technologies* (Washington, D.C.: Institute for Local Self-Reliance, 1986), 2.

37. Russell, "Environmental Racism," 23–26; Gottlieb, *Forcing the Spring,* 189–90; Blumberg and Gottlieb, *War on Waste,* 58–60.

38. K. A. Godfrey Jr., "Municipal Refuse: Is Burning Best?" *Civil Engineering* 55 (Apr. 1985): 54–55; James E. McCarthy, "Incinerating Trash," *Congressional Research Service Review* 7 (Apr. 1986): 19; Institute for Local Self-Reliance, *Environmental Review of Incineration Technologies,* 8; Seldman, "Mass Burn Is Dying," 42; Allen Hershkowitz, "Burning Trash," *Technology Review* (July 1987): 26, 30; Melosi, "Down in the Dumps," 111.

39. Seldman, "Mass Burn Is Dying," 42; U.S. Congress, Office of Technology Assessment, *Facing America's Trash,* 222.

40. John H. Skinner, "The Consequences of New Environmental Requirements," paper presented at Centre Jacques Cartier, Lyon, France, Dec. 1993, 6–7; Margaret Ann Charles, "New Trends in Waste-to-Energy," *Waste Age* 24 (Nov. 1993): 59–60. "Integrated waste management" is a relatively commonsense notion adopted by the EPA that stresses a reliance on "a hierarchy of options from most desirable to least desirable," with source reduction on the high end and with the sanitary landfill on the low end. See James R. Pfafflin and Edward N. Ziegler, eds., *Encyclopedia of Environmental Science and Engineering,* vol. 2 (Philadelphia: Gordon and Breach Science, 1992), 704–5.

41. Charles, "New Trends in Waste-to-Energy," 59–60; Berenyi and Gould, "Municipal Waste Combustion," 51–52; Melosi, "Equity, Eco-racism and Environmental History," 4–11.

42. Schwab, "Garbage In, Garbage Out," 7.

43. Hershkowitz, "Burning Trash," 27; McCarthy, "Incinerating Trash," 19–20.

44. Institute for Local Self-Reliance, *Environmental Review of Incineration Technologies*, 8.

45. David Naguib Pellow, *Garbage Wars* (Cambridge, Mass.: MIT Press, 2002), 9–10.

46. EPA, *Municipal Solid Waste in the United States*, 137–38.

47. Quoted in Long, *Problem of Waste Disposal*, 17.

48. Seldman, "Waste Management," 43–44.

49. "Municipal Solid Waste Management," 6.

50. NSWMA, *Solid Waste Disposal Overview* (Washington, D.C.: NSWMA, 1988), 2; Cynthia Pollock, "There's Gold in Garbage," *Across the Board* (Mar. 1987): 37; "Municipal Solid Waste Management," 7; Debi Kimball, *Recycling in America* (Santa Barbara, Calif.: ABC-Clio, 1992), 3, 5–6, 22–24; EPA, *Municipal Solid Waste in the United States*, 1, 5–6.

51. John Tierney, "Recycling Is Garbage," *New York Times Magazine*, June 30, 1996, 24–26.

52. John T. Aquino, "A Recycling Pilgrim's Progress," *Waste Age* 28 (May 1997): 220, 222, 224, 226, 228, 230–32.

53. Seldman, "Waste Management," 43–44; Schwab, "Garbage In, Garbage Out," 9; EPA, *Recycling Works!* (Washington, D.C.: GPO, Jan. 1989); Nicholas Basta, "A Renaissance in Recycling," *High Technology* 5 (Oct. 1985): 32–39; Barbara Goldoftas, "Recycling: Coming of Age," *Technology Review* (Nov./Dec. 1987): 30–35, 71; Anne Magnuson, "Recycling Gains Ground," *American City and County/Resource Recovery* (1988), RR10; Debra L. Strong, *Recycling in America*, 2nd ed. (Santa Barbara, Calif.: ABC-CLIO, 1997), 1–20.

54. Chaz Miller, "Source Separation Programs," *NCRR Bulletin: Journal of Resource Recovery* 10 (Dec. 1980): 82–83; EPA, *Municipal Solid Waste in the United States*, 13; Jim Glenn, "Curbside Recycling Reaches 40 Million," *BioCycle* 31 (July 1990): 30–31; Susan J. Smith and Kathleen M. Hopkins, "Curbside Recycling in the Top 50 Cities," *Resource Recycling* 11 (Mar. 1992): 101–2; "State of Garbage in America" (1991), 36–37; "Reduce, Reuse, and Recycle" (Dec. 12, 2000), www.epa.gov/epaoswer/non-hw/muncpl/reduce.htm; U.S. Census Bureau, *Statistical Abstract of the United States: 2001*, 218; Lund, *McGraw-Hill Recycling Handbook*, 2.2; James E. McCarthy, "Bottle Bills and Curbside Recycling: Are They Compatible?" *Congressional Research Service Report* (Jan. 27, 1993), 9.

55. Kirsten U. Oldenburg and Joel S. Hirschhorn, "Waste Reduction," *Environment* 29 (Mar. 1987): 17–20, 39–45.

56. "Garbage at the Crossroads," *Chicago Tribune*, Feb. 1, 1984.

57. Savas, "Intracity Competition," 47–48.

58. Armstrong et al., *History of Public Works*, 446–47.

59. Eileen Brettler Berenyi, "Contracting Out Refuse Collection," *Urban Interest* 3 (1981): 30–42; E. S. Savas, "Solid Waste Collection in Metropolitan Areas," in *The Delivery of Urban Services*, ed. Elinor Ostrom (Beverly Hills, Calif.: Sage, 1976), 211–13, 219–21, 228; John N. Collins and Bryant T. Downes, "The Effect of

Size on the Provision of Public Services," *Urban Affairs Quarterly* 12 (Mar. 1977): 345.

60. NSWMA, *Privatizing Municipal Waste Services* (Washington, D.C.: NSWMA, 1988), 1–5.

61. Anne Hartman, "The Solid Waste Control Industry," *Waste Age* 4 (July/ Aug. 1973): 54.

62. Burck, "There's Big Business in All That Garbage," 107–8.

63. *Facing America's Trash* (New York: Van Nostrand Reinhold, 1992), 53; "The Politics of Waste Disposal," *Wall Street Journal*, Sept. 5, 1989; Nancy Shute, "The Selling of Waste Management," *Amicus Journal* 7 (Summer 1985): 8–15; James Cook, "Waste Management Cleans Up," *Forbes*, Nov. 18, 1985, reprint; Bob Sablatura, "BFI, Waste Management Face Probe," *Houston Chronicle*, July 6, 1988; Richard Asinof, "The Nation's Dumpster," *Environmental Action* 17 (May/June 1986): 13–16; Janet Novack, "A New Top Broom," *Forbes*, Nov. 28, 1988, 200, 202.

64. John T. Aquino, "Yanks Abroad," *Waste Age* 29 (Apr. 1998): 84–93; Bethany Barber and John T. Aquino, "The Waste Age 100," *Waste Age* 28 (Sept. 1997): 37.

65. John T. Aquino, "The Future Is (Almost) Now," *Waste Age* 27 (Dec. 1996): 52–53.

66. Scott Jones, "The Latest Moves in Waste Industry Consolidations," *Waste Age* 28 (May 1997): 180, 184.

67. USA Waste Services, Inc., *Hoover's Company Capsules* (Austin, Tex.: Hoover's, 1998); John T. Aquino, "Waste Age 100," *Waste Age* 29 (Sept. 1998): 83–84; "Experts Predict Busier 1998," *Waste News* 3 (Jan. 12, 1998): 1; Cheryl L. Dunson, "Consolidation: Rearranging the Pieces," *Waste Age* (July 1, 1999), http:// wasteage.com.

68. Barnaby J. Feder, "'Mr. Clean' Takes on the Garbage Mess," *New York Times*, Mar. 11, 1990.

69. Harold Crooks, *Dirty Business: The Inside Story of the New Garbage Agglomerates* (Toronto: James Lorimer, 1983), 8.

70. Harold Crooks, *Giants of Garbage* (Toronto: James Lorimer, 1993), 55.

71. John Vickers and George Yarrow, *Privatization* (Cambridge, Mass.: MIT Press, 1988), 41.

72. EPA, *Solid Waste Dilemma: An Agenda for Action*, 2.

73. Melosi, "Waste Management," 7–8; *Facing America's Trash*, 299.

74. Cathy Dombrowski, "Reilly Predicts More Regs and Higher Disposal Costs," *World Wastes* 32 (May 1989): 39.

75. Kovacs, "Legislation and Involved Agencies," 19.

76. Menell, "Beyond the Throwaway Society," 674.

77. Melosi, "Waste Management," 7; Degler, *Federal Pollution Control Programs*, 37–38; Cristofano and Foster, *Management of Local Public Works*, 318; McClain, *U.S. Environmental Laws*, 3-1; William H. Rodgers Jr., *Environmental Law: Pesticides and Toxic Substances* (St. Paul, Minn.: West, 1988), 528–29; Grad, "Role of the Federal and State Governments," 169–83; Kovacs, "Legislation and Involved Agencies," 9; Berry and Horton, *Urban Environmental Management*, 361–62.

78. Rodgers, *Environmental Law*, 530–31; Cristofano and Foster, *Management of Local Public Works*, 318; McClain, *U.S. Environmental Laws*, 3-1–2.

79. Kovacs, "Legislation and Involved Agencies," 10, 12–18.

80. "Waste and the Environment," 8; Melosi, "Historic Development of Sanitary Landfills," 20–24; Repa and Blakey, "Municipal Solid Waste Disposal Trends," 46; Geoffrey F. Segal and Adrian T. Moore, *Privatizing Landfills*, Policy Study No. 267 (May 2000), http://www.reason.org/ps267.html.

BIBLIOGRAPHIC ESSAY

In researching a book of this kind, a wide array of research materials are essential. Some of the most valuable sources are contemporary technical and
popular periodicals that present a vast array of data, from statistics to case
studies of specific municipal practices. They also reflect a variety of viewpoints, including those of government officials, technical experts, medical
officers, sanitarians, and the public. Some periodicals provide excellent
general background about the application of technologies of sanitation to
city purposes, especially *American City*. The transactions, proceedings,
and journals of various engineering societies, especially *Transactions of the
American Society of Civil Engineers*, *Proceedings of the American Society
of Civil Engineers*, and *Journal of the American Water Works Association*,
offer a wide range of information. Particularly useful in these publications are statistical data and impressive analyses of the development of key
technical processes, fiscal policy, and engineering techniques. For detailed
technical information about the systems, consistently the most valuable periodicals are *Civil Engineering, Engineering News and Record, Municipal
Journal and Engineer, Public Works, Scientific American, Sewage Works
Journal, Sewage and Industrial Waste, Water and Sewage Works*, and *Waste
Age*. These periodicals must be used carefully, however, because they often represent issues from the vantage point of the waste industry, the public
works community, or the municipal government. For contemporary health
and environmental issues, few periodicals are better than the *American
Journal of Public Health*. And a wide array of municipal periodicals, such
as the *Survey* and *Municipal Affairs*, brought to light the work of reform
groups on behalf of sanitation. Articles in periodicals for more general consumption, such as *Business Week, Newsweek, Outlook, Popular Science
Monthly, Time*, and *U.S. News and World Report*, often identify issues of
national significance.

Contemporary engineering texts are valuable for an understanding of the
evolution of many technologies, technology transfer, and engineering styles
and procedures. Particularly in the nineteenth and early twentieth centuries, these texts provided some of the best narratives about the history of

sanitary services. Among the most important of these are Rudolph Hering and Samuel A. Greeley, *Collection and Disposal of Municipal Refuse* (New York: McGraw-Hill, 1921); Langdon Pearse, ed., *Modern Sewage Disposal* (New York: Federation of Sewage and Industrial Waste Associations, 1938); and F. E. Turneaure and H. L. Russell, *Public Water-Supplies* (New York: John Wiley and Sons, 1911, 1948). Other significant texts are Harold E. Babbitt and James J. Doland, *Water Supply Engineering,* 4th ed. (New York: McGraw-Hill, 1949); W. H. Corfield, *The Treatment and Utilisation of Sewage,* 3rd ed. (London: Macmillan, 1887); George W. Fuller, *Sewage Disposal* (New York: McGraw-Hill, 1912); William Paul Gerhard, *Sanitation and Sanitary Engineering* (New York: author, 1909); Allen Hazen, *The Filtration of Public Water-Supplies* (New York: John Wiley and Sons, 1905); William P. Mason, *Water-Supply* (New York: John Wiley and Sons, 1897); Leonard Metcalf and Harrison P. Eddy, *American Sewerage Practice,* vol. 1 (New York: McGraw-Hill, 1914); William F. Morse, *The Disposal of Refuse and Garbage* (New York: J. J. O'Brien and Sons, 1899); George W. Rafter and M. N. Baker, *Sewage Disposal in the United States* (New York: D. Van Nostrand, 1894); Samuel Rideal and Eric K. Rideal, *Water Supplies* (London: Crosby Lockwood and Son, 1914); George A. Soper, *Modern Methods of Street Cleaning* (New York: Engineering News, 1907); and Donald C. Stone, *The Management of Municipal Public Works* (Chicago: Public Administration Service, 1939). M. N. Baker, *The Quest for Pure Water: The History of Water Purification from the Earliest Records to the Twentieth Century* (1948; reprint, New York: AWWA, 1981), is written by an eminent sanitarian and is a treasure trove of information, especially on water purification technologies.

Other primary sources were essential. Government reports and studies are plentiful, with useful information on regulations, ordinances, new projects, and planning studies. Every major city has board of health records, as well as records of engineering and public works departments. Federal census material covered many basic municipal statistics and information on fiscal matters, but the researcher will find maddening the practice of changing categories of data from one year to the next.

Some specific reports were key turning points in the history of sanitation and the development of sanitary services. These include Edwin Chadwick, *Report on the Sanitary Condition of the Labouring Population of Great Britain,* ed. with an introduction by M. W. Flinn (Edinburgh: University Press, 1965), which generated the "sanitary idea"; Ellis S. Chesbrough, *Chicago Sewerage* (Chicago: Board of Sewage Commissioners, 1858), which influenced the development of the Chicago Sanitary District; National Council on Public Works Improvement, *Fragile Foundations* (Washington, D.C.: GPO, Feb. 1988), representing the array of reports highlighting the emergence of an infrastructure crisis in the United States; Franklin Associates, Ltd., *Analysis of Trends in Municipal Solid Waste Generation, 1972 to 1987* (Proctor and Gamble, Browning-Ferris Industries, General Mills, and Sears, Jan. 1992), which stimulated national discussion about solid-waste problems; James P. Kirkwood, *Report on the Filtration of River Waters* (New York: Van Nostrand, 1869), which introduced the idea of filtration to American cities; and Lemuel Shattuck, *Report of the Sanitary Commission of Massachusetts, 1850* (Cambridge, Mass.: Harvard University Press, 1948), which relied upon the Chadwick report to make a similar case for sanitation in the United States.

The secondary literature is uneven, more heavily weighted toward the nineteenth and early twentieth centuries than the later twentieth century. (Aside from key works used in this study, the listing herein also incoporates relevant books and articles that were published after the cloth edition of *The Sanitary City* [2000] was prepared. The notes in this abridged edition incorporate many of the works consulted for the original study, but I direct readers to the 2000 edition for a more detailed listing of primary and secondary materials.) Of the general histories of cities, the most directly applicable to the study of infrastructure include several works by Jon Teaford: *The Metropolitan Revolution* (New York: Columbia University Press, 2006); *The Municipal Revolution in America* (Chicago: University of Chicago Press, 1975); *The Rough Road to Renaissance* (Baltimore: Johns Hopkins University Press, 1990); *The Twentieth-Century American City* (Baltimore: Johns Hopkins University Press, 1986); and *The Unheralded Triumph* (Baltimore: Johns Hopkins University Press, 1984). Also useful are Kenneth Fox, *Better City Government* (Philadelphia: Temple University Press, 1977); David R. Goldfield and Blaine A. Brownell, *Urban America,* 2nd ed. (Boston: Houghton Mifflin, 1990); Eric H. Monkkonen, *America Becomes Urban* (Berkeley: University of California Press, 1988); Stanley K. Schultz, *Constructing Urban Culture* (Philadelphia: Temple University Press, 1989); and Sam Bass Warner Jr., *The Urban Wilderness* (New York: Harper and Row, 1972). See also Owen D. Gutfreund, *Twentieth Century Sprawl* (New York: Oxford, 2004); Robert M. Fogelson, *Downtown: Its Rise and Fall, 1880–1950* (New Haven, Conn.: Yale University Press, 2001).

Ernest S. Griffith's somewhat dated volumes on the American city nevertheless provided extraordinary material on city government and fiscal policy: *A History of American City Government: The Conspicuous Failure, 1870–1900* (Washington, D.C.: University Press of America, 1974); *A History of American City Government: The Progressive Years and Their Aftermath, 1900–1920* (1974; reprint, Washington, D.C.: University Press of America, 1983); and Griffith and Charles R. Adrian, *A History of American City Government, 1775–1870: The Formation of Traditions* (1976; reprint, Washington, D.C.: University Press of America, 1983). On planning and infrastructure, see Jon A. Peterson, *The Birth of City Planning in the United States, 1840–1917* (Baltimore: Johns Hopkins University Press, 2003).

A few monographs blended more general treatments of urban themes with important case studies on sanitary services, none better than Olivier Zunz, *The Changing Face of Inequality* (Chicago: University of Chicago Press, 1982). There are few general studies of urban infrastructure development. Those worth consulting include Ellis Armstrong, Michael Robinson, and Suellen Hoy, eds., *History of Public Works in the United States* (Chicago: APWA, 1976), which includes a series of commissioned essays based largely on published public documents; Josef W. Konvitz, *The Urban Millennium* (Carbondale: Southern Illinois University Press, 1985), which focuses on Europe; and Joel A. Tarr and Gabriel Dupuy, eds., *Technology and the Rise of the Networked City in Europe and America* (Philadelphia: Temple University Press, 1988), which includes scholarly articles dealing with a wide array of services but over limited time periods. On a theoretical level, especially dealing with questions of municipal responsibility for services, see the excellent book by Charles D. Jacobson, *Ties That Bind* (Pittsburgh: University of Pittsburgh Press, 2000). See also Gail Radford, "From Municipal Socialism to Public Authorities," *Journal of American History* 90 (Dec. 2003): 863–90.

The best treatment of environmental regulation is Richard N. L. Andrews, *Managing the Environment, Managing Ourselves* (New Haven, Conn.: Yale University Press, 2006). See also Christoph Bernhardt and Genevieve Massard-Guilbaud, eds., *The Modern Demon* (Clermont-Ferrand, France: Presses Universitaires Blaise-Pascalk, 2002); William L. Andreen, "The Evolution of Water Pollution Control in the United States," *Stanford Environmental Law Journal* 22 (Jan. 2003): 145–200; Paul Charles Milazzo, *Unlikely Environmentalists* (Lawrence: University Press of Kansas, 2006); Christine Meisner Rosen, "'Knowing' Industrial Pollution," *Environmental History* 8 (Oct. 2003): 565–97; Joel A. Tarr, "Industrial Waste Disposal in the United States as a Historical Problem," *Ambix* 49 (Mar. 2002): 4–20.

For an understanding of public health issues, the work of John Duffy is essential, especially *A History of Public Health in New York City, 1866–1966* (New York: Russell Sage Foundation, 1974) and *The Sanitarians* (Urbana: University of Illinois Press, 1990). Also important are John H. Ellis, *Yellow Fever and Public Health in the New South* (Lexington: University Press of Kentucky, 1992); Christopher Hamlin, *Public Health and Social Justice in the Age of Chadwick* (Cambridge: Cambridge University Press, 1998); Suellen Hoy, *Chasing Dirt* (New York: Oxford University Press, 1995); Judith Walzer Leavitt, ed., *Women and Health in America* (Madison: University of Wisconsin Press, 1984); Michael P. McCarthy, *Typhoid and the Politics of Public Health in Nineteenth-Century Philadelphia* (Philadelphia: American Philosophical Society, 1987); Charles E. Rosenberg, *The Cholera Years* (1962; reprint, Chicago: University of Chicago Press, 1987); Barbara Gutmann Rosenkrantz, *Public Health and the State* (Cambridge, Mass.: Harvard University Press, 1971); Nancy Tomes, *The Gospel of Germs* (Cambridge, Mass.: Harvard University Press, 1998). See also Sarah S. Elkind, "Public Works and Public Health," *Essays in Public Works History* 19 (Kansas City, Mo.: Public Works Historical Society, Dec. 1999); Steven J. Hoffman, "Progressive Public Health Administration in the Jim Crow South," *Journal of Social History* 35 (2001): 177–94; Margaret Humphreys, *Malaria* (Baltimore: Johns Hopkins University Press, 2003); Steven Johnson, *The Ghost Map: The Story of London's Deadliest Epidemic* (New York: Penguin, 2006); Gerald Markowitz and David Rosner, *Deceit and Denial* (Berkeley: University of California Press, 2002); Gregg Mitman, "In Search of Health: Landscape and Disease in American Environmental History," *Environmental History* 10 (Apr. 2005): 184–210; Harold L. Platt, "'Clever Microbes': Bacteriology and Sanitary Technology in Manchester and Chicago during the Progressive Age," *Osiris* 19 (2004): 149–66; Sally Sheard and Helen Power, eds., *Body and City: Histories of Urban Public Health* (Aldershot, U.K.: Ashgate, 2001); John Welshman, *Municipal Medicine: Public Health in Twentieth-Century Britain* (Oxford, U.K.: Peter Lang, 2000); Louis P. Cain and Elyce J. Rotella, "Death and Spending," *Annales de Demographie Historique* 45 (2002): 139–54.

For the environmental movement, see Robert Gottlieb, *Forcing the Spring* (Washington, D.C.: Island Press, 1993); Adam Rome, *The Bulldozer in the Countryside* (Cambridge: Cambridge University Press, 2001); Mark Dowie, *Losing Ground* (Cambridge, Mass.: MIT Press, 1997); Samuel P. Hays, *Beauty, Health, and Permanence* (Cambridge: Cambridge University Press, 1987).

Municipal, sanitary, and environmental engineering, unfortunately, have not fared as well as public health and medicine in attracting scholarly work. A few studies worth consulting include Terry S. Reynolds, ed., *The Engineer in America* (Chicago: University of Chicago Press, 1991); Peter C. Schulze, ed., *Engineering within Ecological Constraints* (Washington, D.C.: National Academy Press, 1996); Larry D. Lankton, "The 'Practicable' Engineer: John B. Jervis and the Old Croton Aqueduct," *Essays in Public Works History* 5 (Chicago: Public Works Historical Society, 1977); Martin V. Melosi, "Pragmatic Environmentalist: Sanitary Engineer George E. Waring Jr.," *Essays in Public Works History* 4 (Washington, D.C.: Public Works Historical Society, 1977); and Stanley K. Schultz and Clay McShane, "To Engineer the Metropolis," *Journal of American History* 65 (Sept. 1978): 389–411.

Because of its significance to the growth and development of cities, urban water supply has a relatively large bibliography of secondary historical literature, but it is uneven in quality and in the distribution of topics. The best-known work was written forty years ago: Nelson Manfred Blake, *Water for the Cities* (Syracuse, N.Y: Syracuse University Press, 1956). Never published but impressive is Letty Donaldson Anderson, "The Diffusion of Technology in the Nineteenth-Century American City: Municipal Water Supply Investments" (Ph.D. diss., Northwestern University, 1980), and a derivative article, "Hard Choices: Supplying Water to New England," *Journal of Interdisciplinary History* 15 (Autumn 1984): 211–34. See also Stuart Galishoff, "Triumph and Failure: The American Response to the Urban Water Supply Problem, 1860–1923," in *Pollution and Reform in American Cities,* ed. Martin V. Melosi (Austin: University of Texas Press, 1980), 35–57; Sarah S. Elkind, *Bay Cities and Water Politics* (Lawrence: University Press of Kansas, 1998); Elmer W. Becker, *A Century of Milwaukee Water* (Milwaukee: Milwaukee Water Works, 1974); Abraham Hoffman, *Vision or Villainy: Origins of the Owens Valley–Los Angeles Water Controversy* (College Station: Texas A&M University Press, 1981); William L. Kahrl, *Water and Power* (Berkeley: University of California Press, 1982); Fern L. Nesson, *Great Waters* (Hanover, N.H.: University Press of New England, 1983); Howard Rosen and Ann Durkin Keating, eds., *Water and the City* (Chicago: Public Works Historical Society, 1991); Joseph W. Barnes, "Water Works History: Comparison of Albany, Utica, Syracuse, and Rochester," *Rochester History* 39 (July 1977): 1–22; John B. Blake, "Lemuel Shattuck and the Boston Water Supply," *Bulletin of the History of Medicine* 29 (1955): 554–62; Gary A. Donaldson, "Bringing Water to the Crescent City," *Louisiana History* 28 (Fall 1987): 381–96; John H. Ellis and Stuart Galishoff, "Atlanta's Water Supply, 1865–1918," *Maryland Historian* 8 (Spring 1977): 5–22; Gregg R. Hennessey, "The Politics of Water in San Diego, 1895–1897," *Journal of San Diego History* 24 (Summer 1978): 367–83; Carol Hoffecker, "Water and Sewage Works in Wilmington, Delaware, 1810–1910," *Essays in Public Works History* 2 (Chicago: Public Works Historical Society, 1981); Bruce Jordan, "Origins of the Milwaukee Water Works," *Milwaukee History* 9 (Spring 1986): 2–16; Jacob Judd, "Water for Brooklyn," *New York History* 47 (Oct. 1966): 362–71; Michal McMahon, "Makeshift Technology: Water and Politics in Nineteenth-Century Philadelphia," *Environmental Review* 12 (Winter 1988): 21–37; James C. O'Connell, "Chicago's Quest for Pure Water," *Essays in Public Works History* 1 (Washington, D.C.: Public Works Historical Society, 1976); Maureen Ogle, "Redefining 'Public' Water Supplies, 1870–1890: A Study of Three Iowa Cities," *Annals of Iowa* 50 (Spring 1990): 507–30; Terry S. Reynolds, "Cisterns and Fires: Shreveport, Louisiana, as a Case Study of the Emergence of Public Water Supply Systems in the South," *Louisiana History* 22 (Fall 1981): 337–67; Todd A. Shallat, "Fresno's Water Rivalry," *Essays in Public Works History* 8 (1979): 9–13; and Mark J. Tierno, "The Search for Pure Water in Pittsburgh,"

Western Pennsylvania Historical Magazine 60 (Jan. 1977): 23–36. Christopher Hamlin has published several key studies on water purification dealing with England. No similar works are available for the United States. See his *A Science of Impurity* (Berkeley: University of California Press, 1990), and *What Becomes of Pollution?* (New York: Garland, 1987).

Recent studies on water include K. Foss-Mollan, *Hard Water: Politics and Water Supply in Milwaukee* (West Lafayette, Ind.: Purdue University Press, 2001); Robert Jerome Glennon, *Water Follies: Groundwater Pumping and the Fate of America's Fresh Waters* (Washington, D.C.: Island Press, 2002); John Graham-Leigh, *London's Water Wars* (London: Francis Boutle, 2000); Charles Hardy III, "The Watering of Philadelphia," *Pennsylvania Heritage* 30 (Spring 2004): 26–35; John Hassan, *A History of Water in Modern England and Wales* (Manchester: Manchester University Press, 1998); Gerard T. Koeppel, *Water for Gotham* (Princeton, N.J.: Princeton University Press, 2000); Douglas E. Kupel, *Fuel for Growth: Water and Arizona's Urban Environment* (Tucson: University of Arizona Press, 2003); Char Miller, "Running Dry," *Journal of the West* 44 (Summer 2005): 44–51; Catherine Mulholland, *William Mulholland and the Rise of Los Angeles* (Berkeley: University of California Press, 2000); Maureen Ogle, "Water Supply, Waste Disposal, and the Culture of Privatism in the Mid-Nineteenth Century American City," *Journal of Urban History* 25 (1999): 321–47; Jared Orsi, "Reclaiming the City: Water History in the Urban North American West," *Journal of the West* 44 (Summer 2005): 8–11; Jouni Paavola, "Water Quality as Property," *Environment and History* 8 (2002): 295–318; Michael Rawson, "The Nature of Water," *Environmental History* 9 (July 2004): 411–35; Christopher Sellers, "The Artificial Nature of Fluoridated Water," *Osiris* 19 (2004): 182–200; Carolyn G. Shapiro-Shapin, "Filtering the City's Image: Progressivism, Local Control, and the St. Louis Water Supply, 1890–1906," *Journal of the History of Medicine and Applied Sciences* 54 (July 1999): 387–412; Marienka Sokol, "Reclaiming the City: Water and the Urban Landscape in Phoenix and Las Vegas," *Journal of the West* 44 (Summer 2005): 52–61; Werner Troesken, "Race, Disease, and the Provision of Water in American Cities, 1889–1921," *Journal of Economic History* 61 (2001): 750–76; Werner Troesken, "Typhoid Rates and the Public Acquisition of Private Waterworks, 1880–1925," *Journal of Economic History* 59 (1999): 927–48; Werner Troesken, *Water, Race, and Disease* (Cambridge, Mass.: MIT Press, 2004).

The literature on the history of wastewater systems is much smaller than that on water supply, but some is more sophisticated in presentation. It is impossible to think about this topic without encountering the work of Joel A. Tarr. See his "From City to Farm: Urban Wastes and the American Farmer," *Agricultural History* 49 (Oct. 1975): 598–612; "Industrial Wastes and Public Health," *American Journal of Public Health* 75 (Sept. 1985): 1059–67; "The

Separate vs. Combined Sewer Problem," *Journal of Urban History* 5 (May 1979): 308–39; *The Search for the Ultimate Sink* (Akron, Ohio: University of Akron Press, 1996); Joel A. Tarr, James McCurley, and Terry E. Yosie, "The Development and Impact of Urban Wastewater Technology," in Melosi, *Pollution and Reform in American Cities,* 59–82. Also important for Chicago are works by Louis P. Cain, such as *Sanitation Strategy for a Lakefront Metropolis* (De Kalb: Northern Illinois University Press, 1978); "The Creation of Chicago's Sanitary District and Construction of the Sanitary and Ship Canal," *Chicago History* 8 (Summer 1979): 98–110; and "Raising and Watering a City: Ellis Sylvester Chesbrough and Chicago's First Sanitation System," *Technology and Culture* 13 (July 1972): 353–72. See also Jamie Benidickson, *The Culture of Flushing: A Social and Legal History of Sewage* (Vancouver: UBC Press, 2007).

Broader in scope and significance than just wastewater issues for understanding sanitary issues in key cities include Douglas Brinkley, *The Great Deluge* (New York: William Morrow, 2006); Craig E. Colten, *An Unnatural Metropolis* (Baton Rouge: Louisiana State University Press, 2005); William Deverell and Greg Hise, eds. *Land of Sunshine* (Pittsburgh: University of Pittsburgh Press, 2005); Eric Jay Dolin, *Political Waters* (Amherst: University of Massachusetts Press, 2004); Matthew Gandy, *Concrete and Clay* (Cambridge, Mass.: MIT Press, 2002); Blake Gumprecht, *The Los Angeles River* (Baltimore: Johns Hopkins University Press, 1999); Stephen Halliday, *The Great Stink of London* (Stroud: Sutton, 2001); Ari Kelman, *A River and Its City* (Berkeley: University of California Press, 2003); Jared Orsi, *Hazardous Metropolis* (Berkeley: University of California Press, 2004); David L. Pike, *Subterranean Cities* (Ithaca, N.Y.: Cornell University Press, 2005); Harold L. Platt, *Shock Cities* (Chicago: University of Chicago Press, 2005); Joel A. Tarr, ed., *Devastation and Renewal* (Pittsburgh: University of Pittsburgh Press, 2003); and Gavin Weightman, *London's Thames* (New York: St. Martin's Press, 2005).

The first book-length study of the sewage problem is Joanne Abel Goldman, *Building New York's Sewers* (West Lafayette, Ind.: Purdue University Press, 1997). Although the chronology is too narrow for a comprehensive treatment, Goldman does a good job in introducing politics as a key factor in planning a sewerage system. See also Stuart Galishoff, "Drainage, Disease, Comfort, and Class: A History of Newark's Sewers," *Societas* 6 (Spring 1976): 121–38; Christopher Hamlin, "Edwin Chadwick and the Engineers, 1842–1854: Systems and Antisystems in the Pipe-and-Brick Sewers War," *Technology and Culture* 33 (Oct. 1992): 680–709; and Schultz and McShane, "To Engineer the Metropolis."

On solid waste, the literature has been weak, but is getting much richer in recent years. The only volume dealing broadly with the question in the

United States, however, is Martin V. Melosi, *Garbage in the Cities*, rev. ed. (Pittsburgh: University of Pittsburgh Press, 2005). See also my "Down in the Dumps: Is There a Garbage Crisis in America?" in *Urban Public Policy*, ed. Martin V. Melosi (University Park: Pennsylvania State University Press, 1993), 100–127; "Hazardous Waste and Environmental Liability," *Houston Law Review* 25 (July 1988): 1–39; "Historic Development of Sanitary Landfills and Subtitle D," *Energy Laboratory Newsletter* 31 (1994): 20–24; "Sanitary Services and Decision Making in Houston, 1876–1945," *Journal of Urban History* 20 (May 1994): 365–406; "The Viability of Incineration as a Disposal Option," *Public Works Management and Policy* 1 (July 1996): 40–51; "Waste Management: The Cleaning of America," *Environment* 23 (Oct. 1981): 6–13, 41–44; "The Fresno Sanitary Landfill in American Cultural Context," *Public Historian* 24 (Summer 2002): 17–35; and, more generally, *Effluent America* (Pittsburgh: University of Pittsburgh Press, 2001). See also Craig E. Colten, "Chicago's Waste Lands," *Journal of Historical Geography* 20 (1994): 133–34; Larry S. Luton, *The Politics of Garbage* (Pittsburgh: University of Pittsburgh Press, 1996); William Rathje and Cullen Murphy, *Rubbish!* (New York: Harper-Collins, 1992); Patricia Ard, "Garbage in the Garden State," *Public Historian* 27 (Summer 2005): 57–66; Daniel Eli Burnstein, *Next to Godliness* (Urbana: University of Illinois Press, 2006); William A. Cohen and Ryan Johnson, eds., *Filth: Dirt, Disgust, and Modern Life* (Minneapolis: University of Minnesota Press, 2005); Mira Engler, *Designing America's Waste Landscapes* (Baltimore: Johns Hopkins University Press, 2004); Clay McShane and Joel A. Tarr, *The Horse in the City* (Baltimore: Johns Hopkins University Press, 2007) (which deals with much more than waste issues); Benjamin Miller, *Fat of the Land* (New York: Four Windows Eight Walls, 2000); Heather Rogers, *Gone Tomorrow: The Hidden Life of Garbage* (New York: New Press, 2005); Elizabeth Royte, *Garbage Land* (New York: Little, Brown, 2005); Susan Strasser, *Waste and Want* (New York: Metropolitan Books, 1999); Hans Tammemagi, *The Waste Crisis* (New York: Oxford University Press, 1999); Christopher J. Preston and Steven H. Corey, "Public Health and Environmentalism: Adding Garbage to the History of Environmental Ethics," *Environmental Ethics* 27 (Spring 2005): 3–21; Carl Zimring, *Cash for Your Trash* (New Brunswick, N.J.: Rutgers University Press, 2005).

Some useful recent works on environmental justice pertinent to this study include Andrew Szasz, *Ecopopulism* (Minneapolis: University of Minnesota Press, 1994); Barbara L. Allen, *Uneasy Alchemy: Citizens and Experts in Louisiana's Chemical Corridor Disputes* (Cambridge, Mass.: MIT Press, 2003); Robert D. Bullard, Glenn S. Johnson, and Angel O. Torres, *Sprawl City* (Washington, D.C.: Island Press, 2000); Luke Cole and Sheila Foster, *From the Ground Up: Environmental Racism and the Rise of the*

Environmental Justice Movement (New York: NYU Press, 2001); M. Egan, "Subaltern Environmentalism in the United States," *Environment and History* 8 (2002): 21–41; Diane D. Glave and Mark Stoll, eds., *"To Love the Wind and the Rain": African Americans and Environmental History* (Pittsburgh: University of Pittsburgh Press, 2006); Dolores Greenberg, "Reconstructing Race and Protest," *Environmental History* 5 (Apr. 2000): 223–50; Steven J. Hoffman, *Race, Class and Power in the Building of Richmond* (Jefferson, N.C.: McFarland, 2004); Eric J. Krieg, "Race and Environmental Justice in Buffalo, New York," *Society and Natural Resources* 18 (Jan. 2005): 199–213; Steve Lerner, *Diamond: A Struggle for Environmental Justice in Louisiana's Chemical Corridor* (Cambridge, Mass.: MIT Press, 2005); Eileen McGurty, *Transforming Environmentalism* (New Brunswick, N.J.: Rutgers University Press, 2007); David Naguib Pellow, *Garbage Wars* (Cambridge, Mass.: MIT Press, 2002); David Naguib Pellow and Robert J. Brulle, eds., *Power, Justice, and the Environment* (Cambridge, Mass.: MIT Press, 2005); Ellen Stroud, "Troubled Waters in Ecotopia," *Radical History Review* 74 (Spring 1999): 65–95; Julie Sze, *Noxious New York: The Racial Politics of Urban Health and Environmental Justice* (Cambridge, Mass.: MIT Press, 2007); and Sylvia Hood Washington, *Packing Them In* (Lanham, Md.: Lexington Books, 2004).

INDEX

Adams, Julius W., 64
African Americans: disease and, 41–42,
   99; as environmentalists, 219–20; poor
   sanitation and water supply for, 55, 99, 183;
   suburbanization and, 131, 173. *See also*
   race
agriculture, 89, 232–33. *See also* animals;
   fertilizer, sewage as; irrigation
air pollution, 243; from refuse incineration,
   165, 206, 248
Alabama, 240
algae: copper sulfate to control, 94
Allied Waste Industries, 254–55
Altona, Germany, 93
America in Ruins (Choate and Walter), 215
American City surveys, 94–95
American Medical Association (AMA), 45,
   183
American Public Health Association
   (APHA), 46, 78; on refuse collection and
   disposal, 117, 122–23, 126
American Public Works Association
   (APWA), 166–67, 203, 206–8
American Society for Municipal Improve-
   ments (ASMI), 123, 126
American Society of Civil Engineers (ASCE),
   48, 155
American Society of Municipal Engineers
   (ASME), 123, 193, 204–5, 216
American Water Works Association
   (AWWA), 90, 144, 180, 187
Andrews, Richard N. L., 222
animals, 12; garbage fed to, 116, 127, 163, 165,
   206; waste produced by, 26, 113, 115, 128
anthrax, 64
anticontagionism, 31, 42, 49, 77, 100, 116
aqueducts: from Catskill Watershed, for New
   York, 88; Cochituate Aqueduct, for Boston,
   54, 57; from Jamaica Pond to Boston, 22,
   53; Old Croton Aqueduct, for New York,
   22, 53, 56–57, 87; from Owens Valley to LA,
   89, 138–39, 227; Washington Aqueduct, 57
Arizona, 182, 232
Arnott, Neil, 30
ashes: at household level, 113–14, 121, 160
Asia, 108
Atlanta, Georgia, 55, 136, 149–50

Atlantic Ocean: NYC dumping waste into,
   116, 121–22, 126, 162
Augsburg, Germany, 25
Austria, 94
automobiles, 128, 130, 159, 173
aviation, 216

bacteriological theory of disease, 2; effects of,
   72, 76–78, 104; evaluation of water quality
   and, 91–92; focus on cure *vs.* prevention,
   71; focus on water pollution, 96, 104;
   growing acceptance of, 42
bacteriology: disease prevention using,
   94–95; effects on sanitary services, 260–61;
   laboratories for, 79; monitoring bacteria in
   waterways, 105; New Public Health and,
   80–81; sewage treatment using, 108–10
Bahamas, 240
Baker, M. N., 92–95
Baltimore, Maryland, 54, 103, 136, 213
Baltimore Harbor: pollution of, 97–98
Barton, Bruce, 159
bathtubs: in water consumption, 89
Bazalgette, Joseph William, 35–36, 64
Belgium, 94
Belize, 240
Bentham, Jeremy, 29
Benthamite, Chadwick as, 29–31
biochemical oxygen demand (BOD), 96
Birmingham, Alabama, 89
Birmingham, England, 35
Black, William M., 142
boards of health, 78; Chadwick and, 30, 32;
   establishment of, 46–47; in New York City,
   45–46; on refuse collection and disposal,
   116, 126. *See also* health departments
boosterism: by local governments, 51–52, 58,
   83, 88–89
Boston, Massachusetts, 136; health and
   disease in, 43–44, 46; sanitary regulations
   of, 14, 26; sanitary survey of, 43–44; sewer
   systems for, 26, 62, 64–65; solid waste in,
   114, 118; water system for, 22, 53, 57
Boston Harbor: pollution of, 64
Brantford, Ontario, 183
Brewster, NY, 110
Bridge Canyon (on Colorado River), 186

WWII, 201–2; U.S. changing from producer to, 130

convenience: as goal of sanitary services, 5

Coolidge, Calvin, 146

Council of National Public Works Improvement, 215–16

Council on Environmental Quality (CEQ), 217–18, 226

crime: alleged in solid-waste management companies, 204, 255

Crooks, Harold, 255

Croton Aqueduct and Reservoir (for NYC), 22, 53, 87

Cuyahoga River, 187

Dallas, Texas, 172, 185–86, 213

Davenport, Iowa, 162

Davis, Joseph P., 65

death rates: blamed on bad sanitation, 36, 64, 99–100; from cholera, 65–66; falling, 47, 96, 140

Delaware River, 144

Denver, Colorado, 72, 126, 203

detergents, synthetic, 194, 197–98

Detroit, Michigan, 86, 172, 203

developers: expansion of sewer systems by, 98, 103, 182, 193; limited by moratoriums on sewer systems, 228; sprawl and, 130–31, 173

Dibdin, William Joseph, 109

diphtheria, 12–13

disease: causation of, 77, 113, 115, 232; epidemics as motivation to develop sewer systems, 65–66, 98; epidemics as motivation to develop water systems, 15, 18, 53, 93, 98; epidemics of, 11–14, 25, 79, 93, 99; fatalism about, 28, 30; in garbage-fed animals, 116, 163, 206; limited understanding of, 5, 30–31, 40–44, 57; links between poverty and, 30, 75; as moral dilemma, 41–42, 46; race's relation to, 41–42; relation to poor sanitation, 62, 71; responses to epidemics, 30–31, 35, 46–47, 99; waterborne, 57–58, 91–92, 96, 104, 140, 179, 186, 230, 239.

disease prevention, 12, 107, 120; better sanitation for, 4–5, 25, 31, 78; in filth theory of disease, 44; obstacles to, 14–15; sewer systems for, 62, 99, 110–11, 260; understanding causation of cholera, 37–38; vs. cure, 71; water treatment for, 24, 27, 38, 94–96, 139, 259–60

doctors, 46; role in public health, 30–31, 43–45, 78

Domenici, Pete, 238

Dr. Strangelove (Kubrick film), 184

drainage, 62, 65, 100–101. *See also* storm-water

Drinking Water Standards, 187

droughts: effects of, 88, 163, 184–85

dual disposal: of sewage and ground garbage, 153

Duluth, Minnesota, 118

dumping: refuse disposal by, 124–25, 165; on land, 115–16, 126–27, 161–62, 204, 240–41; into water, 115–16, 121–22, 126, 145, 162, 206

dysentery, 30, 65–66

Earth Day, 217

East Coast garbage truck inspection blitz, 246

Eaton, Fred, 88

Echo Park (on Green River), 186

ecofeminism, 222

ecological perspective/environmental paradigm, 154, 176–77

ecological risks, 222

ecology, 222

Eddy, Harrison P., 155

Emergency Relief and Construction Act (1932), 132

energy/power: from refuse incineration, 124–25, 165, 205, 248–49; use of water in generating, 186, 226, 231–32

engineering, 102; civil, 34, 40, 47–48, 178, 223; ecological *vs.* environmental, 223; environmentalism and, 74, 140, 172, 178; municipal, 79–80, 120, 122; role in sanitary services, 3–4, 72, 111–12, 143; sanitary, 80, 122–23, 143, 154–55, 178, 223; in sewer systems, 63, 101–2, 105, 111–12, 195; in solid-waste management, 126; in water systems, 55–56, 59

England, 115; Chadwick's influence on, 29–30; control of public works in, 32–33; sanitary movement in, 28–33, 38; sewer systems in, 26, 63; U.S. and, 38, 52, 59, 110; water closets in, 24, 62; water pollution in, 36–37; water systems in, 16–18. *See also* Great Britain; London

English "sanitary idea," 4–5, 27–28, 40–41

"English system" of water filtering, 18

enterprise zones, 214

Environmental Defense Fund, 177, 238

environmental impact statements (EISs), 217

environmental justice movement, 219–20, 246–47, 250

environmental movement, 171–72, 210, 217–18; environmental justice movement *vs.*, 219–20; legislation from, 177–78; participants in, 177, 221–22; on solid waste issue, 200, 209; on water pollution, 155, 189, 190

environmental problems: relation to sanitary services, 259, 261–63

environmental protection: balancing economic development against, 218–19; federal government's role in, 217–18, 222. *See also* ecological perspective/environmental paradigm; water-pollution control

Environmental Protection Agency (EPA), 221; budget for, 218, 229; Clean Water Act and, 234–36, 238–39; enforcement role of, 208, 234, 236; Office of Ground Water Protection of, 236; roles of, 217–18, 234–35; sewer systems and, 229, 233–34; solid-waste management and, 208, 245, 251, 255–57; tarnished image of, 218–19; water pollution and, 230–33, 236; water supplies and, 229–31, 237–38

environmental reforms: Progressive, 73–75

environmental risks: to water supplies, 230

environmental sanitation, 28, 43, 49, 62, 142; belief in, 52, 54, 71; ecological perspective vs., 176–77, 261; effects of bacteriology on, 72, 76–78; endurance of movement, 71–72; individual health vs., 42, 154; motives for, 259–60; as municipal housekeeping, 113, 117–18; sanitary services in, 60, 113, 259–61

environmental science, 222

Europe, 72; Americans studying sanitary systems of, 58, 65, 67, 102, 117; sanitary practices in, 11–12, 24, 44; sewer systems in, 65, 67, 102, 108; solid-waste management in, 115, 117, 126, 166–67; water systems in, 15, 58–59, 227

Evanston, IL, 183

Exner, Frederick B., 184

Federal Emergency Relief Administration (FERA), 132, 137, 142

Federal Water Pollution Control Act. See Clean Water Act (1972)

Federal Water Pollution Control Act Amendments (1961), 189

Federal Water Pollution Control Act of 1948, 145

Federal Water Pollution Control Administration (FWPCA), 190

Federal Works Agency (FWA), 133

fertilizer, 122; sewage as, 25, 31–33, 36, 108

filth theory of disease, 2, 4–5, 31, 44, 46; belief in danger of sewer gases in, 100, 104; discredited, 77–78, 104; refuse collection in, 120, 158–59; in U.S., 40, 42; water supply in, 57, 60

fire control, 183; central distribution model for, 89–90; in motivations for water systems, 22, 24, 27, 53, 55; water supply for, 16–18

Flanagan, Maureen A., 118

flooding, 137, 195

Florida, 226, 240, 251

fluoridation: of water, 139, 183–84, 230–31

Food, Drug and Cosmetic Act of 1938, 145

Ford, Gerald, 213, 238

Fowler, Gilbert J., 111

Fragile Foundations (Council of National Public Works Improvement), 216

France, 28

Fresh Kills Landfill, 244

Fresno, California, 85, 162–63

Fuller, George W., 105–6, 155, 158

garbage: disposal of, 14, 162; extracting oil from, 125–26; fed to animals, 116, 163, 165, 206; as health risk, 113; increasing production of, 159–60; in-home disposal/grinding, 203; in separation of solid waste, 121. See also refuse collection; refuse disposal; solid waste

Gary, Indiana, 153

Gascoigne, George B., 152

gentrification, 211

Georgia, 85–86

Gerhard, William Paul, 61, 80

Germany, 93, 115, 196

germ theory of disease. See bacteriological theory of disease

Gibb, John, 18

Gibbs, Lois Marie, 219

Glasgow, Scotland, 18

Glen Canyon (on Colorado River), 186

Goldfield, David R., 210

Gorsuch, Anne, 218

government: relations among levels of, 13–14, 51, 143, 176, 189, 255; role of activist, 29, 31, 42, 132–33; role in public health, 44, 46; role in sanitary services, 12–13, 15, 24–25, 38–39, 45; role in water systems, 17, 86

government, city, 74; belief in progress through technology, 79–80; corruption in, 99, 120–21; dealing with growth issues, 73, 172; effects of structure of, 52–53; home rule for, 72, 75–76, 119; politics in, 48–49, 53, 73; reform of boss system in, 73; taxation by, 51–52; waterworks and, 20–21, 85–86. See also cities

government, county, 194

government, federal: in environmental problems and protection, 177–78, 217–18, 221–22; in public health, 47, 79, 179; public works and, 173, 214–16; relations with cities, 6, 129–33, 149–50, 174–76, 212–14; in solid-waste management, 207–9, 255–57; in water-pollution control, 145, 188–90, 198–99, 226, 233–34, 237; in water quality management, 187, 189; water systems and, 137–38, 186, 229

government, local, 79, 175; central vs., 31–32; EPA's reluctance to enforce against, 237–38; flow-control ordinances by, 245–46; New Federalism giving authority to, 213; role in sanitation systems, 1, 14, 38–39, 40, 229; state vs., 48, 79

government, regional, 136, 151–52

government, state: aid to cities, 150, 213–14, 229; cities and, 48, 53, 75–76, 84–86, 99, 192, 212; limiting cities, 21, 175–76, 228; in public health, 79; in solid-waste management, 204, 208–9, 245–46; wastewater systems and, 156, 188, 192, 229; water pollution and, 96, 106–7, 143–45, 156–57, 189, 190, 199; water quality and, 199, 234, 237–38; water systems and, 87–88, 188, 229

Grand Rapids, Michigan, 139, 183

gravity: in water supply system, 89

Great Britain, 11, 74; engineering in, 33–34; refuse disposal in, 124–25, 162–65; sewer systems in, 35–37, 108–11, 196; U.S. vs., 124–25; water pollution in, 35–37; water systems in, 104, 228. See also English "sanitary idea"

Great Depression, 143, 163; effects of, 130, 135, 159; effects on cities' finances, 131, 160; effects on federal/city relations, 131–32; effects on sanitary services, 6, 133–34, 142, 153, 160; effects on sewer systems, 148–49, 153

Great Society program, 175, 190, 199, 213

Great Stink of 1858 (London), 36

Greeley, Samuel A., 26, 118, 122, 158, 160–61

Greenpeace, 250

Greensboro, NC, 149

Griscom, John H., 43

Guiliani, Rudolph W., 244

Hamburg, Germany, 93

Hamilton, Cynthia, 220

Hamlin, Christopher, 34–35, 109

Hansen, Paul, 153

Hartman, Anne, 253–54

Hazardous and Solid Waste Amendments, 256–57

hazardous wastes, 208, 216, 246; Citizens Clearinghouse for Hazardous Wastes, 246–47; Council of National Public Works Improvement's rating of, 216; environmental justice movement and, 219–21; federal role in solid-waste management limited to, 255–57

Hazen, Allen, 94

health, 106, 118, 140; in ecological perspective, 176–77; individual vs. public, 154, 178–79. See also disease; disease prevention; health hazards

health departments, 78, 142; refuse collection and, 120, 122; state, 107, 156, 188. See also boards of health

health hazards, 142, 187; ecological risks broadening focus of, 222; solid waste as, 162, 200

Health of Towns Commission (England), 33

Hering, Rudolph, 98, 105, 110; on refuse collection, 122; on sewer systems, 102–3

Hetch Hetchy (Sierra Nevada area), 186

highways, 225, 245

Hollis, Mark, 178

home rule: for cities, 72, 75–76, 119

homes, 205; ownership of, 131, 172–73; for war-plant workers, 138

Hoover, Herbert, 131–33, 146

Hoskins, J. K., 197

Housing and Community Development Act (1974), 213

Houston, Texas, 213; composting program in, 206–7; expansion of, 172, 182; sewer system of, 148, 153

HUD's Community and Economic Development Task Force, 215

Hyperion Activated Sludge Plant (Los Angeles), 196–97

Ickes, Harold, 150

Illinois, 185, 245

Illinois River, 107, 156–57

Imhoff, Karl, 110

Imhoff tank, 110, 152

immigrants, 72–73, 114; diseases blamed on, 41, 43–44; public health and, 44–45, 78–79

incineration: refuse disposal by, 120, 123–25, 165–66, 205–6, 247–48; opposition to, 246, 249–50; reduction vs., 126–27

Indianapolis, Indiana, 46, 153, 172

industrialization, 72–75

Industrial Revolution, 14, 16

industry, 41; extent of water pollution from, 140–42, 197, 231; products of, 159; response to environmental regulations, 145, 218, 235–36; treatment of wastes from, 187, 229; types of water pollution from, 141–42, 156, 198; wastes dumped into municipal sewer systems, 139–40, 146, 155–56; water-pollution control measures for, 107, 144–45, 153, 189, 234; water pollution from, 36–37, 96, 130, 139, 141, 146, 152, 154, 187, 206, 228; water pollution from sewer overflows and, 62, 195–96; water use by, 51, 55, 62, 89, 181, 186–87, 231–32; waterworks as major, 181; during WWII, 133, 138. See also business

influenza pandemic, 79

infrastructure crisis, 6

inoculation: in Chadwick's plan, 31

Insecticide Act of 1910, 145

Institute for Local Self-Reliance, 250

irrigation: in San Fernando Valley, 88–89; sewage disposal in, 108; water consumed by, 181, 226, 231–32; water projects for, 137, 186. See also fertilizer, sewage as

Izaak Walton League, 155, 190

Jackson, Kenneth, 173

Jacksonville, Florida, 172

Jacobs, Nathan B., 166–67

Jarvis, Edward, 46

Johnson, George A., 94

Johnson, Lyndon, 175, 199, 207–8

Judd, Dennis R., 171

Juvenile Street Cleaning League (NYC), 121

Kansas City, Kansas, 89

Kay, James, 30

Newark, New Jersey, 115
Newburgh, New York, 139, 183–84
New Croton Adqueduct, for New York City, 87
New Deal, 160, 188; federal government's relations with cities in, 132–33; sewer projects in, 148–50; water projects in, 135, 137–38, 142, 146, 149
New Ecology, 6, 171–72, 177, 185–86, 222, 226, 243, 261
New Federalism, 213–14
New Haven, Connecticut, 203
New Jersey, 144–45, 203, 245–46, 251
New Orleans, 72, 126, 150, 162; disease in, 42, 46–47, 238; water system for, 23, 98, 238
New Public Health, 78–81
New York City, 48, 73, 172, 212; boards of health in, 45–46; dumping of solid waste by, 116, 126, 162; incineration of solid waste, 125, 166; landfills for, 162–63, 166, 244; sanitary conditions in, 43, 117; sanitary surveys of, 43, 45; sewer systems for, 62, 67, 149, 194–95; solid-waste management in, 114, 118, 240; street cleaning in, 120–21; water supplies for, 21–22, 56–57, 138; water systems for, 18, 21–22, 53, 85, 89
New York state: solid-waste management and, 203–4, 245, 251; water pollution and, 144–45; water supplies and, 85–88
NIMBYism, 209, 246; evolving into NIABY-ism, 219–21; opposition to incinerators and, 165–66
Nixon, Richard, 212–13, 217
Noble, David, 79
North Carolina, 240
noxious trades, 12, 14
nuisance: filth as, 12, property rights vs. regulation of, 14–15; regulation of, 14–15; smells as, 36, 77; solid waste as, 33, 113, 117, 200, 258; waste disposal and, 24, 156
Nuisance Removal Act of 1855 (England), 35

Oakland, California, 136, 185
Oberstar, James L., 236
Odum, H. T., 223
Office of Solid Waste (of EPA), 256
Ogallala aquifer, 232
Ohio Plan, 136
Ohio River, 96, 141, 154, 197
oil industry, 146, 218
Oil Pollution Control Act of 1924, 145–46
Okun, Daniel, 237
Old Croton Aqueduct for NYC, 56–57
Olmsted, Frederick Law, 1
Oregon, 251–52
Oregon State University, 208
Ottawa, Canada, 93
Owens Valley, LA water from, 88–89, 138–39, 227

Pacific Ocean, 162
packaging: as solid waste, 201–2, 242
Panic of 1873, 47, 51
paper: in waste stream, 201, 207, 242–43
Paris, France, 15–16, 25
Pasteur, Louis, 42, 77
Pataki, George E., 244
path dependence, 4, 262
Pennsylvania, 107, 144, 151–52, 194, 245, 251
permanence: sanitary services' search for, 161, 261–62; lack of, 171, 194–95
personal hygiene vs. environmental sanitation, 42
pesticides: water pollution from, 232, 233, 236
Peterson, Jon, 3
Phelps, Earle B., 142
Philadelphia, Pennsylvania, 41, 118, 213; sewer systems for, 99, 194, 203; water systems for, 19–20, 23, 50, 93–95, 259
Phillip II, King (of France), 25
Philosophical Radicals, 29
Phoenix, Arizona, 172, 213
Pinchot, Gifford, 88–89
Pittsburgh, Pennsylvania, 93, 99, 114–15
Plymouth, Pennsylvania, 93, 107
politics, 116, 150, 174; cities', 48–49, 53, 73; federal government's relations with cities in, 131–32, 175; in garbage crisis, 252–53; over sewer systems, 65, 67, 103; over water systems, 53–54; Progressive reforms of, 73, 86; in public health reform, 44–45
pollution, 203, 261–62; biological, 130; effects of, in ecological perspective, 176–77; faith in science and technology to solve, 178, 223; relation to race, 219–21; solid waste as, 200, 241. See also under industry; water pollution
Poor Laws (England), 29–30
population density, 15
population growth, 72; cities trying to keep services up with, 50–51, 147–48; increasing solid waste from, 115, 201; rapid U.S., 40–41; sewer systems and, 25, 147–48; water consumption and, 136, 226; water systems and, 20, 50–51, 82
Portland, Oregon, 152
Potomac River, 57
Poughkeepsie, New York, 58, 93
poverty. See class, social
power. See energy/power
practical vital theory, 42
preservation vs. development, 177
private companies, 40, 135, 214; cities push-ing out of water systems, 51, 54–55, 83–85; consolidation of solid-waste management, 204, 245, 253–55; not interested in sewerage systems, 62, 97–98, 148; refuse collection by, 119–20, 160–61, 243, 253; sewer systems by, 148; solid-waste management by, 160,

164, 203, 253–54, 257; water systems by, 59, 61, 84–86, 90, 227–28
privy vaults, 24–25, 62, 97, 100
Progressive Era, conservation perspective in, 154, 178
Progressive reforms, 86, 117; effects on city finances, 76, 85; environmental, 71–75
property, taxation based, 52, 133
property rights, 14–15, 106
public health, 42; Britain's leadership in, 28; changes in field of, 78, 178–79; expertise in, 44, 80; failures in, 39, 99; as focus of sanitary services, 80–81, 176–77; in France, 28; government role in, 13, 29–30, 44, 46–47, 79; as municipal responsibility, 51–52, 98; participants in, 43–46, 143; politics in reform of, 44–45; priorities of, 72; refuse disposal and, 158–59; sewer systems and, 99, 108; vs. common law, 106. See also disease prevention
Public Health Act of 1848 (England), 32–33
Public Health Act of 1875 (England), 32–33
public health officials: acceptance of bacteriology, 77; engineers vs., 111–12, 155; environmentalism of, 74, 172; on sewage treatment, 63, 111–12, 155; on water pollution, 140–42; on water treatment, 59, 142, 155
public transportation, 130
public utilities, 83, 85, 122, 176; regulation of, 84–86
Public Utility Regulatory Policies Act (PURPA), 248–49
public works: decision making and expertise in, 180, 223, 226–27; deficiencies of, 180, 262; federal involvement with cities', 132–33, 216; infrastructure crisis in, 214–15, 239, 261–62; infrastructure of, 150, 262; investment in, 131, 214–16; politics over, 48–49, 103; during WWII, 133, 175
Public Works Administration (PWA), 132, 137–38, 149–50
public works departments, 120, 126
Public Works Improvement Act of 1984, 215–16
Pucinski, Roman, 252–53
Pullman, Illinois, 108
Pure Food and Drug Act of 1906, 145

quarantine: as response to epidemics, 13–14, 47, 99

race: environmental justice movement and, 219–21, 250; relation to disease, 41–42, 179; sanitary services and, 55, 203, 250; suburbanization and, 173–74, 182–83, 211
Ramapo Water Company, 85
Rampart (on Yukon), 186
Rathje, William, 241
Reagan, Ronald, 214, 218, 229, 235
Reclamation Project Act (1939), 138

Reconstruction Finance Corporation (RFC), 132, 137, 149
recycling: in refuse disposal, 127, 206–7, 241–43, 246, 250–52
reduction, in refuse disposal, 124, 126, 163–65
Refuse Act of 1899 (Rivers and Harbors Act), 145
refuse collection, 122, 123; combined vs. separate systems for, 123; compaction in, 202–3; cost of, 123–24, 160, 202–3, 243; dissatisfactions with, 100, 116–17, 124; environmental sanitation and, 113, 260; European systems praised, 166–67; improvements in, 115, 160–61; as municipal responsibility, 119–22, 128, 203–4; by private companies, 119–20, 161, 203–4, 243, 253; problems in, 202–3, 243–44; recycling and, 250–52; service charges for, 151, 160, 202; transfer stations in, 161, 203; Waring in charge of NY's, 120–22. See also solid-waste management
refuse disposal: beyond individual responsibility, 113, 119; beyond local governments' capabilities, 207; calls for more attention to, 116–18, 158–59, 166–67; centralization of, 166; citizens' groups in, 118–19, 246–47; costs of, 123, 126, 162, 202, 245, 247, 257; dual (sewage and ground garbage), 153; environmental racism in, 220; European systems praised, 166–67; evaluation of methods of, 123, 160, 205, 251–52; exporting garbage to other states, 244–45; flow-control ordinances in, 245–46; garbage crisis in, 240–41, 244; by incineration vs. reduction, 126–27; in-home garbage disposals, 166; methods of, 160–62, 247–48; as municipal responsibility, 122, 128; private companies in, 253–54; in sanitary services, 1; technology in, 134; Waring's plan for, 121–22; as weakness of environmental sanitation, 261. See also solid-waste management; specific methods, i.e., incineration
Reilly, William K., 218–19, 221
Report Card for America's Infrastructure (ASCE's), 216
"Report on the Filtration of River Water" (Kirkwood), 58, 92
Report on the Sanitary Condition of the Labouring Population of Great Britain (Chadwick), 4, 30
Resource Recovery Act (1970), 208, 256–57
Resources for the Future, 177
revenue-sharing programs, 213–14
Revolutionary War, 13
Reynolds, Terry, 47
Rhode Island, 251
Ricardo, David, 29
Richmond, Virginia, 58

River Pollution Prevention Act (1876)
(England), 37
Rodgers, William H., Jr., 235
Roe, John, 31–32, 34
Romney, George, 182–83
Roosevelt, Franklin D., 132–33
Roosevelt, Theodore, 120
Rosenberg, Charles E., 41
Royal Sewage Commission (England), 37
rubbish: production of, 159–60; separation
of, 121, 164. *See also* refuse disposal;
solid-waste management
Ruckelshaus, William D., 200–201, 218
Rudolfs, Willem, 152

Sabine River, 185
Safe Drinking Water Act (1974), 237–39
Salvato, Joseph A., 193
San Diego, California, 148, 172, 182, 208
San Francisco, California, 85, 89, 162–63,
182, 185
sanitarians, 63, 77, 142; in decision-making
process, 3–4, 49; environmentalism of, 74,
172; on health issues, 113, 140
Sanitary Act of 1866 (England), 32
*The Sanitary Condition of the Laboring
Population of New York* (Griscom), 43
sanitary conventions, 45–46
sanitary engineering. *See* engineering,
sanitary
sanitary landfills, 127, 162–63, 165–66,
204–5, 246. *See also* landfills
sanitary movement: in England, 28–33; in
U.S., 40, 47, 50, 117
sanitary reforms. *See* environmental reforms
Sanitary Report (Chadwick), 4
sanitary services: bio-physico-chemical
perspective in, 152, 154; Chadwick's
plan for, 31–33; decision making in, 3–4,
38–39, 79, 129, 143, 154, 262; deficiencies
of, 180, 239, 247, 261–62; definition of, 1;
effectiveness in disease prevention, 30–31,
99–100, 140; effects of bacteriology theory
on, 260–61; effects of Great Depression
and WWII on, 133–34; England's influence
on U.S., 27–28, 38; in environmental
sanitation paradigm, 259–61; expansion of,
40, 131, 171, 194, 199, 262–63; government
role in, 12–13, 31, 138; impermanence in, 4,
6, 161, 171, 261–62; increasing complexity
of, 210; infrastructure for, 176, 216–17;
levels of, 15, 83, 199, 259, 262–63; priorities
in, 158–60; public cooperation with, 45,
121, 167; regulations in, 45, 99; relation
to environmental problems, 259, 261–63;
responsibility for, 12–13, 15, 31, 46; role of
private companies in, 38–39; technologies
of, 80–81. *See also* specific services
sanitary surveys: of Boston, 43–44; of
Massachusetts, 44–46; of Memphis, 99; of
NYC, 43, 45

San Jose, California, 85, 172
San Marcos, Texas, 111
SCA Services, 204
scavengers: animals in cities as, 12; refuse
collection by, 115, 161; in utilization of
refuse, 121, 163–65; waste disposal by,
25–26, 114, 119
Schneider, Franz, Jr., 115
science, 36, 79, 222, 261; faith in, 75, 127;
hope for environmental solutions from,
178, 217; in water treatment, 91, 104. *See
also* technology
Seattle, Washington, 152, 162
Sedgwick, William T., 92
Service Delivery in the City, 1
sewage: as biological, not chemical, threat,
104; considered nuisance, 24; constitution
of, 25–26, 197; contaminating water
supplies, 17, 92; in disease causation,
37–38, 58, 92; sewer systems and, 97–98,
104; uses of, 25, 36, 108, 228–29. *See also*
water pollution, from sewage
sewage disposal, 261; British methods of,
33, 108–11; effects on water supplies, 87;
as health problem, 25, 106; at household
level, 24, 26, 97–98, 193–94; on land, 108;
pretreatment in, 105–6, 112; regulation of,
25–26, 188; responsibility for, 119–20; in
sanitary services, 1, 158; through dilution,
67, 105, 111–12, 157, 197. *See also* water
pollution, from sewage disposal
sewage farms, 108
sewage treatment, 157, 187, 196; aeration in,
111; biological *vs.* chemical, 153; chemical
precipitation as, 108, 153; by cities,
156, 199; contact filtration as, 109–10;
disinfection of, 94, 110–11; effectiveness of,
228, 239; effects of synthetic detergents on,
197–98; electrolytic process in, 111; federal
funding for, 137, 153–54, 188–89, 198–99,
239; growing understanding of biological
process of, 36, 105; intermittent filtration
as, 108; percentage of systems including,
152–53; primary *vs.* secondary, 110, 152,
196; public support for, 105–6, 155; secondary treatment in, 236; in septic tanks, 110,
193–94; technologies in, 107–10, 196–97,
228; trickling filtration as, 109–10, 152;
water treatment *vs.*, 106, 111–12, 139,
152–54, 236. *See also* sludge
sewerage districts: metropolitan, 136, 152,
194
sewer gases: believed dangerous, 100, 104
sewer systems: access to, 24, 103; benefits of
underground, 62, 97; as capital intensive,
98–99; in Chadwick's plan, 31–32, 34;
chemical pollution causing explosions
in, 141, 156; Chicago's, 65–66; combined,
63–64, 195–96; combined *vs.* separate,
63–64, 66, 99, 101–2, 102–3, 104, 147–48,
195–96; construction of, 196, 198; cost

of, 63, 65, 103, 148, 176, 192–94, 196, 198; decision making in, 102–4, 223, 229–30; deficiencies in, 180, 192, 194–95, 228; effectiveness in disease prevention, 5, 36; effects of piped-in water on, 62, 97, 260; expansion of, 66–67, 98–99, 147–48, 148–49, 193, 228; federal funding for, 137–38, 142, 160; federal role in, 234; funding for, 66, 189, 194, 229; ground garbage disposal into, 166, 203; industry's right to use, 139–40, 155–56; interceptors in, 64, 66; London's, 34–36; Memphis's, 100–102; numbers of, 98, 147–48, 152–53; as open ditches, 24, 26, 62; outlets for, 33, 35–36, 101; overflows of, 195–96, 233–34, 237; private companies in, 227; quantities of effluent in, 147–48; ratings of, 71, 216, 225; regionalization of, 34, 151–52, 194, 228; regulation of disposal, 14; responsibility for, 24, 97–98, 103, 106; service charges for, 149–50, 150–51, 193–94, 202; slowness in development of, 61–62; stormwater and, 34–35, 237; technology in, 101, 153, 195, 234; used for drainage *vs.* sewage, 24–26; water systems not integrated with, 67, 97–98, 136–37, 226, 260
Shattuck, Lemuel, 43–46
Sheibley, Edward G., 143
Sherman, Texas, 118–19
Sierra Club, 177, 217
Silent Spring (Carson), 177
sludge: activated sludge plants, 149, 152, 153; in sewage treatment processes, 109–11, 196–97; spread on land, 228–29
slums, 29–30, 41, 174
Small, William E., 200
smallpox, 12–13
smells: from dumps, 116; from garbage reduction plants, 126, 164; from refuse incineration, 165–66; responses to, 77, 117; from sewage-polluted water, 36, 97–98; in water, 91, 139
Smith, Thomas Southwood, 29–30
Snow, John, 37–38, 57
soil erosion, 187
soil pollution, 99–100
solid waste: composition of, 201–2, 207, 241–43; efforts to reduce, 241, 246–47; in filth *vs.* germ theories of disease, 5, 42, 67; quantities of, 113–15, 127, 159–60, 166–67, 201, 203, 241; as third pollution, 5–6, 200, 207, 209, 241; utilization of, 163–65. *See also* hazardous wastes; refuse collection; refuse disposal
Solid Waste Disposal Act: in Clean Air Act, 207–8, 256
solid-waste management, 216; attempts to change citizen behavior and, 128; cost of, 202–3, 243; as environmental issue, 200, 209; federal government in, 207–9; garbage crisis in, 257–58; integrated systems of,

255–56; level of government entities in, 207, 255; as local responsibility, 129, 160, 207; private companies in, 203–4, 245, 253–55; research on, 207–8; separation in, 121, 161, 164, 204, 243–44; states in, 208–9. *See also* refuse collection
Southeast Coastal Plain, 232
South Water Filtration Plant (Chicago), 137
St. Louis, Missouri: suing Chicago over pollution of Mississippi River, 107, 156–57; water systems for, 22–23, 54, 58, 89
St. Louis Water Works, 54
St. Paul, Minnesota, 126
St. Petersburg, Florida, 206–7
State and Local Assistance Act of 1972, 213
steam, for water pumping, 15, 19–20
Stein, Albert, 58
Stine, Jeffrey K., 223
Stockman, David, 218
stormwater, 63; sewer systems and, 25–26, 34–35, 100–101, 148, 195–96; water pollution from runoff, 233, 237
Strasser, Susan, 114
Stream Pollution Investigation Station, 96
stream purification, 96
street cleaning, 25, 119, 122
Streeter, H. W., 142
Streeter, Harold W., 154
streets, 25, 100; belief in increased cleanliness from cars, 127–28; as "commons," 26
Strong, William L., 120, 122
suburbs, 210, 213; demand for services in, 173–74, 176; growth of, 41, 129–31, 171–72; race and class in, 173, 182–83, 212; relations with cities, 174–75, 182, 194, 210–12; sanitary services for, 5–6, 194; sewer systems for, 193–94; water systems for, 86, 90, 181–82
Sun Belt migration, 172
Szasz, Andrew, 219–20

Tarr, Joel A., 25, 111–12
taxes: changing basis for, 51–52; in city finances, 131, 173, 212–14; funding sanitary systems, 84, 98; on property values, 76, 133; in urban crisis, 173, 175–76
Teaford, Jon C., 73, 210–11
technology, 2–3, 261; cities' networked, 79–80; faith in, 79–80, 223; hope for environmental solutions from, 178, 217, 235–36, 249–50; of refuse incineration, 124–25, 205–6, 248–50; of sanitation, 80–81; in sewage treatment, 107–10, 196–97, 228; in sewer systems, 195, 234; in solid-waste management, 134, 160–61, 203, 208; in water systems, 82, 89, 95, 139, 187. *See also* science
technology forcing, 234–37
Texas, 226, 232
Thames River, 37, 109
Thom, Robert, 18

184, 186, 226, 233, 238; criticisms of, 22,
187; drinking-water standards, 187, 237–39;
efforts to protect, 236; evaluation of, 17,
57, 91, 96, 99, 139–40, 142, 176, 187, 216;
federal government in management of,
189, 226, 234–35; improvements in, 57–58,
71, 92; purity vs. safety in, 91; standards
for, 142, 226; steps to assure, 50, 57, 90; of
surface vs. groundwater, 231–32, 236. *See
also* water pollution; water treatment
Water Quality Act of 1965, 189, 199, 234
Water Quality Improvement Act (1970), 234
water sources, 181; changing, 18–19;
expansion of, 59–60; groundwater as,
186–87, 227–28, 232, 236, 238; groundwater
vs. surface water, 182, 231–32; ponds and
lakes, 54; search for new, 87–89; streams
for, 87, 185; wells and cisterns, 55, 87, 100,
185, 232
water supplies: access to, 23, 59; adequacy of,
71, 216, 225–26; in Chadwick's sanitation
plan, 33; deficiencies in, 180, 182; depletion
of, 184–85; effects of contamination of,
5, 15, 17–18, 30–31, 37, 57–58, 79, 230;
federal studies on, 186; for fire control,
16–18; generating revenue for cities, 176;
governmental entities' relations over,
136, 182; government role in, 17–19, 129,
146; increased usage of, 24, 62, 83, 97, 260;
municipal responsibility for, 38, 83, 143;
planning for needs, 50, 83, 186; protection
of, 96, 230, 236, 238; public vs. private
responsibility for, 17, 21–22, 51; in sanitary
services, 1, 158; shortages in, 86, 100,
184–87, 190–91; wastewater issues and, 62,
97, 226, 260
water systems: aqueducts in, 15–16, 22,
53–54, 56–57, 87–89, 138–39, 227; cities
developing, 21–23, 53–55, 59–60, 65–66, 85;
cities pushing private companies out of,
51, 54, 83–85; costs of, 90, 181–83, 187, 215;
dams in, 186; decision making in, 226–27;
deficiencies of, 180–81, 225; distribution
in, 15–16, 22–23, 31, 57, 59, 89–90, 183,
260; effects of local circumstances on,
226–27; expansion of, 90, 182; federal role
in, 137–38, 142, 160, 189, 234; funding for,
138, 142, 160, 189, 229; improvements
in, 57; inequities in, 86, 183; metering in,
86–87, 193; need to address new types of
pollution, 262; in New Deal, 142, 149, 160;
not included in "infrastructure crisis"
predictions, 225; not integrated with
sewer systems, 97–98, 136–37, 260; older
protosystems for, 82–83; ownership of,
83–85, 181, 226; pipes in, 15, 19–20, 20, 23,
59–60, 90; planning for, 97–98, 186; private
companies in, 15, 16–17, 21–22, 59, 61–62,
84–86, 90, 227–28; proposals to increase

efficiency of, 185–86; pumping in, 23,
59–60, 89; rates for, 84–85; regionalization
of, 227–28; research on, 189; reservoirs
in, 23, 185–86; sewer systems and, 67,
192; standardization of, 90; technological
innovations in, 59–60, 82, 89; tunnels and
conduits in, 138; use of electricity in, 82,
89
water treatment, 142, 188; as alternative
to new sources, 87, 91–92; chlorination
in, 93–95, 139, 183, 230; deficiencies in,
187, 197, 225; effectiveness of, 94–96,
141–42, 186, 237; eras of, 92–94, 153; faith
in dilution as, 141, 197; fluoridation in,
139, 183–84, 230–31; increasing use of, 83,
139; sewage treatment vs., 105–6, 111–12,
152–54, 236; softening, 139; water filtration
and, 104, 183
waterworks: in Chadwick's plan, 31–32; cost
of, 136–37; development by cities, 50–53,
82; expansion of, 135, 146; motivations for,
24, 27, 93; municipal ownership of, 20–21,
96, 100, 135; number of, 19, 50–51, 82, 135,
181; prototypes for, 19–20, 22–24, 50, 57;
as public enterprise, 259–60; regulation
of, 85–86
Watt, James, 218
Wayne, Pennsylvania, 95
welfare statism, vs. sanitary reforms, 74
wells: in municipal water systems, 22, 51;
water quality from, 22; yields of, 232
Whipple, George C., 78, 91, 104, 155
White Wings (NYC street cleaners), 121
Wilderness Act (1964), 177–78
Wilmington, Delaware, 98
Wilson and Company, 22–23
Wisconsin, 85–86
Wolman, Abel, 140, 143, 178
women: in environmental movement,
221–22; in refuse reform, 117–18
Women in Toxics Organizing Conference,
222
Works Progress Administration (WPA), 132,
137–38, 149
World War I, 78–79, 163–65
World War II: effects of, 160, 175; effects on
health departments, 142–43; effects on
sanitary services, 6, 133–34, 138, 148, 153;
federal involvement with cities during,
133, 138; throwaway culture after, 201–2;
utilization of refuse during, 163–65

yellow fever, 12–13, 23, 42; causes of, 42, 58;
epidemics of, 19–20, 46–47, 99; sanitation
systems as response to, 19–20, 98
Youngstown, Ohio, 136

zoning: of noxious trades, 14